Contents

MEMOIRS
of the
American Mathematical Society

Volume 244 • Number 1154 (third of 4 numbers) • November 2016

Proof of the 1-Factorization and Hamilton Decomposition Conjectures

Béla Csaba
Daniela Kühn
Allan Lo
Deryk Osthus
Andrew Treglown

ISSN 0065-9266 (print) ISSN 1947-6221 (online)

American Mathematical Society
Providence, Rhode Island

Library of Congress Cataloging-in-Publication Data

Names: Csaba, Béla, 1968–

Title: Proof of the 1-factorization and Hamilton decomposition conjectures / Béla Csaba [and four others].

Description: Providence, Rhode Island : American Mathematical Society, 2016. — Series: Memoirs of the American Mathematical Society, ISSN 0065-9266 ; volume 244, number 1154 — Includes bibliographical references.

Identifiers: LCCN 2016031065 (print) — LCCN 2016037506 (ebook) — ISBN 9781470420253 (alk. paper) — ISBN 9781470435080 (ebook)

Subjects: LCSH: Factorization (Mathematics) — Decomposition (Mathematics)

Classification: LCC QA161.F3 P76 2016 (print) — LCC QA161.F3 (ebook) — DDC 512.9/23–dc23 LC record available at `https://lccn.loc.gov/2016031065`

DOI: `http://dx.doi.org/10.1090/memo/1154`

Memoirs of the American Mathematical Society

This journal is devoted entirely to research in pure and applied mathematics.

Subscription information. Beginning with the January 2010 issue, *Memoirs* is accessible from `www.ams.org/journals`. The 2016 subscription begins with volume 239 and consists of six mailings, each containing one or more numbers. Subscription prices for 2016 are as follows: for paper delivery, US$890 list, US$712.00 institutional member; for electronic delivery, US$784 list, US$627.20 institutional member. Upon request, subscribers to paper delivery of this journal are also entitled to receive electronic delivery. If ordering the paper version, add US$10 for delivery within the United States; US$69 for outside the United States. Subscription renewals are subject to late fees. See `www.ams.org/help-faq` for more journal subscription information. Each number may be ordered separately; *please specify number* when ordering an individual number.

Back number information. For back issues see `www.ams.org/backvols`.

Subscriptions and orders should be addressed to the American Mathematical Society, P. O. Box 845904, Boston, MA 02284-5904 USA. *All orders must be accompanied by payment.* Other correspondence should be addressed to 201 Charles Street, Providence, RI 02904-2294 USA.

Memoirs of the American Mathematical Society (ISSN 0065-9266 (print); 1947-6221 (online)) is published bimonthly (each volume consisting usually of more than one number) by the American Mathematical Society at 201 Charles Street, Providence, RI 02904-2294 USA. Periodicals postage paid at Providence, RI. Postmaster: Send address changes to Memoirs, American Mathematical Society, 201 Charles Street, Providence, RI 02904-2294 USA.

This publication is indexed in *Mathematical Reviews®*, *Zentralblatt MATH*, *Science Citation Index®*, *Science Citation Index™-Expanded*, *ISI Alerting Services℠*, *SciSearch®*, *Research Alert®*, *CompuMath Citation Index®*, *Current Contents®/Physical, Chemical & Earth Sciences*. This publication is archived in *Portico* and *CLOCKSS*.

Printed in the United States of America.

∞ The paper used in this book is acid-free and falls within the guidelines established to ensure permanence and durability.

Visit the AMS home page at `http://www.ams.org/`

10 9 8 7 6 5 4 3 2 1 21 20 19 18 17 16

Abstract

In this paper we prove the following results (via a unified approach) for all sufficiently large n:

(i) [*1-factorization conjecture*] Suppose that n is even and $D \geq 2\lceil n/4 \rceil - 1$. Then every D-regular graph G on n vertices has a decomposition into perfect matchings. Equivalently, $\chi'(G) = D$.

(ii) [*Hamilton decomposition conjecture*] Suppose that $D \geq \lfloor n/2 \rfloor$. Then every D-regular graph G on n vertices has a decomposition into Hamilton cycles and at most one perfect matching.

(iii) [*Optimal packings of Hamilton cycles*] Suppose that G is a graph on n vertices with minimum degree $\delta \geq n/2$. Then G contains at least $\mathrm{reg}_{\mathrm{even}}(n,\delta)/2 \geq (n-2)/8$ edge-disjoint Hamilton cycles. Here $\mathrm{reg}_{\mathrm{even}}(n,\delta)$ denotes the degree of the largest even-regular spanning subgraph one can guarantee in a graph on n vertices with minimum degree δ.

(i) was first explicitly stated by Chetwynd and Hilton. (ii) and the special case $\delta = \lceil n/2 \rceil$ of (iii) answer questions of Nash-Williams from 1970. All of the above bounds are best possible.

Received by the editor August 13, 2013 and, in revised form, June 13, 2014 and October 20, 2014.

Article electronically published on June 21, 2016.

DOI: http://dx.doi.org/10.1090/memo/1154

2010 *Mathematics Subject Classification.* Primary 05C70, 05C45.

Key words and phrases. 1-factorization, Hamilton cycle, Hamilton decomposition.

The research leading to these results was partially supported by the European Research Council under the European Union's Seventh Framework Programme (FP/2007–2013) / ERC Grant Agreement no. 258345 (B. Csaba, D. Kühn and A. Lo), 306349 (D. Osthus) and 259385 (A. Treglown). The research was also partially supported by the EPSRC, grant no. EP/J008087/1 (D. Kühn and D. Osthus).

CHAPTER 1

Introduction

1.1. Introduction

In this paper we provide a unified approach towards proving three long-standing conjectures for all sufficiently large graphs. Firstly, the 1-factorization conjecture, which can be formulated as an edge-colouring problem; secondly, the Hamilton decomposition conjecture, which provides a far-reaching generalization of Walecki's result [**26**] that every complete graph of odd order has a Hamilton decomposition and thirdly, a best possible result on packing edge-disjoint Hamilton cycles in Dirac graphs. The latter two problems were raised by Nash-Williams [**28**–**30**] in 1970.

1.1.1. The 1-factorization Conjecture. Vizing's theorem states that for any graph G of maximum degree Δ, its edge-chromatic number $\chi'(G)$ is either Δ or $\Delta + 1$. However, the problem of determining the precise value of $\chi'(G)$ for an arbitrary graph G is NP-complete [**12**]. Thus, it is of interest to determine classes of graphs G that attain the (trivial) lower bound Δ – much of the recent book [**34**] is devoted to the subject. For regular graphs G, $\chi'(G) = \Delta(G)$ is equivalent to the existence of a 1-factorization: a 1-*factorization* of a graph G consists of a set of edge-disjoint perfect matchings covering all edges of G. The long-standing 1-factorization conjecture states that every regular graph of sufficiently high degree has a 1-factorization. It was first stated explicitly by Chetwynd and Hilton [**3**, **5**] (who also proved partial results). However, they state that according to Dirac, it was already discussed in the 1950s. Here we prove the conjecture for large graphs.

THEOREM 1.1.1. *There exists an $n_0 \in \mathbb{N}$ such that the following holds. Let $n, D \in \mathbb{N}$ be such that $n \geq n_0$ is even and $D \geq 2\lceil n/4 \rceil - 1$. Then every D-regular graph G on n vertices has a 1-factorization. Equivalently, $\chi'(G) = D$.*

The bound on the minimum degree in Theorem 1.1.1 is best possible. To see this, suppose first that $n = 2 \pmod 4$. Consider the graph which is the disjoint union of two cliques of order $n/2$ (which is odd). If $n = 0 \pmod 4$, consider the graph obtained from the disjoint union of cliques of orders $n/2 - 1$ and $n/2 + 1$ (both odd) by deleting a Hamilton cycle in the larger clique.

Note that Theorem 1.1.1 implies that for every regular graph G on an even number of vertices, either G or its complement has a 1-factorization. Also, Theorem 1.1.1 has an interpretation in terms of scheduling round-robin tournaments (where n players play all of each other in $n - 1$ rounds): one can schedule the first half of the rounds arbitrarily before one needs to plan the remainder of the tournament.

The best previous result towards Theorem 1.1.1 is due to Perkovic and Reed [**32**], who proved an approximate version, i.e. they assumed that $D \geq n/2 + \varepsilon n$. This was generalized by Vaughan [**35**] to multigraphs of bounded multiplicity. Indeed, he proved an approximate form of the following multigraph version of the 1-factorization conjecture which was raised by Plantholt and Tipnis [**33**]: Let G be a regular multigraph of even order n with multiplicity at most r. If the degree of G is at least $rn/2$ then G is 1-factorizable.

In 1986, Chetwynd and Hilton [**4**] made the following 'overfull subgraph' conjecture. Roughly speaking, this says that a dense graph satisfies $\chi'(G) = \Delta(G)$ unless there is a trivial obstruction in the form of a dense subgraph H on an odd number of vertices. Formally, we say that a subgraph H of G is *overfull* if $e(H) > \Delta(G)\lfloor |H|/2 \rfloor$ (note this requires $|H|$ to be odd).

CONJECTURE 1.1.2. *A graph G on n vertices with $\Delta(G) \geq n/3$ satisfies $\chi'(G) = \Delta(G)$ if and only if G contains no overfull subgraph.*

It is easy to see that this generalizes the 1-factorization conjecture (see e.g. [**2**] for the details). The overfull subgraph conjecture is still wide open – partial results are discussed in [**34**], which also discusses further results and questions related to the 1-factorization conjecture.

1.1.2. The Hamilton Decomposition Conjecture. Rather than asking for a 1-factorization, Nash-Williams [**28, 30**] raised the more difficult problem of finding a Hamilton decomposition in an even-regular graph. Here, a *Hamilton decomposition* of a graph G consists of a set of edge-disjoint Hamilton cycles covering all edges of G. A natural extension of this to regular graphs G of odd degree is to ask for a decomposition into Hamilton cycles and one perfect matching (i.e. one perfect matching M in G together with a Hamilton decomposition of $G - M$). The following result solves the problem of Nash-Williams for all large graphs.

THEOREM 1.1.3. *There exists an $n_0 \in \mathbb{N}$ such that the following holds. Let $n, D \in \mathbb{N}$ be such that $n \geq n_0$ and $D \geq \lfloor n/2 \rfloor$. Then every D-regular graph G on n vertices has a decomposition into Hamilton cycles and at most one perfect matching.*

Again, the bound on the degree in Theorem 1.1.3 is best possible. Indeed, Proposition 1.3.1 shows that a smaller degree bound would not even ensure connectivity. Previous results include the following: Nash-Williams [**27**] showed that the degree bound in Theorem 1.1.3 ensures a single Hamilton cycle. Jackson [**13**] showed that one can ensure close to $D/2 - n/6$ edge-disjoint Hamilton cycles. Christofides, Kühn and Osthus [**6**] obtained an approximate decomposition under the assumption that $D \geq n/2 + \varepsilon n$. Under the same assumption, Kühn and Osthus [**22**] obtained an exact decomposition (as a consequence of the main result in [**21**] on Hamilton decompositions of robustly expanding graphs).

Note that Theorem 1.1.3 does not quite imply Theorem 1.1.1, as the degree threshold in the former result is slightly higher.

A natural question is whether one can extend Theorem 1.1.3 to sparser (quasi)-random graphs. Indeed, for random regular graphs of bounded degree this was proved by Kim and Wormald [**16**] and for (quasi-)random regular graphs of linear degree this was proved in [**22**] as a consequence of the main result in [**21**]. However, the intermediate range remains open.

1.1.3. Packing Hamilton Cycles in Graphs of Large Minimum Degree. Although Dirac's theorem is best possible in the sense that the minimum degree condition $\delta \geq n/2$ is best possible, the conclusion can be strengthened considerably: a remarkable result of Nash-Williams [**29**] states that every graph G on n vertices with minimum degree $\delta(G) \geq n/2$ contains $\lfloor 5n/224 \rfloor$ edge-disjoint Hamilton cycles. He raised the question of finding the best possible bound, which we answer in Corollary 1.1.5 below.

We actually answer a more general form of this question: what is the number of edge-disjoint Hamilton cycles one can guarantee in a graph G of minimum degree δ?

A natural upper bound is obtained by considering the largest degree of an even-regular spanning subgraph of G. Let $\mathrm{reg}_{\mathrm{even}}(G)$ be the largest degree of an even-regular spanning subgraph of G. Then let

$$\mathrm{reg}_{\mathrm{even}}(n,\delta) := \min\{\mathrm{reg}_{\mathrm{even}}(G) : |G| = n,\ \delta(G) = \delta\}.$$

Clearly, in general we cannot guarantee more than $\mathrm{reg}_{\mathrm{even}}(n,\delta)/2$ edge-disjoint Hamilton cycles in a graph of order n and minimum degree δ. The next result shows that this bound is best possible (if $\delta < n/2$, then $\mathrm{reg}_{\mathrm{even}}(n,\delta) = 0$).

THEOREM 1.1.4. *There exists an $n_0 \in \mathbb{N}$ such that the following holds. Suppose that G is a graph on $n \geq n_0$ vertices with minimum degree $\delta \geq n/2$. Then G contains at least* $\mathrm{reg}_{\mathrm{even}}(n,\delta)/2$ *edge-disjoint Hamilton cycles.*

The main result of Kühn, Lapinskas and Osthus [**19**] proves Theorem 1.1.4 unless G is close to one of the extremal graphs for Dirac's theorem. This will allow us to restrict our attention to the latter situation (i.e. when G is close to the complete balanced bipartite graph or close to the union of two disjoint copies of a clique).

An approximate version of Theorem 1.1.4 for $\delta \geq n/2 + \varepsilon n$ was obtained earlier by Christofides, Kühn and Osthus [**6**]. Hartke and Seacrest [**11**] gave a simpler argument with improved error bounds.

Precise estimates for $\mathrm{reg}_{\mathrm{even}}(n,\delta)$ (which yield either one or two possible values for any n, δ) are proved in [**6, 10**] using Tutte's theorem: Suppose that $n, \delta \in \mathbb{N}$ and $n/2 \leq \delta < n$. Then the bounds in [**10**] imply that

$$\frac{\delta + \sqrt{n(2\delta - n) + 8}}{2} - \varepsilon \leq \mathrm{reg}_{\mathrm{even}}(n,\delta) \leq \frac{\delta + \sqrt{n(2\delta - n)}}{2} + 1, \tag{1.1.1}$$

where $0 < \varepsilon \leq 2$ is chosen to make the left hand side of (1.1.1) an even integer. Note that (1.1.1) determines $\mathrm{reg}_{\mathrm{even}}(n, n/2)$ exactly (the upper bound in this case was already proved by Katerinis [**15**]). Moreover, (1.1.1) implies that if $\delta \geq n/2$ then $\mathrm{reg}_{\mathrm{even}}(n,\delta) \geq (n-2)/4$. So we obtain the following immediate corollary of Theorem 1.1.4, which answers a question of Nash-Williams [**28–30**].

COROLLARY 1.1.5. *There exists an $n_0 \in \mathbb{N}$ such that the following holds. Suppose that G is a graph on $n \geq n_0$ vertices with minimum degree $\delta \geq n/2$. Then G contains at least $(n-2)/8$ edge-disjoint Hamilton cycles.*

The following construction (which is based on a construction of Babai, see [**28**]) shows that the bound in Corollary 1.1.5 is best possible for $n = 8k+2$, where $k \in \mathbb{N}$. Consider the graph G consisting of one empty vertex class A of size $4k$, one vertex class B of size $4k+2$ containing a perfect matching and no other edges, and all possible edges between A and B. Thus G has order $n = 8k+2$ and minimum degree $4k+1 = n/2$. Any Hamilton cycle in G must contain at least two edges of the perfect matching in B, so G contains at most $\lfloor |B|/4 \rfloor = k = (n-2)/8$ edge-disjoint Hamilton cycles. The lower bound on $\mathrm{reg}_{\mathrm{even}}(n,\delta)$ in (1.1.1) follows from a generalization of this construction.

The following conjecture from [**19**] would be a common generalization of both Theorems 1.1.3 and 1.1.4 (apart from the fact that the degree threshold in Theorem 1.1.3 is slightly lower). It would provide a result which is best possible for every graph G (rather than the class of graphs with minimum degree at least δ).

CONJECTURE 1.1.6. *Suppose that G is a graph on n vertices with minimum degree $\delta(G) \geq n/2$. Then G contains $\mathrm{reg}_{\mathrm{even}}(G)/2$ edge-disjoint Hamilton cycles.*

For $\delta \geq (2-\sqrt{2}+\varepsilon)n$, this conjecture was proved in [**22**], based on the main result of [**21**]. Recently, Ferber, Krivelevich and Sudakov [**7**] were able to obtain an approximate version of Conjecture 1.1.6, i.e. a set of $(1-\varepsilon)\mathrm{reg}_{\mathrm{even}}(G)/2$ edge-disjoint Hamilton cycles under the assumption that $\delta(G) \geq (1+\varepsilon)n/2$. It also makes sense to consider a directed version of Conjecture 1.1.6. Some related questions for digraphs are discussed in [**22**].

It is natural to ask for which other graphs one can obtain similar results. One such instance is the binomial random graph $G_{n,p}$: for any p, asymptotically almost surely it contains $\lfloor \delta(G_{n,p})/2 \rfloor$ edge-disjoint Hamilton cycles, which is clearly optimal. This follows from the main result of Krivelevich and Samotij [**18**] combined with that of Knox, Kühn and Osthus [**17**] (which builds on a number of previous results). The problem of packing edge-disjoint Hamilton cycles in hypergraphs has been considered in [**8**]. Further questions in the area are discussed in the recent survey [**23**].

1.1.4. Overall Structure of the Argument. For all three of our main results, we split the argument according to the structure of the graph G under consideration:

(i) G is close to the complete balanced bipartite graph $K_{n/2,n/2}$;
(ii) G is close to the union of two disjoint copies of a clique $K_{n/2}$;
(iii) G is a 'robust expander'.

Roughly speaking, G is a robust expander if for every set S of vertices, its neighbourhood is at least a little larger than $|S|$, even if we delete a small proportion of the vertices and edges of G. The main result of [**21**] states that every dense regular robust expander has a Hamilton decomposition (see Theorem 1.3.4). This immediately implies Theorems 1.1.1 and 1.1.3 in Case (iii). For Theorem 1.1.4, Case (iii) is proved in [**19**] using a more involved argument, but also based on the main result of [**21**] (see Theorem 1.3.7).

Case (i) is proved in Chapter 4 whilst Chapter 2 tackles Case (ii). We defer the proof of some of the key lemmas needed for Case (ii) until Chapter 3. (These lemmas provide a suitable decomposition of the set of 'exceptional edges' – these include the edges between the two almost complete graphs induced by G.) Case (ii) is by far the hardest case for Theorems 1.1.1 and 1.1.3, as the extremal examples are all close to the union of two cliques. On the other hand, the proof of Theorem 1.1.4 is comparatively simple in this case, as for this result, the extremal construction is close to the complete balanced bipartite graph.

The arguments in Cases (i) and (ii) make use of an 'approximate' decomposition result. We defer the proof of this result until Chapter 5. The arguments for both

(i) and (ii) use the main lemma from [**21**] (the 'robust decomposition lemma') when transforming this approximate decomposition into an exact one.

In Section 1.3, we derive Theorems 1.1.1, 1.1.3 and 1.1.4 from the structural results covering Cases (i)–(iii).

The main proof in [**21**] (but not the proof of the robust decomposition lemma) makes use of Szemerédi's regularity lemma. So due to Case (iii) the bounds on n_0 in our results are very large (of tower type). However, the case of Theorem 1.1.1 when both $\delta \geq n/2$ and (iii) hold was proved by Perkovic and Reed [**32**] using 'elementary' methods, i.e. with a much better bound on n_0. Since the arguments for Cases (i) and (ii) do not rely on the regularity lemma, this means that if we assume that $\delta \geq n/2$, we get much better bounds on n_0 in our 1-factorization result (Theorem 1.1.1).

1.2. Notation

Unless stated otherwise, all the graphs and digraphs considered in this paper are simple and do not contain loops. So in a digraph G, we allow up to two edges between any two vertices, at most one edge in each direction. Given a graph or digraph G, we write $V(G)$ for its vertex set, $E(G)$ for its edge set, $e(G) := |E(G)|$ for the number of edges in G and $|G| := |V(G)|$ for the number of vertices in G. We denote the complement of G by $\overline{G}$.

Suppose that G is an undirected graph. We write $\delta(G)$ for the minimum degree of G, $\Delta(G)$ for its maximum degree and $\chi'(G)$ for the edge-chromatic number of G. Given a vertex v of G, we write $N_G(v)$ for the set of all neighbours of v in G. Given a set $A \subseteq V(G)$, we write $d_G(v, A)$ for the number of neighbours of v in G which lie in A. Given $A, B \subseteq V(G)$, we write $E_G(A)$ for the set of edges of G which have both endvertices in A and $E_G(A, B)$ for the set of edges of G which have one endvertex in A and its other endvertex in B. We also call the edges in $E_G(A, B)$ *AB-edges* of G. We let $e_G(A) := |E_G(A)|$ and $e_G(A, B) := |E_G(A, B)|$. We denote by $G[A]$ the subgraph of G with vertex set A and edge set $E_G(A)$. If $A \cap B = \emptyset$, we denote by $G[A, B]$ the bipartite subgraph of G with vertex classes A and B and edge set $E_G(A, B)$. If $A = B$ we define $G[A, B] := G[A]$. We often omit the index G if the graph G is clear from the context. An *AB-path* in G is a path with one endpoint in A and the other in B. A spanning subgraph H of G is an *r-factor* of G if the degree of every vertex of H is r.

Given a vertex set V and two multigraphs G and H with $V(G), V(H) \subseteq V$, we write $G + H$ for the multigraph whose vertex set is $V(G) \cup V(H)$ and in which the multiplicity of xy in $G + H$ is the sum of the multiplicities of xy in G and in H (for all $x, y \in V(G) \cup V(H)$). Similarly, if $\mathcal{H} := \{H_1, \dots, H_\ell\}$ is a set of graphs, we define $G + \mathcal{H} := G + H_1 + \cdots + H_\ell$. If G and H are simple graphs, we write $G \cup H$ for the (simple) graph whose vertex set is $V(G) \cup V(H)$ and whose edge set is $E(G) \cup E(H)$. We write $G - H$ for the subgraph of G which is obtained from G by deleting all the edges in $E(G) \cap E(H)$. Given $A \subseteq V(G)$, we write $G - A$ for the graph obtained from G by deleting all vertices in A.

We say that a graph or digraph G has a *decomposition* into $H_1, \dots, H_r$ if $G = H_1 + \cdots + H_r$ and the H_i are pairwise edge-disjoint.

A *path system* is a graph Q which is the union of vertex-disjoint paths (some of them might be trivial). We say that P is a *path in Q* if P is a component of Q and, abusing the notation, sometimes write $P \in Q$ for this. A *path sequence* is a digraph which is the union of vertex-disjoint directed paths (some of them might be trivial). We often view a matching M as a graph (in which every vertex has degree precisely one).

If G is a digraph, we write xy for an edge directed from x to y. If $xy \in E(G)$, we say that y is an *outneighbour* of x and x is an *inneighbour* of y. A digraph G is an *oriented graph* if there are no $x, y \in V(G)$ such that $xy, yx \in E(G)$. Unless stated otherwise, when we refer to paths and cycles in digraphs, we mean directed paths and cycles, i.e. the edges on these paths/cycles are oriented consistently. If x is a vertex of a digraph G, then $N_G^+(x)$ denotes the *outneighbourhood* of x, i.e. the set of all those vertices y for which $xy \in E(G)$. Similarly, $N_G^-(x)$ denotes the *inneighbourhood* of x, i.e. the set of all those vertices y for which $yx \in E(G)$. The *outdegree* of x is $d_G^+(x) := |N_G^+(x)|$ and the *indegree* of x is $d_G^-(x) := |N_G^-(x)|$. We write $d_G^+(x, A)$ for the number of outneighbours of x lying inside A and define $d_G^-(x, A)$ similarly. We denote the minimum outdegree of G by $\delta^+(G)$ and the minimum indegree by $\delta^-(G)$. We write $\delta(G)$ and $\Delta(G)$ for the minimum and maximum degrees of the underlying simple undirected graph of G respectively.

Given a digraph G and $A, B \subseteq V(G)$, an *AB-edge* is an edge with initial vertex in A and final vertex in B, and $e_G(A, B)$ denotes the number of these edges in G. If $A \cap B = \emptyset$, we denote by $G[A, B]$ the bipartite subdigraph of G whose vertex classes are A and B and whose edges are all AB-edges of G. By a bipartite digraph $G = G[A, B]$ we mean a digraph which only contains AB-edges. A spanning subdigraph H of G is an *r-factor* of G if the outdegree and the indegree of every vertex of H is r.

If P is a path and $x, y \in V(P)$, we write xPy for the subpath of P whose endvertices are x and y. We define xPy similarly if P is a directed path and x precedes y on P.

Let $V_1, \dots, V_k$ be pairwise disjoint sets of vertices and let $C = V_1 \dots V_k$ be a directed cycle on these sets. We say that an edge xy of a digraph R *winds around* C if there is some i such that $x \in V_i$ and $y \in V_{i+1}$. In particular, we say that R *winds around* C if all edges of R wind around C.

In order to simplify the presentation, we omit floors and ceilings and treat large numbers as integers whenever this does not affect the argument. The constants in the hierarchies used to state our results have to be chosen from right to left. More precisely, if we claim that a result holds whenever $0 < 1/n \ll a \ll b \ll c \leq 1$ (where n is the order of the graph or digraph), then this means that there are non-decreasing functions $f : (0, 1] \to (0, 1]$, $g : (0, 1] \to (0, 1]$ and $h : (0, 1] \to (0, 1]$ such that the result holds for all $0 < a, b, c \leq 1$ and all $n \in \mathbb{N}$ with $b \leq f(c)$, $a \leq g(b)$ and $1/n \leq h(a)$. We will not calculate these functions explicitly. Hierarchies with more constants are defined in a similar way. We will write $a = b \pm c$ as shorthand for $b - c \leq a \leq b + c$.

1.3. Derivation of Theorems 1.1.1, 1.1.3, 1.1.4 from the Main Structural Results

In this section, we combine the main auxiliary results of this paper (together with results from [**22**] and [**19**]) to derive Theorems 1.1.1, 1.1.3 and 1.1.4. Before this, we first show that the bound on the minimum degree in Theorem 1.1.3 is best possible.

PROPOSITION 1.3.1. *For every $n \geq 6$, let $D^* := \lfloor n/2 \rfloor - 1$. Unless both D^* and n are odd, there is a disconnected D^*-regular graph G on n vertices. If both D^* and n are odd, there is a disconnected $(D^* - 1)$-regular graph G on n vertices.*

Note that if both D^* and n are odd, no D^*-regular graph exists.

Proof. If n is even, take G to be the disjoint union of two cliques of order $n/2$. Suppose that n is odd and D^* is even. This implies $n = 3 \pmod 4$. Let G be the graph obtained from the disjoint union of cliques of orders $\lfloor n/2 \rfloor$ and $\lceil n/2 \rceil$ by deleting a perfect matching in the bigger clique. Finally, suppose that n and D^* are both odd. This implies that $n = 1 \pmod 4$. In this case, take G to be the graph obtained from the disjoint union of cliques of orders $\lfloor n/2 \rfloor - 1$ and $\lceil n/2 \rceil + 1$ by deleting a 3-factor in the bigger clique. □

1.3.1. Deriving Theorems 1.1.1 and 1.1.3. As indicated in Section 1.1, in the proofs of our main results we will distinguish the cases when our given graph G is close to the union of two disjoint copies of $K_{n/2}$, close to a complete bipartite graph $K_{n/2,n/2}$ or a robust expander. We will start by defining these concepts.

We say that a graph G on n vertices is *ε-close to the union of two disjoint copies of $K_{n/2}$* if there exists $A \subseteq V(G)$ with $|A| = \lfloor n/2 \rfloor$ and such that $e(A, V(G) \setminus A) \leq \varepsilon n^2$. We say that G is *ε-close to $K_{n/2,n/2}$* if there exists $A \subseteq V(G)$ with $|A| = \lfloor n/2 \rfloor$ and such that $e(A) \leq \varepsilon n^2$. We say that G is *ε-bipartite* if there exists $A \subseteq V(G)$ with $|A| = \lfloor n/2 \rfloor$ such that $e(A), e(V(G) \setminus A) \leq \varepsilon n^2$. So every ε-bipartite graph is ε-close to $K_{n/2,n/2}$. Conversely, if $1/n \ll \varepsilon$ and G is a regular graph on n vertices which ε-close to $K_{n/2,n/2}$, then G is 2ε-bipartite.

Given $0 < \nu \leq \tau < 1$, we say that a graph G on n vertices is a *robust (ν, τ)-expander*, if for all $S \subseteq V(G)$ with $\tau n \leq |S| \leq (1 - \tau)n$ the number of vertices that have at least νn neighbours in S is at least $|S| + \nu n$.

The following observation from [**19**] implies that we can split the proofs of Theorems 1.1.1 and 1.1.3 into three cases.

LEMMA 1.3.2. *Suppose that $0 < 1/n \ll \kappa \ll \nu \ll \tau, \varepsilon < 1$. Let G be a graph on n vertices of minimum degree $\delta := \delta(G) \geq (1/2 - \kappa)n$. Then G satisfies one of the following properties:*

(i) *G is ε-close to $K_{n/2,n/2}$;*
(ii) *G is ε-close to the union of two disjoint copies of $K_{n/2}$;*
(iii) *G is a robust (ν, τ)-expander.*

Recall that in Chapter 2 we prove Theorems 1.1.1 and 1.1.3 in Case (ii) when our given graph G is ε-close to the union of two disjoint copies of $K_{n/2}$. The following result is sufficiently general to imply both Theorems 1.1.1 and 1.1.3 in this case. We will prove it in Section 2.10.

THEOREM 1.3.3. *For every $\varepsilon_{\rm ex} > 0$ there exists an $n_0 \in \mathbb{N}$ such that the following holds for all $n \geq n_0$. Suppose that $D \geq n - 2\lfloor n/4 \rfloor - 1$ and that G is a D-regular graph on n vertices which is $\varepsilon_{\rm ex}$-close to the union of two disjoint copies of $K_{n/2}$. Let F be the size of a minimum cut in G. Then G can be decomposed into $\lfloor \min\{D, F\}/2 \rfloor$ Hamilton cycles and $D - 2\lfloor \min\{D, F\}/2 \rfloor$ perfect matchings.*

Note that Theorem 1.3.3 provides structural insight into the extremal graphs for Theorem 1.1.3 – they are those with a cut of size less than D.

Throughout this paper, we will use the following fact.

$$n - 2\lfloor n/4 \rfloor - 1 = \begin{cases} n/2 - 1 & \text{if } n = 0 \pmod 4, \\ (n-1)/2 & \text{if } n = 1 \pmod 4, \\ n/2 & \text{if } n = 2 \pmod 4, \\ (n+1)/2 & \text{if } n = 3 \pmod 4. \end{cases} \tag{1.3.1}$$

The next result from **[22]** (derived from the main result of **[21]**) shows that every even-regular robust expander of linear degree has a Hamilton decomposition. It will be used to prove Theorems 1.1.1 and 1.1.3 in the case when our given graph G is a robust expander.

THEOREM 1.3.4. *For every $\alpha > 0$ there exists $\tau > 0$ such that for every $\nu > 0$ there exists $n_0 = n_0(\alpha, \nu, \tau)$ for which the following holds. Suppose that*

(i) *G is an r-regular graph on $n \geq n_0$ vertices, where $r \geq \alpha n$ is even;*
(ii) *G is a robust (ν, τ)-expander.*

Then G has a Hamilton decomposition.

The following result implies Theorems 1.1.1 and 1.1.3 in the case when our given graph is ε-close to $K_{n/2,n/2}$. Note that unlike the case when G is ε-close to the union of two disjoint copies of $K_{n/2}$, we have room to spare in the lower bound on D.

THEOREM 1.3.5. *There are $\varepsilon_{\rm ex} > 0$ and $n_0 \in \mathbb{N}$ such that the following holds. Let $n \geq n_0$ and suppose that $D \geq (1/2 - \varepsilon_{\rm ex})n$ is even. Suppose that G is a D-regular graph on n vertices which is $\varepsilon_{\rm ex}$-bipartite. Then G has a Hamilton decomposition.*

Theorem 1.3.5 is one of the two main results proven in Chapter 4. The following result is an easy consequence of Tutte's theorem and gives the degree threshold for a single perfect matching in a regular graph. Note the condition on D is the same as in Theorem 1.1.1.

PROPOSITION 1.3.6. *Suppose that $D \geq 2\lceil n/4 \rceil - 1$ and n is even. Then every D-regular graph G on n vertices has a perfect matching.*

Proof. If $D \geq n/2$ then G has a Hamilton cycle (and thus a perfect matching) by Dirac's theorem. So we may assume that $D = n/2 - 1$ and so $n = 0 \pmod 4$. In this case, we will use Tutte's theorem which states that a graph G has a perfect matching if for every set $S \subseteq V(G)$ the graph $G - S$ has at most $|S|$ odd components (i.e. components on an odd number of vertices). The latter condition holds if $|S| \leq 1$ and if $|S| \geq n/2$.

If $|S| = n/2 - 1$ and $G - S$ has more than $|S|$ odd components, then $G - S$ consists of isolated vertices. But this implies that each vertex outside S is joined to all vertices in S, contradicting the $(n/2 - 1)$-regularity of G.

If $2 \leq |S| \leq n/2 - 2$, then every component of $G - S$ has at least $n/2 - |S|$ vertices and so $G - S$ has at most $\lfloor (n - |S|)/(n/2 - |S|) \rfloor$ components. But $\lfloor (n - |S|)/(n/2 - |S|) \rfloor \leq |S|$ unless $n = 8$ and $|S| = 2$. (Indeed, note that $(n - |S|)/(n/2 - |S|) \leq |S|$ if and only if $n + |S|^2 - (n/2 + 1)|S| \leq 0$. The latter holds for $|S| = 3$ and $|S| = n/2 - 2$, and so for all values in between. The case $|S| = 2$ can be checked separately.) If $n = 8$ and $|S| = 2$, it is easy to see that $G - S$ has at most two odd components.

□

Proof of Theorem 1.1.1. Let $\tau = \tau(1/3)$ be the constant returned by Theorem 1.3.4 for $\alpha := 1/3$. Choose $n_0 \in \mathbb{N}$ and constants $\nu, \varepsilon_{\rm ex}$ such that $1/n_0 \ll \nu \ll \tau, \varepsilon_{\rm ex}$ and $\varepsilon_{\rm ex} \ll 1$. Let $n \geq n_0$ and let G be a D-regular graph as in Theorem 1.1.1. Lemma 1.3.2 implies that G satisfies one of the following properties:

(i) G is $\varepsilon_{\rm ex}$-close to $K_{n/2,n/2}$;
(ii) G is $\varepsilon_{\rm ex}$-close to the union of two disjoint copies of $K_{n/2}$;
(iii) G is a robust (ν, τ)-expander.

If (i) holds and D is even, then as observed at the beginning of this subsection, this implies that G is $2\varepsilon_{\rm ex}$-bipartite. So Theorem 1.3.5 implies that G has a Hamilton decomposition and thus also a 1-factorization (as n is even and so every Hamilton cycle can be decomposed into two perfect matchings). Suppose that (i) holds and D is odd. Then Proposition 1.3.6 implies that G contains a perfect matching M. Now $G - M$ is still $\varepsilon_{\rm ex}$-close to $K_{n/2,n/2}$ and so Theorem 1.3.5 implies that $G - M$ has a Hamilton decomposition. Thus G has a 1-factorization. If (ii) holds, then Theorem 1.3.3 and (1.3.1) imply that G has a 1-factorization. If (iii) holds and D is odd, we use Proposition 1.3.6 to choose a perfect matching M in G and let $G' := G - M$. If D is even, let $G' := G$. In both cases, $G' - M$ is still a robust $(\nu/2, \tau)$-expander. So Theorem 1.3.4 gives a Hamilton decomposition of G'. So G has a 1-factorization. □

The proof of Theorem 1.1.3 is similar to that of Theorem 1.1.1.

Proof of Theorem 1.1.3. Choose $n_0 \in \mathbb{N}$ and constants $\tau, \nu, \varepsilon_{\rm ex}$ as in the proof of Theorem 1.1.1. Let $n \geq n_0$ and let G be a D-regular graph as in Theorem 1.1.3. As before, Lemma 1.3.2 implies that G satisfies one of (i)–(iii). Suppose first that (i) holds. If D is odd, n must be even and so $D \geq n/2$. Choose a perfect matching M in G (e.g. by applying Dirac's theorem) and let $G' := G - M$. If D is even, let $G' := G$. Note that in both cases G' is $\varepsilon_{\rm ex}$-close to $K_{n/2,n/2}$ and so $2\varepsilon_{\rm ex}$-bipartite. Thus Theorem 1.3.5 implies that G' has a Hamilton decomposition.

Suppose next that (ii) holds. Note that by (1.3.1), $D \geq n - 2\lfloor n/4 \rfloor - 1$ unless $n = 3 \pmod 4$ and $D = \lfloor n/2 \rfloor$. But the latter would mean that both n and D are odd, which is impossible. So the conditions of Theorem 1.3.3 are satisfied. Moreover, since $D \geq \lfloor n/2 \rfloor$, Proposition 2.2.1(ii) implies that the size of a minimum cut in G is at least D. Thus Theorem 1.3.3 implies that G has a decomposition into Hamilton cycles and at most one perfect matching.

Finally, suppose that (iii) holds. If D is odd (and thus n is even), we can apply Proposition 1.3.6 again to find a perfect matching M in G and let $G' := G - M$. If D is even, let $G' := G$. In both cases, G' is still a robust $(\nu/2, \tau)$-expander. So Theorem 1.3.4 gives a Hamilton decomposition of G'. □

1.3.2. Deriving Theorem 1.1.4. The derivation of Theorem 1.1.4 is similar to that of the previous two results. We will replace the use of Lemma 1.3.2 and Theorem 1.3.4 with the following result, which is an immediate consequence of the two main results in [**19**].

THEOREM 1.3.7. *For every $\varepsilon_{\rm ex} > 0$ there exists an $n_0 \in \mathbb{N}$ such that the following holds. Suppose that G is a graph on $n \geq n_0$ vertices with $\delta(G) \geq n/2$. Then G satisfies one of the following properties:*

(i) *G is $\varepsilon_{\rm ex}$-close to $K_{n/2,n/2}$;*
(ii) *G is $\varepsilon_{\rm ex}$-close to the union of two disjoint copies of $K_{n/2}$;*
(iii) *G contains $\mathrm{reg}_{\rm even}(n,\delta)/2$ edge-disjoint Hamilton cycles.*

To deal with the near-bipartite case (i), we will apply the following result which we prove in Chapter 4.

THEOREM 1.3.8. *For each $\alpha > 0$ there are $\varepsilon_{\rm ex} > 0$ and $n_0 \in \mathbb{N}$ such that the following holds. Suppose that F is an $\varepsilon_{\rm ex}$-bipartite graph on $n \geq n_0$ vertices with $\delta(F) \geq (1/2 - \varepsilon_{\rm ex})n$. Suppose that F has a D-regular spanning subgraph G such that $n/100 \leq D \leq (1/2 - \alpha)n$ and D is even. Then F contains $D/2$ edge-disjoint Hamilton cycles.*

The next result immediately implies Theorem 1.1.4 in Case (ii) when G is ε-close to the union of two disjoint copies of $K_{n/2}$. We will prove it in Chapter 2 (Section 2.5). Since G is far from extremal in this case, we obtain almost twice as many edge-disjoint Hamilton cycles as needed for Theorem 1.1.4.

THEOREM 1.3.9. *For every $\varepsilon > 0$, there exist $\varepsilon_{\rm ex} > 0$ and $n_0 \in \mathbb{N}$ such that the following holds. Suppose $n \geq n_0$ and G is a graph on n vertices such that G is $\varepsilon_{\rm ex}$-close to the union of two disjoint copies of $K_{n/2}$ and such that $\delta(G) \geq n/2$. Then G has at least $(1/4 - \varepsilon)n$ edge-disjoint Hamilton cycles.*

We will also use the following well-known result of Petersen.

THEOREM 1.3.10. *Every regular graph of positive even degree contains a 2-factor.*

Proof of Theorem 1.1.4. Choose $n_0 \in \mathbb{N}$ and $\varepsilon_{\rm ex}$ such that $1/n_0 \ll \varepsilon_{\rm ex} \ll 1$. In particular, we choose $\varepsilon_{\rm ex} \leq \varepsilon^1_{\rm ex}(1/12)$, where $\varepsilon^1_{\rm ex}(1/12)$ is the constant returned by Theorem 1.3.9 for $\varepsilon := 1/12$, as well as $\varepsilon_{\rm ex} \leq \varepsilon^2_{\rm ex}(1/6)/2$, where $\varepsilon^2_{\rm ex}(1/6)$ is the constant returned by Theorem 1.3.8 for $\alpha := 1/6$. Let G be a graph on $n \geq n_0$ vertices with $\delta := \delta(G) \geq n/2$. Theorem 1.3.7 implies that we may assume that G satisfies either (i) or (ii). Note that in both cases it follows that $\delta(G) \leq (1/2+5\varepsilon_{\rm ex})n$. So (1.1.1) implies that $n/5 \leq \mathrm{reg}_{\rm even}(n,\delta) \leq 3n/10$.

Suppose first that (i) holds. As mentioned above, this implies that G is $2\varepsilon_{\rm ex}$-bipartite. Let G' be a D-regular spanning subgraph of G such that D is even and $D \geq \mathrm{reg}_{\rm even}(n,\delta)$. Petersen's theorem (Theorem 1.3.10) implies that by successively deleting 2-factors of G', if necessary, we may in addition assume that $D \leq n/3$. Then Theorem 1.3.8 (applied with $\alpha := 1/6$) implies that G contains at least $D/2 \geq \mathrm{reg}_{\rm even}(n,\delta)/2$ edge-disjoint Hamilton cycles.

Finally suppose that (ii) holds. Then Theorem 1.3.9 (applied with $\varepsilon := 1/12$) implies that G contains $n/6 \geq \mathrm{reg}_{\rm even}(n,\delta)/2$ edge-disjoint Hamilton cycles. □

1.4. Tools

1.4.1. ε-regularity. If $G = (A, B)$ is an undirected bipartite graph with vertex classes A and B, then the *density* of G is defined as

$$d(A,B) := \frac{e_G(A,B)}{|A||B|}.$$

For any $\varepsilon > 0$, we say that G is *ε-regular* if for any $A' \subseteq A$ and $B' \subseteq B$ with $|A'| \geq \varepsilon|A|$ and $|B'| \geq \varepsilon|B|$ we have $|d(A', B') - d(A, B)| < \varepsilon$. We say that G is *$(\varepsilon, \geq d)$-regular* if it is ε-regular and has density d' for some $d' \geq d - \varepsilon$.

We say that G is *$[\varepsilon, d]$-superregular* if it is ε-regular and $d_G(a) = (d \pm \varepsilon)|B|$ for every $a \in A$ and $d_G(b) = (d \pm \varepsilon)|A|$ for every $b \in B$. G is *$[\varepsilon, \geq d]$-superregular* if it is *$[\varepsilon, d']$-superregular* for some $d' \geq d$.

Given disjoint vertex sets X and Y in a digraph G, recall that $G[X, Y]$ denotes the bipartite subdigraph of G whose vertex classes are X and Y and whose edges are all the edges of G directed from X to Y. We often view $G[X, Y]$ as an undirected bipartite graph. In particular, we say $G[X, Y]$ is *ε-regular*, *$(\varepsilon, \geq d)$-regular*, $[\varepsilon, d]$-superregular or $[\varepsilon, \geq d]$-superregular if this holds when $G[X, Y]$ is viewed as an undirected graph.

The following proposition states that the graph obtained from a superregular pair by removing a small number of edges at every vertex is still superregular (with slightly worse parameters). We omit the proof which follows straightforwardly from the definition of superregularity. A similar argument is for example included in [**21**].

PROPOSITION 1.4.1. *Suppose that $0 < 1/m \ll \varepsilon \leq d' \ll d \leq 1$. Let G be a bipartite graph with vertex classes A and B of size m. Suppose that G' is obtained from G by removing at most $d'm$ vertices from each vertex class and at most $d'm$ edges incident to each vertex from G. If G is $[\varepsilon, d]$-superregular then G' is $[2\sqrt{d'}, d]$-superregular.*

We will also use the following well-known observation, which easily follows from Hall's theorem and the definition of $[\varepsilon, d]$-superregularity.

PROPOSITION 1.4.2. *Suppose that $0 < 1/m \ll \varepsilon \ll d \leq 1$. Suppose that G is an $[\varepsilon, d]$-superregular bipartite graph with vertex classes of size m. Then G contains a perfect matching.*

We will also apply the following simple fact.

FACT 1.4.3. *Let $\varepsilon > 0$. Suppose that G is a bipartite graph with vertex classes of size n such that $\delta(G) \geq (1 - \varepsilon)n$. Then G is $[\sqrt{\varepsilon}, 1]$-superregular.*

1.4.2. A Chernoff-Hoeffding Bound. We will often use the following Chernoff-Hoeffding bound for binomial and hypergeometric distributions (see e.g. [**14**, Corollary 2.3 and Theorem 2.10]). Recall that the binomial random variable with parameters (n, p) is the sum of n independent Bernoulli variables, each taking value 1 with probability p or 0 with probability $1-p$. The hypergeometric random variable X with parameters (n, m, k) is defined as follows. We let N be a set of size n, fix $S \subseteq N$ of size $|S| = m$, pick a uniformly random $T \subseteq N$ of size $|T| = k$, then define $X := |T \cap S|$. Note that $\mathbb{E}X = km/n$.

PROPOSITION 1.4.4. *Suppose X has binomial or hypergeometric distribution and $0 < a < 3/2$. Then $\mathbb{P}(|X - \mathbb{E}X| \geq a\mathbb{E}X) \leq 2e^{-a^2\mathbb{E}X/3}$.*

1.4.3. Other Useful Results. We will need the following fact, which is a simple consequence of Vizing's theorem and was first observed by McDiarmid and independently by de Werra (see e.g. [**37**]).

PROPOSITION 1.4.5. *Let G be a graph with $\chi'(G) \leq m$. Then G has a decomposition into m matchings $M_1, \ldots, M_m$ with $|e(M_i) - e(M_j)| \leq 1$ for all $i, j \leq m$.*

It is also useful to state Proposition 1.4.5 in the following alternative form.

COROLLARY 1.4.6. *Let H be a graph with maximum degree at most Δ. Then $E(H)$ can be decomposed into $\Delta+1$ edge-disjoint matchings $M_1,\dots,M_{\Delta+1}$ such that $|e(M_i)-e(M_j)|\le 1$ for all $i,j\le \Delta+1$.*

The following partition result will also be useful.

LEMMA 1.4.7. *Suppose that $0<1/n\ll \varepsilon,\varepsilon_1\ll\varepsilon_2\ll 1/K\ll 1$, that $r\le 2K$, that $Km\ge n/4$ and that $r,K,n,m\in\mathbb{N}$. Let G and F be graphs on n vertices with $V(G)=V(F)$. Suppose that there is a vertex partition of $V(G)$ into $U,R_1,\dots,R_r$ with the following properties:*

- *$|U|=Km$.*
- *$\delta(G[U])\ge\varepsilon n$ or $\Delta(G[U])\le\varepsilon n$.*
- *For each $j\le r$ we either have $d_G(u,R_j)\le\varepsilon n$ for all $u\in U$ or $d_G(x,U)\ge\varepsilon n$ for all $x\in R_j$.*

Then there exists a partition of U into K parts $U_1,\dots,U_K$ satisfying the following properties:

(i) *$|U_i|=m$ for all $i\le K$.*
(ii) *$d_G(v,U_i)=(d_G(v,U)\pm\varepsilon_1 n)/K$ for all $v\in V(G)$ and all $i\le K$.*
(iii) *$e_G(U_i,U_{i'})=2(e_G(U)\pm\varepsilon_2\max\{n,e_G(U)\})/K^2$ for all $1\le i\ne i'\le K$.*
(iv) *$e_G(U_i)=(e_G(U)\pm\varepsilon_2\max\{n,e_G(U)\})/K^2$ for all $i\le K$.*
(v) *$e_G(U_i,R_j)=(e_G(U,R_j)\pm\varepsilon_2\max\{n,e_G(U,R_j)\})/K$ for all $i\le K$ and $j\le r$.*
(vi) *$d_F(v,U_i)=(d_F(v,U)\pm\varepsilon_1 n)/K$ for all $v\in V(F)$ and all $i\le K$.*

Proof. Consider an equipartition $U_1,\dots,U_K$ of U which is chosen uniformly at random. So (i) holds by definition. Note that for a given vertex $v\in V(G)$, $d_G(v,U_i)$ has the hypergeometric distribution with mean $d_G(v,U)/K$. So if $d_G(v,U)\ge\varepsilon_1 n/K$, Proposition 1.4.4 implies that

$$\mathbb{P}\left(\left|d_G(v,U_i)-\frac{d_G(v,U)}{K}\right|\ge\frac{\varepsilon_1 d_G(v,U)}{K}\right)\le 2\exp\left(-\frac{\varepsilon_1^2 d_G(v,U)}{3K}\right)\le\frac{1}{n^2}.$$

Thus we deduce that for all $v\in V(G)$ and all $i\le K$,

$$\mathbb{P}\left(|d_G(v,U_i)-d_G(v,U)/K|\ge\varepsilon_1 n/K\right)\le 1/n^2.$$

Similarly,

$$\mathbb{P}\left(|d_F(v,U_i)-d_F(v,U)/K|\ge\varepsilon_1 n/K\right)\le 1/n^2.$$

So with probability at least 3/4, both (ii) and (vi) are satisfied.

We now consider (iii) and (iv). Fix $i,i'\le K$. If $i\ne i'$, let $X:=e_G(U_i,U_{i'})$. If $i=i'$, let $X:=2e_G(U_i)$. For an edge $f\in E(G[U])$, let E_f denote the event that $f\in E(U_i,U_{i'})$. So if $f=xy$ and $i\ne i'$, then

$$\mathbb{P}(E_f)=2\mathbb{P}(x\in U_i)\mathbb{P}(y\in U_{i'}\mid x\in U_i)=2\frac{m}{|U|}\cdot\frac{m}{|U|-1}. \tag{1.4.1}$$

Similarly, if f and f' are disjoint (that is, f and f' have no common endpoint) and $i\ne i'$, then

$$\mathbb{P}(E_{f'}\mid E_f)=2\frac{m-1}{|U|-2}\cdot\frac{m-1}{|U|-3}\le 2\frac{m}{|U|}\cdot\frac{m}{|U|-1}=\mathbb{P}(E_{f'}). \tag{1.4.2}$$

By (1.4.1), if $i \neq i'$, we also have

$$\mathbb{E}(X) = 2\frac{e_G(U)}{K^2} \cdot \frac{|U|}{|U|-1} = \left(1 \pm \frac{2}{|U|}\right)\frac{2e_G(U)}{K^2} = (1 \pm \varepsilon_2/4)\,\frac{2e_G(U)}{K^2}. \tag{1.4.3}$$

If $f = xy$ and $i = i'$, then

$$\mathbb{P}(E_f) = \mathbb{P}(x \in U_i)\mathbb{P}(y \in U_i \mid x \in U_i) = \frac{m}{|U|} \cdot \frac{m-1}{|U|-1}. \tag{1.4.4}$$

So if $i = i'$, similarly to (1.4.2) we also obtain $\mathbb{P}(E_{f'} \mid E_f) \leq \mathbb{P}(E_f)$ for disjoint f and f' and we obtain the same bound as in (1.4.3) on $\mathbb{E}(X)$ (recall that $X = 2e_G(U_i)$ in this case).

Note that if $i \neq i'$ then

$$\begin{aligned}
\mathrm{Var}(X) &= \sum_{f \in E(U)} \sum_{f' \in E(U)} (\mathbb{P}(E_f \cap E_{f'}) - \mathbb{P}(E_f)\mathbb{P}(E_{f'})) \\
&= \sum_{f \in E(U)} \mathbb{P}(E_f) \sum_{f' \in E(U)} (\mathbb{P}(E_{f'} \mid E_f) - \mathbb{P}(E_{f'})) \\
&\stackrel{(1.4.2)}{\leq} \sum_{f \in E(U)} \mathbb{P}(E_f) \cdot 2\Delta(G[U]) \stackrel{(1.4.3)}{\leq} \frac{3e_G(U)}{K^2} \cdot 2\Delta(G[U]) \\
&\leq e_G(U)\Delta(G[U]).
\end{aligned}$$

Similarly, if $i = i'$ then

$$\mathrm{Var}(X) = 4 \sum_{f \in E(U)} \sum_{f' \in E(U)} (\mathbb{P}(E_f \cap E_{f'}) - \mathbb{P}(E_f)\mathbb{P}(E_{f'})) \leq e_G(U)\Delta(G[U]).$$

Let $a := e_G(U)\Delta(G[U])$. In both cases, from Chebyshev's inequality, it follows that

$$\mathbb{P}\left(|X - \mathbb{E}(X)| \geq \sqrt{a/\varepsilon^{1/2}}\right) \leq \varepsilon^{1/2}.$$

Suppose that $\Delta(G[U]) \leq \varepsilon n$. If we also have have $e_G(U) \leq n$, then $\sqrt{a/\varepsilon^{1/2}} \leq \varepsilon^{1/4} n \leq \varepsilon_2 n/2K^2$. If $e_G(U) \geq n$, then $\sqrt{a/\varepsilon^{1/2}} \leq \varepsilon^{1/4} e_G(U) \leq \varepsilon_2 e_G(U)/2K^2$.

If we do not have $\Delta(G[U]) \leq \varepsilon n$, then our assumptions imply that $\delta(G[U]) \geq \varepsilon n$. So $\Delta(G[U]) \leq n \leq \varepsilon e_G(G[U])$ with room to spare. This in turn means that $\sqrt{a/\varepsilon^{1/2}} \leq \varepsilon^{1/4} e_G(U) \leq \varepsilon_2 e_G(U)/2K^2$. So in all cases, we have

$$\mathbb{P}\left(|X - \mathbb{E}(X)| \geq \frac{\varepsilon_2 \max\{n, e_G(U)\}}{2K^2}\right) \leq \varepsilon^{1/2}. \tag{1.4.5}$$

Now note that by (1.4.3) we have

$$\left|\mathbb{E}(X) - \frac{2e_G(U)}{K^2}\right| \leq \frac{\varepsilon_2 e_G(U)}{2K^2}. \tag{1.4.6}$$

So (1.4.5) and (1.4.6) together imply that for fixed i, i' the bound in (iii) fails with probability at most $\varepsilon^{1/2}$. The analogue holds for the bound in (iv). By summing over all possible values of $i, i' \leq K$, we have that (iii) and (iv) hold with probability at least $3/4$.

A similar argument shows that for all $i \leq K$ and $j \leq r$, we have

$$\mathbb{P}\left(\left|e_G(U_i, R_j) - \frac{e_G(U, R_j)}{K}\right| \geq \frac{\varepsilon_2 \max\{n, e_G(U, R_j)\}}{K}\right) \leq \varepsilon^{1/2}. \tag{1.4.7}$$

Indeed, fix $i \leq K$, $j \leq r$ and let $X := e_G(U_i, R_j)$. For an edge $f \in G[U, R_j]$, let E_f denote the event that $f \in E(U_i, R_j)$. Then $\mathbb{P}(E_f) = m/|U| = 1/K$ and so $\mathbb{E}(X) = e_G(U, R_j)/K$. The remainder of the argument proceeds as in the previous case (with slightly simpler calculations).

So (v) holds with probability at least 3/4, by summing over all possible values of $i \leq K$ and $j \leq r$ again. So with positive probability, the partition satisfies all requirements. $\square$

CHAPTER 2

The two cliques case

This chapter is concerned with proving Theorems 1.1.1, 1.1.3 and 1.1.4 in the case when our graph is close to the union of two disjoint copies of a clique $K_{n/2}$ (Case (ii)). More precisely, we prove Theorem 1.3.9 (i.e. Case (ii) of Theorem 1.1.4) and Theorem 1.3.3, which is a common generalization of Case (ii) of Theorems 1.1.1 and 1.1.3. In Section 2.1, we give a sketch of the arguments for the 'two cliques' Case (ii) (i.e. the proofs of Theorems 1.3.3 and 1.3.9). Sections 2.2–2.4 (and part of Section 2.5) are common to the proofs of both Theorems 1.3.3 and 1.3.9. Theorem 1.3.9 is proved in Section 2.5. All the subsequent sections of this chapter are devoted to the proof of Theorem 1.3.3.

In this chapter (and Chapter 3) it is convenient to view matchings as graphs (in which every vertex has degree precisely one).

2.1. Overview of the Proofs of Theorems 1.3.3 and 1.3.9

The proof of Theorem 1.3.9 is much simpler than that of Theorems 1.3.3 (mainly because its assertion leaves some leeway – one could probably find a slightly larger set of edge-disjoint Hamilton cycles than guaranteed by Theorem 1.3.9). Moreover, the ideas used in the former all appear in the proof of the latter too.

2.1.1. Proof Overview for Theorem 1.3.9. Let G be a graph on n vertices with $\delta(G) \geq n/2$ which is close to being the union of two disjoint cliques. So there is a vertex partition of G into sets A and B of roughly equal size so that $G[A]$ and $G[B]$ are almost complete. Our aim is to construct almost $n/4$ edge-disjoint Hamilton cycles.

Several techniques have recently been developed which yield approximate decompositions of dense (almost) regular graphs, i.e. a set of Hamilton cycles covering almost all the edges (see e.g. [**6, 7, 9, 24, 31**]). This leads to the following idea: replace $G[A]$ and $G[B]$ by multigraphs G_A and G_B so that any suitable pair of Hamilton cycles C_A and C_B of G_A and G_B respectively corresponds to a single Hamilton cycle C in the original graph G. We will construct G_A and G_B by deleting some edges of G and introducing some 'fictive edges'. (The introduction of these fictive edges is the reason why G_A and G_B are multigraphs.)

We next explain the key concept of these 'fictive edges'. The following graph G provides an instructive example: suppose that $n = 0 \pmod 4$. Let G be obtained from two disjoint cliques induced by sets A and B of size $n/2$ by adding a perfect matching M between A and B. Note that G is $n/2$-regular. Now pair up the edges of M into $n/4$ pairs (e_i, e_{i+1}) for $i = 1, 3, \dots, n/2 - 1$. Write $e_i =: x_i y_i$ with $x_i \in A$ and $y_i \in B$. Next let G_A be the multigraph obtained from $G[A]$ by adding all the edges $x_i x_{i+1}$, where i is odd. Similarly, let G_B be obtained from $G[B]$ by adding all the edges $y_i y_{i+1}$, where i is odd. We call the edges $x_i x_{i+1}$ and $y_i y_{i+1}$ fictive edges.

Note that G_A and G_B are regular multigraphs. Now pair off the fictive edges in G_A with those in G_B, i.e. x_ix_{i+1} is paired off with y_iy_{i+1}. Suppose that C_A is a Hamilton cycle in G_A which contains x_ix_{i+1} (and no other fictive edges) and C_B is a Hamilton cycle in G_B which contains y_iy_{i+1} (and no other fictive edges). Then together, C_A and C_B correspond to a Hamilton cycle C in the original graph G (where fictive edges are replaced by the corresponding matching edges in M again).

So we have reduced the problem of finding many edge-disjoint Hamilton cycles in G to that of finding many edge-disjoint Hamilton cycles in the almost complete graph G_A (and G_B), with the additional requirement that each such Hamilton cycle contains a unique fictive edge. This can be achieved via the 'approximate decomposition result' (see Lemma 2.5.4 which is proved in Chapter 5).

Additional difficulties arise from 'exceptional' vertices, namely those which have high degree into both A and B. (It is easy to see that there cannot be too many of these vertices.) Fictive edges also provide a natural way of 'eliminating' these exceptional vertices. Suppose for example that G' is obtained from the graph G above by adding a vertex a so that a is adjacent to half of the vertices in A and half of the vertices in B. (Note that $\delta(G')$ is a little smaller than $|G'|/2$, but G' is similar to graphs actually occurring in the proof.) Then we can pair off the neighbours of a into pairs within A and introduce a fictive edge f_i between each pair of neighbours. We also introduce fictive edges f_i between pairs of neighbours of a in B. Without loss of generality, we have fictive edges $f_1, f_3, \dots, f_{n/2-1}$ (and recall that $|G'| = n+1$). So we have $V(G'_A) = A$ and $V(G'_B) = B$ again. We then require each pair of Hamilton cycles C_A, C_B of G'_A and G'_B to contain x_ix_{i+1}, y_iy_{i+1} and a fictive edge f_i (which may lie in A or B) where i is odd, see Figure 2.1.1. Then C_A and C_B together correspond to a Hamilton cycle C in G' again. The subgraph J of G' which corresponds to three such fictive edges x_ix_{i+1}, y_iy_{i+1} and f_i of C is called a 'Hamilton exceptional system'. J will always be a path system. So in general, we will first find a sufficient number of edge-disjoint Hamilton exceptional systems J. Then we apply Lemma 2.5.4 to find edge-disjoint Hamilton cycles in G'_A and G'_B, where each pair of cycles contains a suitable set J^* of fictive edges (corresponding to some Hamilton exceptional system J).

For Lemma 2.5.4, we need each of the Hamilton exceptional systems J to be 'localized': given a partition of A and B into clusters, the endpoints of the corresponding set J^* of fictive edges need to be contained in a single cluster of A and of B. The fact that the Hamilton exceptional systems need to be localized is one reason for treating exceptional vertices differently from the others by introducing fictive edges for them.

2.1.2. Proof Overview for Theorem 1.3.3. The main result of this chapter is Theorem 1.3.3. Suppose that G is a D-regular graph satisfying the conditions of that theorem.

Using the approach of the previous subsection, one can obtain an approximate decomposition of G, i.e. a set of edge-disjoint Hamilton cycles covering almost all edges of G. However, one does not have any control over the 'leftover' graph H, which makes a complete decomposition seem infeasible. This problem was overcome in [**21**] by introducing the concept of a 'robustly decomposable graph' G^{rob}. Roughly speaking, this is a sparse regular graph with the following property: given *any* very sparse regular graph H with $V(H) = V(G^{\mathrm{rob}})$ which is edge-disjoint from

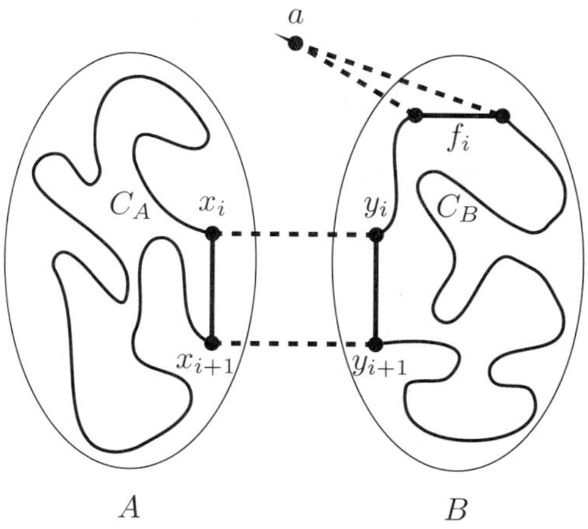

FIGURE 2.1.1. Transforming the problem of finding a Hamilton cycle on $V(G')$ into finding two Hamilton cycles C_A and C_B on A and B respectively.

G^{rob}, one can guarantee that $G^{\mathrm{rob}} \cup H$ has a Hamilton decomposition. This leads to a natural (and very general) strategy to obtain a decomposition of G:

(1) find a (sparse) robustly decomposable graph G^{rob} in G and let G' denote the leftover;
(2) find an approximate Hamilton decomposition of G' and let H denote the (very sparse) leftover;
(3) find a Hamilton decomposition of $G^{\mathrm{rob}} \cup H$.

It is of course far from clear that one can always find such a graph G^{rob}. The main 'robust decomposition lemma' of [**21**] guarantees such a graph G^{rob} in any regular robustly expanding graph of linear degree. Since G is close to the disjoint union of two cliques, we are of course not in this situation. However, a regular almost complete graph is certainly a robust expander, i.e. our assumptions imply that G is close to being the disjoint union of two regular robustly expanding graphs G_A and G_B, with vertex sets A and B.

So very roughly, the strategy is to apply the robust decomposition lemma of [**21**] to G_A and G_B separately, to obtain a Hamilton decomposition of both G_A and G_B. Now we pair up Hamilton cycles of G_A and G_B in this decomposition, so that each such pair corresponds to a single Hamilton cycle of G and so that all edges of G are covered. It turns out that we can achieve this as in the proof of Theorem 1.3.9: we replace all edges of G between A and B by suitable 'fictive edges' in G_A and G_B. We then need to ensure that each Hamilton cycle in G_A and G_B contains a suitable set of fictive edges – and the set-up of the robust decomposition lemma does allow for this.

One significant difficulty compared to the proof of Theorem 1.3.9 is that this time we need a *decomposition* of all the 'exceptional' edges (i.e. those between A and B and those incident to the exceptional vertices) into Hamilton exceptional systems. The nature of the decomposition depends on the structure of the bipartite subgraph $G[A', B']$ of G, where A' is obtained from A by including some subset A_0 of the exceptional vertices, and B' is obtained from B by including the remaining set B_0 of exceptional vertices. We say that G is 'critical' if many edges of $G[A', B']$ are incident to very few (exceptional) vertices. In our decomposition into Hamilton exceptional systems, we will need to distinguish between the critical and non-critical case (when in addition $G[A', B']$ contains many edges) and the case when $G[A', B']$ contains only a few edges. The lemmas guaranteeing this decomposition are stated and discussed in Section 2.7, but their proofs are deferred until Chapter 3.

Finding these localized Hamilton exceptional systems becomes more feasible if we can assume that there are no edges with both endpoints in the exceptional set A_0 or both endpoints in B_0. So in Section 2.6, we find and remove a set of edge-disjoint Hamilton cycles covering all edges in $G[A_0]$ and $G[B_0]$. We can then find the localized Hamilton exceptional systems in Section 2.7. After this, we need to extend and combine them into certain path systems and factors in Section 2.8, before we can use them as an 'input' for the robust decomposition lemma in Section 2.9. Finally, all these steps are combined in Section 2.10 to prove Theorem 1.3.3.

2.2. Partitions and Frameworks

2.2.1. Edges between Partition Classes. Let A', B' be a partition of the vertex set of a graph G. The aim of this subsection is to give some useful bounds on the number $e_G(A', B')$ of edges between A' and B' in G.

PROPOSITION 2.2.1. *Let G be a graph on n vertices with $\delta(G) \geq D$. Let A', B' be a partition of $V(G)$. Then the following properties hold:*

(i) $e_G(A', B') \geq (D - |B'| + 1)|B'|$.
(ii) *If $D \geq n - 2\lfloor n/4 \rfloor - 1$, then $e_G(A', B') \geq D$ unless $n = 0 \pmod 4$, $D = n/2 - 1$ and $|A'| = |B'| = n/2$.*

Proof. Since $\delta(G) \geq D$ we have $d(v, A') \geq D - |B'| + 1$ for all $v \in B'$ and so $e_G(A', B') \geq (D - |B'| + 1)|B'|$, which implies (i). (ii) follows from (1.3.1) and (i). □

PROPOSITION 2.2.2. *Let G be a D-regular graph on n vertices together with a vertex partition A', B'. Then*

(i) *$e_G(A', B')$ is odd if and only if both $|A'|$ and D are odd.*
(ii) $e_G(A', B') = e_{\overline{G}}(A') + e_{\overline{G}}(B') + \frac{(2D+2-n)n}{4} - \frac{(|A'|-|B'|)^2}{4}$.

Proof. Note that $e_G(A', B') = \sum_{v\in A'} d(v, B') = \sum_{v \in A'}(D - d(v, A')) = |A'|D - 2e_G(A')$. Hence (i) follows.

For (ii), note that

$$e_{\overline{G}}(A') = \binom{|A'|}{2} - e_G(A') = \binom{|A'|}{2} - \frac{1}{2}\left(D|A'| - e_G(A', B')\right),$$

and similarly $e_{\overline{G}}(B') = \binom{|B'|}{2} - (D|B'| - e_G(A', B'))/2$. Since $|A'| + |B'| = n$ it follows that

$$\begin{aligned} e_G(A', B') &= e_{\overline{G}}(A') + e_{\overline{G}}(B') - \frac{1}{2}\left(|A'|^2 + |B'|^2 - n\left(D+1\right)\right) \\ &= e_{\overline{G}}(A') + e_{\overline{G}}(B') + \frac{(2D+2-n)n}{4} - \frac{(|A'|-|B'|)^2}{4}, \end{aligned}$$

as required. □

PROPOSITION 2.2.3. *Let G be a D-regular graph on n vertices with $D \geq \lfloor n/2 \rfloor$. Let A', B' be a partition of $V(G)$ with $|A'|, |B'| \geq D/2$ and $\Delta(G[A', B']) \leq D/2$. Then*

$$e_{G-U}(A', B') \geq \begin{cases} D - 28 & \text{if } D \geq n/2, \\ D/2 - 28 & \text{if } D = (n-1)/2 \end{cases}$$

for every $U \subseteq V(G)$ with $|U| \leq 3$.

Proof. Without loss of generality, we may assume that $|A'| \geq |B'|$. Set $G' := G[A', B']$. If $|B'| \leq D - 4$, then $e(G') \geq (D - |B'| + 1)|B'| \geq 5D/2$ by Proposition 2.2.1(i). Since $\Delta(G') \leq D/2$ we have $e(G' - U) \geq e(G') - 3D/2 \geq D$. Thus we may assume that $|B'| \geq D - 3$. For every $v \in B'$, we have

$$d_{G'}(v) = d_G(v, A') = D - d_G(v, B') = D - (|B'| - d_{\overline{G}}(v, B') - 1) \leq d_{\overline{G}}(v, B') + 4,$$

and similarly $d_{G'}(v) \leq d_{\overline{G}}(v, A') + 4$ for all $v \in A'$. Thus

$$\begin{aligned} \sum_{u \in U} d_{G'}(u) &\leq 12 + \sum_{u \in U \cap A'} d_{\overline{G}}(u, A') + \sum_{u \in U \cap B'} d_{\overline{G}}(u, B') \\ &\leq 15 + e_{\overline{G}}(A') + e_{\overline{G}}(B'). \end{aligned} \tag{2.2.1}$$

Note that $|A'| - |B'| \leq 7$ since $|A'| \geq |B'| \geq D - 3 \geq \lfloor n/2 \rfloor - 3$. By Proposition 2.2.2(ii), we have

$$\begin{aligned} e(G' - U) &\geq e(G') - \sum_{u \in U} d_{G'}(u) \\ &\geq e_{\overline{G}}(A') + e_{\overline{G}}(B') + \frac{(2D + 2 - n)n}{4} - \frac{(|A'| - |B'|)^2}{4} - \sum_{u \in U} d_{G'}(u) \\ &\overset{(2.2.1)}{\geq} \frac{(2D + 2 - n)n}{4} - \frac{(|A'| - |B'|)^2}{4} - 15 \geq \frac{(2D + 2 - n)n}{4} - 28. \end{aligned}$$

Hence the proposition follows. □

The following result is an analogue of Proposition 2.2.3 for the case when G is $(n/2 - 1)$-regular with $n = 0 \pmod 4$ and $|A'| = n/2 = |B'|$.

PROPOSITION 2.2.4. *Let G be an $(n/2 - 1)$-regular graph on n vertices with $n = 0 \pmod 4$. Let A', B' be a partition of $V(G)$ with $|A'| = n/2 = |B'|$. Then*

$$e_G(A' \setminus X, B') \geq e_G(X, B') - |X|(|X| - 1)$$

for every vertex set $X \subseteq A'$. Moreover, $\Delta(G[A', B']) \leq e_G(A', B')/2$.

Proof. For every $v \in A'$, we have

$$d_G(v, B') = n/2 - 1 - d_G(v, A') = |A'| - 1 - d_G(v, A') = d_{\overline{G}}(v, A').$$

By summing over all $v \in A'$ we obtain

$$\begin{aligned} e_G(A', B') = 2e_{\overline{G}}(A') &\geq 2\left(\sum_{x \in X} d_{\overline{G}}(x, A') - \binom{|X|}{2}\right) \\ &= 2\sum_{x \in X} d_G(x, B') - |X|(|X| - 1) \\ &= 2e_G(X, B') - |X|(|X| - 1). \end{aligned}$$

Therefore,

$$e_G(A' \setminus X, B') = e_G(A', B') - e_G(X, B') \geq e_G(X, B') - |X|(|X| - 1).$$

In particular, this implies that for each vertex $x \in A'$ we have $e_G(A' \setminus \{x\}, B') \geq e_G(\{x\}, B') = d_G(x, B')$ and so $2d_G(x, B') \leq e_G(A', B')$. By symmetry, for any $y \in B'$ we have $2d(y, A') \leq e_G(A', B')$. Therefore, $\Delta(G[A', B']) \leq e_G(A', B')/2$. □

2.2.2. Frameworks. Throughout this chapter, we will consider partitions into sets A and B of equal size (which induce 'near-cliques') as well as 'exceptional sets' A_0 and B_0. The following definition formalizes this. Given a graph G, we say that (G, A, A_0, B, B_0) is an (ε_0, K)-*framework* if the following holds, where $A' := A_0 \cup A$, $B' := B_0 \cup B$ and $n := |V(G)|$:

(FR1) A, A_0, B, B_0 forms a partition of $V(G)$.
(FR2) $e(A', B') \leq \varepsilon_0 n^2$.
(FR3) $|A| = |B|$ is divisible by K, $|A_0| \geq |B_0|$ and $|A_0| + |B_0| \leq \varepsilon_0 n$.
(FR4) If $v \in A$ then $d(v, B') < \varepsilon_0 n$ and if $v \in B$ then $d(v, A') < \varepsilon_0 n$.

We often write V_0 for $A_0 \cup B_0$ and think of the vertices in V_0 as 'exceptional vertices'. Also, whenever (G, A, A_0, B, B_0) is an (ε_0, K)-framework, we will write $A' := A_0 \cup A$, $B' := B_0 \cup B$.

PROPOSITION 2.2.5. *Let $0 < 1/n \ll \varepsilon_{\rm ex}, 1/K \ll 1$ and $\varepsilon_{\rm ex} \ll \varepsilon_0 \ll 1$. Let G be a graph on n vertices with $\delta(G) = D \geq n - 2\lfloor n/4 \rfloor - 1$ that is $\varepsilon_{\rm ex}$-close to the union of two disjoint copies of $K_{n/2}$. Then there is a partition A, A_0, B, B_0 of $V(G)$ such that (G, A, A_0, B, B_0) is an (ε_0, K)-framework, $d(v, A') \geq d(v)/2$ for all $v \in A'$ and $d(v, B') \geq d(v)/2$ for all $v \in B'$.*

Proof. Write $\varepsilon := \varepsilon_{\rm ex}$. Since G is ε-close to the union of two disjoint copies of $K_{n/2}$, there exists a partition A'', B'' of $V(G)$ such that $|A''| = \lfloor n/2 \rfloor$ and $e(A'', B'') \leq \varepsilon n^2$. If there exists a vertex $v \in A''$ such that $d(v, A'') < d(v, B'')$, then we move v to B''. We still denote the vertex classes thus obtained by A'' and B''. Similarly, if there exists a vertex $v \in B''$ such that $d(v, B'') < d(v, A'')$, then we move v to A''. We repeat this process until $d(v, A'') \geq d(v, B'')$ for all $v \in A''$ and $d(v, B'') \geq d(v, A'')$ for all $v \in B''$. Note that this process must terminate since at each step the value of $e(A'', B'')$ decreases. Let A', B' denote the resulting partition. By relabeling the classes if necessary we may assume that $|A'| \geq |B'|$. By construction, $e(A', B') \leq e(A'', B'') \leq \varepsilon n^2$ and so (FR2) holds. Suppose that $|B'| < (1 - 5\varepsilon)n/2$. Then at some stage in the process we have that $|B''| = (1 - 5\varepsilon)n/2$. But then by Proposition 2.2.1(i),

$$e(A'', B'') \geq (D - |B''| + 1)|B''| > \varepsilon n^2,$$

a contradiction to the definition of ε-closeness (as the number of edges between the partition classes has not increased while moving the vertices). Hence, $|A'| \geq |B'| \geq (1 - 5\varepsilon)n/2$. Let B'_0 be the set of vertices v in B' such that $d(v, A') \geq \sqrt{\varepsilon} n$. Since $\sqrt{\varepsilon} n |B'_0| \leq e(A', B') \leq \varepsilon n^2$ we have $|B'_0| \leq \sqrt{\varepsilon} n$. Note that

$$|B'| - |B'_0| \geq (1 - 5\varepsilon)n/2 - \sqrt{\varepsilon} n \geq (1 - 3\sqrt{\varepsilon})n/2. \tag{2.2.2}$$

Similarly, let A'_0 be the set of vertices v in A' such that $d(v, B') \geq \sqrt{\varepsilon} n$. Thus, $|A'_0| \leq \sqrt{\varepsilon} n$ and $|A'| - |A'_0| \geq n/2 - |A'_0| \geq (1 - 2\sqrt{\varepsilon})n/2$. Let m be the largest integer such that $Km \leq |A'| - |A'_0|, |B'| - |B'_0|$. Let A and B be Km-subsets of

$A' \setminus A'_0$ and $B' \setminus B'_0$ respectively. Set $A_0 := A' \setminus A$ and $B_0 := B' \setminus B$. Note that (2.2.2) and its analogue for A' together imply that $|A_0| + |B_0| \leq 3\sqrt{\varepsilon}n + 2K \leq \varepsilon_0 n$. Therefore, (G, A, A_0, B, B_0) is an (ε_0, K)-framework. □

2.3. Exceptional Systems and (K, m, ε_0)-Partitions

The definitions and observations in this section will enable us to 'reduce' the problem of finding Hamilton cycles in G to that of finding suitable pairs C_A, C_B of cycles with $V(C_A) = A$ and $V(C_B) = B$. In particular, they will enable us to 'ignore' the exceptional set $V_0 = A_0 \cup B_0$. Roughly speaking, for each Hamilton cycle we seek, we find a certain path system J covering V_0 (called an exceptional system). From this, we derive a set J^* of edges whose endvertices lie in $A \cup B$ by replacing paths of J with 'fictive edges' in a suitable way. We can then work with J^* instead of J when constructing our Hamilton cycles (see Proposition 2.3.1 and the explanation preceding it).

Suppose that A, A_0, B, B_0 forms a partition of a vertex set V of size n such that $|A| = |B|$. Let $V_0 := A_0 \cup B_0$. An *exceptional cover* J is a graph which satisfies the following properties:

(EC1) J is a path system with $V_0 \subseteq V(J) \subseteq V$.
(EC2) $d_J(v) = 2$ for every $v \in V_0$ and $d_J(v) \leq 1$ for every $v \in V(J) \setminus V_0$.
(EC3) $e_J(A), e_J(B) = 0$.

We say that J is an *exceptional system with parameter* ε_0, or an *ES* for short, if J satisfies the following properties:

(ES1) J is an exceptional cover.
(ES2) One of the following is satisfied:
  (HES) The number of AB-paths in J is even and positive. In this case we say J is a *Hamilton exceptional system*, or *HES* for short.
  (MES) $e_J(A', B') = 0$. In this case we say J is a *matching exceptional system*, or *MES* for short.
(ES3) J contains at most $\sqrt{\varepsilon_0}n$ AB-paths.

Note that by definition, every AB-path in J is maximal. So the number of AB-paths in J is the number of genuine 'connections' between A and B (and thus between A' and B'). If we want to extend J into a Hamilton cycle using only edges induced by A and edges induced by B, this number clearly has to be even and positive. Hamilton exceptional systems will always be extended into Hamilton cycles and matching exceptional systems will always be extended into two disjoint even cycles which together span all vertices (and thus consist of two edge-disjoint perfect matchings).

Since each maximal path in J has endpoints in $A \cup B$ and internal vertices in V_0, an exceptional system J naturally induces a matching J^*_{AB} on $A \cup B$. More precisely, if $P_1, \dots, P_{\ell'}$ are the non-trivial paths in J and x_i, y_i are the endpoints of P_i, then we define $J^*_{AB} := \{x_i y_i : i \leq \ell'\}$. Thus $e_{J^*_{AB}}(A, B)$ is equal to the number of AB-paths in J. In particular, if J is a matching exceptional system, then $e_{J^*_{AB}}(A, B) = 0$.

Let $x_1 y_1, \dots, x_{2\ell} y_{2\ell}$ be a fixed enumeration of the edges of $J^*_{AB}[A, B]$ with $x_i \in A$ and $y_i \in B$. Define

$$J^*_A := J^*_{AB}[A] \cup \{x_{2i-1} x_{2i} : 1 \leq i \leq \ell\} \quad \text{and} \quad J^*_B := J^*_{AB}[B] \cup \{y_{2i} y_{2i+1} : 1 \leq i \leq \ell\}$$

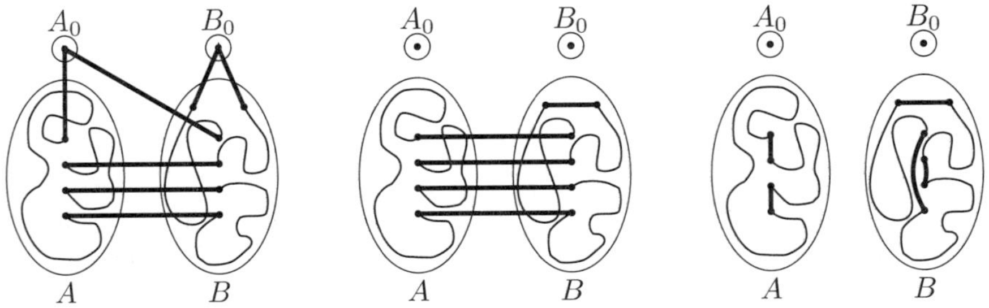

(A) J (B) J^*_{AB} (C) J^*

FIGURE 2.3.1. The thick lines illustrate the edges of J, J^*_{AB} and J^* respectively.

(with indices considered modulo 2ℓ). Let $J^* := J^*_A + J^*_B$, see Figure 2.3.1. Note that J^* is the union of one matching induced by A and another on B, and $e(J^*) = e(J^*_{AB})$. Moreover, by (EC2) we have

$$e(J^*) = e(J^*_{AB}) \leq |V_0| + e_J(A', B') \leq 2\sqrt{\varepsilon_0}n. \tag{2.3.1}$$

We will call the edges in J^* *fictive* edges. Note that if J_1 and J_2 are two edge-disjoint exceptional systems, then J_1^* and J_2^* may not be edge-disjoint. However, we will always view fictive edges as being distinct from each other and from the edges in other graphs. So in particular, whenever J_1 and J_2 are two exceptional systems, we will view J_1^* and J_2^* as being edge-disjoint.

We say that a path P is *consistent with* J^*_A if P contains J^*_A and (there is an orientation of P which) visits the vertices $x_1, \dots, x_{2\ell}$ in this order. A path P is *consistent with* J^*_B if P contains J^*_B and visits the vertices $y_2, \dots, y_{2\ell}, y_1$ in this order. In a similar way we define when a cycle is consistent with J^*_A or J^*_B.

The next result shows that if J is a Hamilton exceptional system and C_A, C_B are two Hamilton cycles on A and B respectively which are consistent with J^*_A and J^*_B, then the graph obtained from $C_A + C_B$ by replacing $J^* = J^*_A + J^*_B$ with J is a Hamilton cycle on V which contains J, see Figure 2.3.1. When choosing our Hamilton cycles, this property will enable us ignore all the vertices in V_0 and to consider the (almost complete) graphs induced by A and by B instead. Similarly, if J is a matching exceptional system and both $|A'|$ and $|B'|$ are even, then the graph obtained from $C_A + C_B$ by replacing J^* with J is the edge-disjoint union of two perfect matchings on V.

PROPOSITION 2.3.1. *Suppose that A, A_0, B, B_0 forms a partition of a vertex set V. Let J be an exceptional system. Let C_A and C_B be two cycles such that*

- *C_A is a Hamilton cycle on A that is consistent with J^*_A;*
- *C_B is a Hamilton cycle on B that is consistent with J^*_B.*

Then the following assertions hold.

(i) *If J is a Hamilton exceptional system, then $C_A + C_B - J^* + J$ is a Hamilton cycle on V.*
(ii) *If J is a matching exceptional system, then $C_A + C_B - J^* + J$ is the union of a Hamilton cycle on A' and a Hamilton cycle on B'. In particular, if*

both $|A'|$ and $|B'|$ are even, then $C_A + C_B - J^* + J$ is the union of two edge-disjoint perfect matchings on V.

Proof. Suppose that J is a Hamilton exceptional system. Let $x_1y_1, \dots, x_{2\ell}y_{2\ell}$ be an enumeration of the edges of $J^*_{AB}[A, B]$ with $x_i \in A$ and $y_i \in B$ and such that $J^*_A = J^*_{AB}[A] \cup \{x_{2i-1}x_{2i} : 1 \leq i \leq \ell\}$ and $J^*_B = J^*_{AB}[B] \cup \{y_{2i}y_{2i+1} : 1 \leq i \leq \ell\}$. Let $P^A_1, \dots, P^A_\ell$ be the paths in $C_A - \{x_{2i-1}x_{2i} : 1 \leq i \leq \ell\}$. Since C_A is consistent with J^*_A, we may assume that P^A_i is a path from x_{2i-2} to x_{2i-1} for all $i \leq \ell$. Similarly, let $P^B_1, \dots, P^B_\ell$ be the paths in $C_B - \{y_{2i}y_{2i+1} : 1 \leq i \leq \ell\}$. Again, we may assume that P^B_i is a path from y_{2i-1} to y_{2i} for all $i \leq \ell$. Define C^* to be the 2-regular graph on $A \cup B$ obtained from concatenating $P^A_1, x_1y_1, P^B_1, y_2x_2, P^A_2, x_3y_3, \dots, P^B_\ell$ and $y_{2\ell}x_{2\ell}$. Together with (HES), the construction implies that C^* is a Hamilton cycle on $A \cup B$ and $C^* = C_A + C_B - J^* + J^*_{AB}$. Thus $C := C^* - J^*_{AB} + J$ is a Hamilton cycle on V. Since $C = C_A + C_B - J^* + J$, (i) holds.

The proof of (ii) is similar to that of (i). Indeed, the previous argument shows that C^* is the union of a Hamilton cycle on A and a Hamilton cycle on B. (MES) now implies that C is the union of a Hamilton cycle on A' and one on B'. □

In general, we construct an exceptional system by first choosing an exceptional system candidate (defined below) and then extending it to an exceptional system. More precisely, suppose that A, A_0, B, B_0 forms a partition of a vertex set V. Let $V_0 := A_0 \cup B_0$. A graph F is called an *exceptional system candidate with parameter* ε_0, or an *ESC* for short, if F satisfies the following properties:

(ESC1) F is a path system with $V_0 \subseteq V(F) \subseteq V$ and such that $e_F(A), e_F(B) = 0$.
(ESC2) $d_F(v) \leq 2$ for all $v \in V_0$ and $d_F(v) = 1$ for all $v \in V(F) \setminus V_0$.
(ESC3) $e_F(A', B') \leq \sqrt{\varepsilon_0}n/2$. In particular, $|V(F) \cap A|, |V(F) \cap B| \leq 2|V_0| + \sqrt{\varepsilon_0}n/2$.
(ESC4) One of the following holds:
- (HESC) Let $b(F)$ be the number of maximal paths in F with one endpoint in A' and the other in B'. Then $b(F)$ is even and $b(F) > 0$. In this case we say that F is a *Hamilton exceptional system candidate*, or *HESC* for short.
- (MESC) $e_F(A', B') = 0$. In this case, F is called a *matching exceptional system candidate* or *MESC* for short.

Note that if $d_F(v) = 2$ for all $v \in V_0$, then F is an exceptional system. Also, if F is a Hamilton exceptional system candidate with $e(F) = 2$, then F consists of two independent $A'B'$-edges. Moreover, note that (EC2) allows an exceptional cover J (and so also an exceptional system J) to contain vertices in $A \cup B$ which are isolated in J. However, (ESC2) does not allow for this in an exceptional system candidate F.

Similarly to condition (HES), in (HESC) the parameter $b(F)$ counts the number of 'connections' between A' and B'. In order to extend a Hamilton exceptional system candidate into a Hamilton cycle without using any additional $A'B'$-edges, it is clearly necessary that $b(F)$ is positive and even.

The next result shows that we can extend an exceptional system candidate into an exceptional system by adding suitable A_0A- and B_0B-edges. In the proof of Lemma 2.6.1 we will use that if G is a D-regular graph with $D \geq n/100$ (say) and (G, A, A_0, B, B_0) is an (ε_0, K)-framework with $\Delta(G[A', B']) \leq D/2$, then conditions (i) and (ii) below are satisfied.

LEMMA 2.3.2. *Suppose that* $0 < 1/n \ll \varepsilon_0 \ll 1$ *and that* $n \in \mathbb{N}$. *Let* G *be a graph on* n *vertices so that*

(i) A, A_0, B, B_0 *forms a partition of* $V(G)$ *with* $|A_0 \cup B_0| \leq \varepsilon_0 n$.
(ii) $d(v, A) \geq \sqrt{\varepsilon_0} n$ *for all* $v \in A_0$ *and* $d(v, B) \geq \sqrt{\varepsilon_0} n$ *for all* $v \in B_0$.

Let F *be an exceptional system candidate with parameter* ε_0. *Then there exists an exceptional system* J *with parameter* ε_0 *such that* $F \subseteq J \subseteq G + F$ *and such that every edge of* $J - F$ *lies in* $G[A_0, A] + G[B_0, B]$. *Moreover, if* F *is a Hamilton exceptional system candidate, then* J *is a Hamilton exceptional system. Otherwise* J *is a matching exceptional system.*

Proof. For each vertex $v \in A_0$, we select $2 - d_F(v)$ edges uv in G with $u \in A \setminus V(F)$. Since $d_G(v, A) \geq \sqrt{\varepsilon_0} n \geq |V(F) \cap A| + 2|V_0|$ by (ESC3), these edges can be chosen such that they have no common endpoint in A. Similarly, for each vertex $v \in B_0$, we select $2 - d_F(v)$ edges uv in G with $u \in B \setminus V(F)$. Again, these edges are chosen such that they have no common endpoint in B. Let J be the graph obtained from F by adding all these edges. Note that J is an exceptional cover such that every edge of $J - F$ lies in $G[A_0, A] + G[B_0, B]$. Furthermore, the number of AB-paths in J is at most $e_F(A', B') \leq \sqrt{\varepsilon_0} n/2$.

Suppose F is a Hamilton exceptional system candidate with parameter ε_0. Our construction of J implies that the number of AB-paths in J equals $b(F)$. So (HES) follows from (HESC). Now suppose F is a matching exceptional system candidate. Then (MES) is satisfied since $e_J(A', B') = e_F(A', B') = 0$ by (MESC). This proves the lemma. □

Let $K, m \in \mathbb{N}$ and $\varepsilon_0 > 0$. A (K, m, ε_0)*-partition* $\mathcal{P}$ of a set V of vertices is a partition of V into sets $A_0, A_1, \dots, A_K$ and $B_0, B_1, \dots, B_K$ such that $|A_i| = |B_i| = m$ for all $i \geq 1$ and $|A_0 \cup B_0| \leq \varepsilon_0 |V|$. The sets $A_1, \dots, A_K$ and $B_1, \dots, B_K$ are called *clusters* of $\mathcal{P}$ and A_0, B_0 are called *exceptional sets*. We often write V_0 for $A_0 \cup B_0$ and think of the vertices in V_0 as 'exceptional vertices'. Unless stated otherwise, whenever $\mathcal{P}$ is a (K, m, ε_0)-partition, we will denote the clusters by $A_1, \dots, A_K$ and $B_1, \dots, B_K$ and the exceptional sets by A_0 and B_0. We will also write $A := A_1 \cup \dots \cup A_K$, $B := B_1 \cup \dots \cup B_K$, $A' := A_0 \cup A_1 \cup \dots \cup A_K$ and $B' := B_0 \cup B_1 \cup \dots \cup B_K$.

Given a (K, m, ε_0)-partition $\mathcal{P}$ and $1 \leq i, i' \leq K$, we say that J is an (i, i')*-localized Hamilton exceptional system* (abbreviated as (i, i')*-HES*) if J is a Hamilton exceptional system and $V(J) \subseteq V_0 \cup A_i \cup B_{i'}$. In a similar way, we define

- (i, i')*-localized matching exceptional systems* ((i, i')*-MES*),
- (i, i')*-localized exceptional systems* ((i, i')*-ES*),
- (i, i')*-localized Hamilton exceptional system candidates* ((i, i')*-HESC*),
- (i, i')*-localized matching exceptional system candidates* ((i, i')*-MESC*),
- (i, i')*-localized exceptional system candidates* ((i, i')*-ESC*).

To make clear with which partition we are working, we sometimes also say that J is an (i, i')-localized Hamilton exceptional system with respect to $\mathcal{P}$ etc.

2.4. Schemes and Exceptional Schemes

It will often be convenient to consider the 'exceptional' and 'non-exceptional' part of a graph G separately. For this, we introduce a 'scheme' (which corresponds to the non-exceptional part and also incorporates a refined partition of G) and an 'exceptional scheme' (which corresponds to the exceptional part and also incorporates a refined partition of G).

Given a graph G and a partition $\mathcal{P}$ of a vertex set V, we call $(G, \mathcal{P})$ a $(K, m, \varepsilon_0, \varepsilon)$-*scheme* if the following properties hold:

(Sch1) $\mathcal{P}$ is a (K, m, ε_0)-partition of V.
(Sch2) $V(G) = A \cup B$ and $e_G(A, B) = 0$.
(Sch3) For all $1 \le i \le K$ and all $v \in A$ we have $d(v, A_i) \ge (1 - \varepsilon)m$. Similarly, for all $1 \le i \le K$ and all $v \in B$ we have $d(v, B_i) \ge (1 - \varepsilon)m$.

The next proposition shows that if $(G, \mathcal{P})$ is a scheme and G' is obtained from G by removing a small number of edges at each vertex, then $(G', \mathcal{P})$ is also a scheme with slightly worse parameters. Its proof is immediate from the definition of a scheme.

PROPOSITION 2.4.1. *Suppose that $0 < 1/m \ll \varepsilon, \varepsilon' \ll 1$ and that $K, m \in \mathbb{N}$. Let $(G, \mathcal{P})$ be a $(K, m, \varepsilon_0, \varepsilon)$-scheme. Let G' be a spanning subgraph of G such that $\Delta(G - G') \le \varepsilon' m$. Then $(G', \mathcal{P})$ is a $(K, m, \varepsilon_0, \varepsilon + \varepsilon')$-scheme.*

Given a graph G on n vertices and a partition $\mathcal{P}$ of $V(G)$ we call $(G, \mathcal{P})$ a $(K, m, \varepsilon_0, \varepsilon)$-*exceptional scheme* if the following properties are satisfied:

(ESch1) $\mathcal{P}$ is a (K, m, ε_0)-partition of $V(G)$.
(ESch2) $e(A), e(B) = 0$.
(ESch3) If $v \in A$ then $d(v, B') < \varepsilon_0 n$ and if $v \in B$ then $d(v, A') < \varepsilon_0 n$.
(ESch4) For all $v \in V(G)$ and all $1 \le i \le K$ we have $d(v, A_i) = (d(v, A) \pm \varepsilon n)/K$ and $d(v, B_i) = (d(v, B) \pm \varepsilon n)/K$.
(ESch5) For all $1 \le i, i' \le K$ we have

$$\begin{aligned} e(A_0, A_i) &= (e(A_0, A) \pm \varepsilon \max\{e(A_0, A), n\})/K, \\ e(B_0, A_i) &= (e(B_0, A) \pm \varepsilon \max\{e(B_0, A), n\})/K, \\ e(A_0, B_i) &= (e(A_0, B) \pm \varepsilon \max\{e(A_0, B), n\})/K, \\ e(B_0, B_i) &= (e(B_0, B) \pm \varepsilon \max\{e(B_0, B), n\})/K, \\ e(A_i, B_{i'}) &= (e(A, B) \pm \varepsilon \max\{e(A, B), n\})/K^2. \end{aligned}$$

Suppose that (G, A, A_0, B, B_0) is an (ε_0, K)-framework. The next lemma shows that there is a refinement of the vertex partition A, A_0, B, B_0 of $V(G)$ into a (K, m, ε_0)-partition $\mathcal{P}$ such that $(G[A] + G[B], \mathcal{P})$ is a scheme and $(G - G[A] - G[B], \mathcal{P})$ is an exceptional scheme.

LEMMA 2.4.2. *Suppose that $0 < 1/n \ll \varepsilon_0 \ll 1/K \ll 1$, that $\varepsilon_0 \ll \varepsilon_1 \le \varepsilon_2 \ll 1$, that $1/n \ll \mu \ll \varepsilon_2$ and that $n, K, m \in \mathbb{N}$. Let G be a graph on n vertices such that $\delta(G) \ge (1 - \mu)n/2$. Let (G, A, A_0, B, B_0) be an (ε_0, K)-framework with $|A| = |B| = Km$. Then there are partitions $A_1, \dots, A_K$ of A and $B_1, \dots, B_K$ of B which satisfy the following properties:*

(i) *The partition $\mathcal{P}$ formed by A_0, B_0 and all these $2K$ clusters is a (K, m, ε_0)-partition of $V(G)$.*
(ii) *$(G[A] + G[B], \mathcal{P})$ is a $(K, m, \varepsilon_0, \varepsilon_2)$-scheme.*
(iii) *$(G - G[A] - G[B], \mathcal{P})$ is a $(K, m, \varepsilon_0, \varepsilon_1)$-exceptional scheme.*
(iv) *For all $v \in V(G)$ and all $1 \le i \le K$ we have $d_G(v, A_i) = (d_G(v, A) \pm \varepsilon_0 n)/K$ and $d_G(v, B_i) = (d_G(v, B) \pm \varepsilon_0 n)/K$.*

Proof. Define a new constant ε_1' such that $\varepsilon_0 \ll \varepsilon_1' \ll \varepsilon_1, 1/K$. In order to find the required partitions $A_1, \dots, A_K$ of A and $B_1, \dots, B_K$ of B we will apply Lemma 1.4.7 twice, as follows.

In our first application of Lemma 1.4.7 we let $F := G$, $U := A$ and let A_0, B_0, B play the roles of R_1, R_2, R_3. Note that $\delta(G[A]) \geq \delta(G) - |A_0| - \varepsilon_0 n \geq \varepsilon_0 n$ (with room to spare) by (FR3), (FR4) and that $d(a, R_j) \leq |R_j| \leq \varepsilon_0 n$ for all $a \in A$ and $j = 1, 2$ by (FR3). Moreover, (FR4) implies that $d(a, R_3) \leq d(a, B') \leq \varepsilon_0 n$ for all $a \in A$. Thus we can apply Lemma 1.4.7 with $\varepsilon_0, \varepsilon_0$ and ε_1' playing the roles of $\varepsilon, \varepsilon_1$ and ε_2 to obtain a partition of A into K clusters $A_1, \dots, A_K$, each of size m. Then by Lemma 1.4.7(ii) for all $v \in V(G)$ and all $1 \leq i \leq K$ we have

$$d_G(v, A_i) = (d_G(v, A) \pm \varepsilon_0 n)/K. \tag{2.4.1}$$

Moreover, Lemma 1.4.7(v) implies that the first two equalities in (ESch5) hold with respect to ε_1' (for G and thus also for $G - G[A] - G[B]$). Furthermore,

$$e_G(A_i, B) = (e_G(A, B) \pm \varepsilon_1' \max\{n, e_G(A, B)\})/K. \tag{2.4.2}$$

For the second application of Lemma 1.4.7 we let $F := G$, $U := B$ and let $B_0, A_0, A_1, \dots, A_K$ play the roles of $R_1, \dots, R_{K+2}$. As before, $\delta(G[B]) \geq \varepsilon_0 n$ by (FR3), (FR4) and $d(b, R_j) \leq |R_j| \leq \varepsilon_0 n$ for all $b \in B$ and $j = 1, 2$ by (FR3). Moreover, (FR4) implies that $d(b, R_j) \leq d(b, A') \leq \varepsilon_0 n$ for all $b \in B$ and all $j = 3, \dots, K+2$. Thus we can apply Lemma 1.4.7 with $\varepsilon_0, \varepsilon_0$ and ε_1' playing the roles of $\varepsilon, \varepsilon_1$ and ε_2 to obtain a partition of B into K clusters $B_1, \dots, B_K$, each of size m. Similarly as before one can show that for all $v \in V(G)$ and all $1 \leq i \leq K$ we have

$$d_G(v, B_i) = (d_G(v, B) \pm \varepsilon_0 n)/K, \tag{2.4.3}$$

and that the third and the fourth equalities in (ESch5) hold with respect to ε_1' (for G and thus also for $G - G[A] - G[B]$). Moreover, Lemma 1.4.7(v) implies that for all $1 \leq i' \leq K$ we have

$$\begin{aligned} e_G(A_i, B_{i'}) &= (e_G(A_i, B) \pm \varepsilon_1' \max\{n, e_G(A_i, B)\})/K \\ &\overset{(2.4.2)}{=} \frac{e_G(A, B) \pm \varepsilon_1' \max\{n, e_G(A, B)\} \pm K\varepsilon_1' \max\{n, e_G(A_i, B)\}}{K^2} \\ &= (e_G(A, B) \pm \varepsilon_1 \max\{n, e_G(A, B)\})/K^2, \end{aligned}$$

i.e. the last equality in (ESch5) holds too. Let $\mathcal{P}$ be the partition formed by $A_0, A_1, \dots, A_K$ and $B_0, B_1, \dots, B_K$. Then (i) holds.

Let us now verify (ii). Clearly $(G[A] + G[B], \mathcal{P})$ satisfies (Sch1) and (Sch2). In order to check (Sch3), let $G_1 := G[A] + G[B]$ and note that for all $v \in A$ and all $1 \leq i \leq K$ we have

$$\begin{aligned} d_{G_1}(v, A_i) = d_G(v, A_i) &\overset{(2.4.1)}{\geq} (d_G(v, A) - \varepsilon_0 n)/K \overset{(\text{FR4})}{\geq} (\delta(G) - |A_0| - 2\varepsilon_0 n)/K \\ &\overset{(\text{FR3})}{\geq} ((1-\mu)n/2 - 3\varepsilon_0 n)/K \geq (1 - \varepsilon_2)m. \end{aligned}$$

Similarly one can use (2.4.3) to show that $d_{G_1}(v, B_i) \geq (1 - \varepsilon_2)m$ for all $v \in B$ and all $1 \leq i \leq K$. This implies (Sch3) and thus (ii).

Note that (iv) follows from (2.4.1) and (2.4.3). Thus it remains to check (iii). Clearly $(G - G[A] - G[B], \mathcal{P})$ satisfies (ESch1), (ESch2) and we have already verified (ESch5). (ESch3) follows from (FR4) and (ESch4) follows from (2.4.1) and (2.4.3). □

2.5. Proof of Theorem 1.3.9

An important tool in the proof of Theorem 1.3.9 is Lemma 2.5.4, which guarantees an 'approximate' Hamilton decomposition of a graph G, provided that G is close to the union of two disjoint copies of $K_{n/2}$. This yields the required number of Hamilton cycles for Theorem 1.3.9. As an 'input', Lemma 2.5.4 requires an appropriate number of localized Hamilton exceptional systems.

To find these, we proceed as follows: the next lemma (Lemma 2.5.1) guarantees many edge-disjoint Hamilton exceptional systems in a given framework. We will apply it to 'localized subgraphs' (obtained from Lemma 2.5.2) of the original graph to ensure that the exceptional systems guaranteed by Lemma 2.5.1 are also localized. These can then be used as the required input for Lemma 2.5.4.

LEMMA 2.5.1. *Suppose that $0 < 1/n \ll \varepsilon_0 \ll \varepsilon \ll \alpha \ll 1$ and that $n, \alpha n \in \mathbb{N}$. Let G be a graph on n vertices. Suppose that (G, A, A_0, B, B_0) is an (ε_0, K)-framework which satisfies the following conditions:*

(a) $e_G(A', B') \geq 2(\alpha + \varepsilon)n$.
(b) $e_{G-v}(A', B') \geq \alpha n$ *for all* $v \in A_0 \cup B_0$.
(c) $d(v) \geq 2(\alpha + \varepsilon)n$ *for all* $v \in A_0 \cup B_0$.
(d) $d(v, A') \geq d(v, B') - \varepsilon n$ *for all* $v \in A_0$ *and* $d(v, B') \geq d(v, A') - \varepsilon n$ *for all* $v \in B_0$.

Then there exist αn edge-disjoint Hamilton exceptional systems with parameter ε_0 in G.

Proof. First we will find αn edge-disjoint matchings of size 2 in $G[A', B']$. If $\Delta(G[A', B']) \leq (\alpha + \varepsilon/2)n$, then by (a) and Proposition 1.4.5 we can find such matchings. So suppose that $\Delta(G[A', B']) \geq (\alpha + \varepsilon/2)n$ and let v be a vertex such that $d_{G[A',B']}(v) \geq (\alpha + \varepsilon/2)n$. Thus $v \in A_0 \cup B_0$ by (FR4). By (b) there are αn edges $e_1, \dots, e_{\alpha n}$ in $G[A', B'] - v$. Since $d_{G[A',B']}(v) \geq (\alpha + \varepsilon/2)n$, for each e_s in turn we can find an edge e'_s incident to v in $G[A', B']$ such that e'_s is vertex-disjoint from e_s and such that the e'_s are distinct for different indices $s \leq \alpha n$. Then the matchings consisting of e_s and e'_s are as required. Thus in both cases we can find edge-disjoint matchings $M_1, \dots, M_{\alpha n}$ of size 2 in $G[A', B']$.

Our aim is to extend each M_s into a Hamilton exceptional system J_s such that all these J_s are pairwise edge-disjoint. Initially, we set $F_s := M_s$ for all $s \leq \alpha n$. So each F_s is a Hamilton exceptional system candidate. For each $v \in V_0$ in turn, we are going to assign at most two edges joining v to $A \cup B$ to each of $F_1, \dots, F_{\alpha n}$ in such a way that now each F_s is a Hamilton exceptional system candidate with $d_{F_s}(v) = 2$. Thus after we have carried out these assignments for all $v \in V_0$, every F_s will be a Hamilton exceptional system with parameter ε_0.

So consider any $v \in V_0$. Without loss of generality we may assume that $v \in A_0$. Moreover, by relabelling the F_s if necessary, we may assume that there exists an integer $0 \leq r \leq \alpha n$ such that $d_{F_s}(v) = 1$ for all $s \leq r$ and $d_{F_s}(v) = 0$ for $r < s \leq \alpha n$. For each $s \leq r$ our aim is to assign some edge vw_s between v and A to F_s such that $w_s \notin V(F_s)$ and such that the vertices w_s are distinct for different $s \leq r$. To check that such an assignment of edges is possible, note that $|V(F_s) \cap A|, |V(F_s) \cap B| \leq 2|V_0| + 2 \leq 3\varepsilon_0 n$. Together with (c) and (d) this implies that

$$d(v, A) \geq d(v, A') - |A_0| \geq (\alpha + \varepsilon/2 - \varepsilon_0)n > r + |V(F_s) \cap A|.$$

Thus for all $s \leq r$ we can assign an edge vw_s to F_s as required.

It remains to assign two edges at v to each of $F_{r+1}, \dots, F_{\alpha n}$. We will do this for each $s = r+1, \dots, \alpha n$ in turn and for each such s we will either assign two edges between v and A to F_s or two edges between v and B. (This will ensure that we still have $b(F_s) = 2$, where $b(F_s)$ is the number of vertex-disjoint $A'B'$-paths in the path system F_s.) So suppose that for some $r < s \leq \alpha n$ we have already assigned two edges at v to each of $F_{r+1}, \dots, F_{s-1}$. Set $G_s := G - \sum_{s'=1}^{\alpha n} F_{s'}$. The fact that v has degree at most two in each $F_{s'}$ and (c) together imply that $d_{G_s}(v) \geq d_G(v) - 2\alpha n \geq 10\varepsilon_0 n$. So either $d_{G_s}(v, A') \geq 5\varepsilon_0 n$ or $d_{G_s}(v, B') \geq 5\varepsilon_0 n$. If the former holds then

$$d_{G_s}(v, A) \geq d_{G_s}(v, A') - |A_0| \geq 4\varepsilon_0 n \geq |V(F_s) \cap A| + 2$$

and so we can assign two edges vw and vw' of G_s to F_s such that $w, w' \in A \setminus V(F_s)$. Similarly if $d_{G_s}(v, B') \geq 5\varepsilon_0 n$ then we can assign two edges vw and vw' in G_s to F_s such that $w, w' \in B \setminus V(F_s)$. This shows that to each of $F_{r+1}, \dots, F_{\alpha n}$ we can assign two suitable edges at v.

Let $J_1, \dots, J_{\alpha n}$ be the graphs obtained after carrying out these assignments for all $v \in V_0$. Then the J_s are pairwise edge-disjoint and it is easy to check that each J_s is a Hamilton exceptional system with parameter ε_0. (Note that (ES2) and (ES3) hold since $b(J_s) = 2$ and so the number of AB-paths is two.) □

The next lemma guarantees a decomposition of an exceptional scheme $(G, \mathcal{P})$ into suitable 'localized slices' $G(i, i')$ whose edges are induced by A_0, B_0 and two clusters of $\mathcal{P}$. We will use it again in Chapter 3.

LEMMA 2.5.2. *Suppose that $0 < 1/n \ll \varepsilon_0 \ll \varepsilon \ll 1/K \ll 1$ and that $n, K, m \in \mathbb{N}$. Let $(G, \mathcal{P})$ be a $(K, m, \varepsilon_0, \varepsilon)$-exceptional scheme with $|G| = n$ and $e_G(A_0), e_G(B_0) = 0$. Then G can be decomposed into edge-disjoint spanning subgraphs $H(i, i')$ and $H'(i, i')$ of G (for all $i, i' \leq K$) such that the following properties hold, where $G(i, i') := H(i, i') + H'(i, i')$:*

- (a_1) *Each $H(i, i')$ contains only A_0A_i-edges and $B_0B_{i'}$-edges.*
- (a_2) *All edges of $H'(i, i')$ lie in $G[A_0 \cup A_i, B_0 \cup B_{i'}]$.*
- (a_3) $e(H'(i, i')) = (e_G(A', B') \pm 4\varepsilon \max\{n, e_G(A', B')\})/K^2$.
- (a_4) $d_{H'(i,i')}(v) = (d_{G[A',B']}(v) \pm 2\varepsilon n)/K^2$ *for all* $v \in V_0$.
- (a_5) $d_{G(i,i')}(v) = (d_G(v) \pm 4\varepsilon n)/K^2$ *for all* $v \in V_0$.

Proof. First we decompose G into K^2 'random' edge-disjoint spanning subgraphs $G(i, i')$ (one for all $i, i' \leq K$) as follows:

- Initially set $V(G(i, i')) := V(G)$ and $E(G(i, i')) := \emptyset$ for all $i, i' \leq K$.
- Add all the $A_iB_{i'}$-edges of G to $G(i, i')$.
- Choose a partition of $E(A_0, B_0)$ into K^2 sets $U_{i,i'}$ (one for all $i, i' \leq K$) whose sizes are as equal as possible. Add the edges in $U_{i,i'}$ to $G(i, i')$.
- For all $i \leq K$, choose a random partition of $E(A_0, A_i)$ into K sets $U'_{i'}$ of equal size (one for each $i' \leq K$) and add the edges in $U'_{i'}$ to $G(i, i')$. (If $e(A_0, A_i)$ is not divisible by K, first distribute up to $K-1$ edges arbitrarily among the $U'_{i'}$ to achieve divisibility.) For all $i' \leq K$ proceed similarly to distribute each edge in $E(B_0, B_{i'})$ to $G(i, i')$ for some $i \leq K$.
- For all $i' \leq K$, choose a random partition of $E(A_0, B_{i'})$ into K sets U''_i of equal size (one for each $i \leq K$) and add the edges in U''_i to $G(i, i')$. (If $e(A_0, B_{i'})$ is not divisible by K, first distribute up to $K - 1$ edges

arbitrarily among the U_i'' to achieve divisibility.) For all $i \le K$ proceed similarly to distribute each edge in $E(B_0, A_i)$ to $G(i,i')$ for some $i' \le K$.

Thus every edge of G is added to precisely one of the subgraphs $G(i,i')$. Set $H(i,i') := G(i,i')[A'] + G(i,i')[B']$ and $H'(i,i') := G(i,i')[A', B']$. So conditions ($a_1$) and ($a_2$) hold. Fix any $i, i' \le K$ and set $H := H(i,i')$ and $H' := H'(i,i')$. To verify (a_3), note that

$$\begin{aligned} e(H') &= e_{H'}(A_i, B_{i'}) + e_{H'}(A_0, B_0) + e_{H'}(A_0, B_{i'}) + e_{H'}(B_0, A_i) \\ &= e_G(A_i, B_{i'}) + e_G(A_0, B_0)/K^2 + e_G(A_0, B_{i'})/K + e_G(B_0, A_i)/K \pm 3 \\ &= \frac{e_G(A,B) + e_G(A_0,B_0) + e_G(A_0,B) + e_G(B_0,A) \pm 3\varepsilon \max\{e_G(A',B'), n\}}{K^2} \pm 3 \\ &= \frac{e_G(A',B') \pm 4\varepsilon \max\{e_G(A',B'), n\}}{K^2}. \end{aligned}$$

Here the third equality follows from (ESch5).

To prove (a_4), suppose first that $v \in A_0$. If $d_G(v, B_{i'}) \le \varepsilon n/K^2$ then clearly $0 \le d_{H'(i,i')}(v) \le \varepsilon n/K^2 + |V_0| \le 2\varepsilon n/K^2$. Further by (ESch4) we have $d_G(v,B) \le K d_G(v, B_{i'}) + \varepsilon n = \varepsilon n/K + \varepsilon n$. So $d_G(v, B') \le 2\varepsilon n$. Together this shows that (a_4) is satisfied.

So assume that $d_G(v, B_{i'}) \ge \varepsilon n/K^2$. Proposition 1.4.4 implies that with probability at least $1 - \mathrm{e}^{-\sqrt{n}}$ (with room to spare) we have

$$d_{G(i,i')}(v, B_{i'}) = (d_G(v, B_{i'}) \pm \varepsilon n/2K)/K \overset{\text{(ESch4)}}{=} (d_G(v,B) \pm 3\varepsilon n/2)/K^2. \tag{2.5.1}$$

Since

$$\begin{aligned} d_{H'(i,i')}(v) &= d_{G(i,i')}(v, B_{i'}) + d_{G(i,i')}(v, B_0) = d_{G(i,i')}(v, B_{i'}) \pm \varepsilon_0 n \\ &\overset{(2.5.1)}{=} (d_G(v, B') \pm 2\varepsilon n)/K^2, \end{aligned}$$

it follows that v satisfies (a_4). The argument for the case when $v \in B_0$ is similar. Thus (a_4) holds with probability at least $1 - n\mathrm{e}^{-\sqrt{n}}$.

Similarly as (2.5.1) one can show that with probability at least $1 - n\mathrm{e}^{-\sqrt{n}}$ we have $d_{G(i,i')}(v, A_i) = (d_G(v,A) \pm 3\varepsilon n/2)/K^2$ for all $v \in A_0$ and $d_{G(i,i')}(v, B_{i'}) = (d_G(v,B) \pm 3\varepsilon n/2)/K^2$ for all $v \in B_0$. Together with the fact that $e_G(A_0), e_G(B_0) = 0$ and (a_4) this now implies (a_5). □

The next lemma first applies the previous one to construct localized subgraphs $G(i,i')$ and then applies Lemma 2.5.1 to find many Hamilton exceptional systems within each of the localized slices $G(i,i')$. Altogether, this yields many localized Hamilton exceptional systems in G.

LEMMA 2.5.3. *Suppose that $0 < 1/n \ll \varepsilon_0 \ll \varepsilon \ll \phi, 1/K \ll 1$ and that $n, K, m, (1/4-\phi)n/K^2 \in \mathbb{N}$. Suppose that (G, A, A_0, B, B_0) is an (ε_0, K)-framework with $|G| = n$, $\delta(G) \ge n/2$ and such that $d_G(v, A') \ge d_G(v)/2$ for all $v \in A'$ and $d_G(v, B') \ge d_G(v)/2$ for all $v \in B'$. Suppose that $\mathcal{P} = \{A_0, A_1, \dots, A_K, B_0, B_1, \dots, B_K\}$ is a refinement of the partition A, A_0, B, B_0 such that $(G - G[A] - G[B], \mathcal{P})$ is a $(K, m, \varepsilon_0, \varepsilon)$-exceptional scheme. Then there is a set $\mathcal{J}$ of $(1/4-\phi)n$ edge-disjoint Hamilton exceptional systems with parameter ε_0 in G such that, for each $i, i' \le K$, $\mathcal{J}$ contains precisely $(1/4-\phi)n/K^2$ (i,i')-HES.*

Proof. Let $\alpha := (1/4-\phi)/K^2$ and choose a new constant ε' such that $\varepsilon \ll \varepsilon' \ll \phi, 1/K$. Note that (FR3) implies that $|A'| \geq |B'|$. If $|B'| < n/2$, then Proposition 2.2.1(i) implies that $e_G(A', B') \geq 2|B'| \geq (1-\varepsilon_0)n \geq 3K^2\alpha n$ (where the second inequality follows from (FR3) and there is room to spare in the final inequality). Since $d_{G[A',B']}(v) \leq n/2$ for every vertex $v \in V(G)$, it follows that $e_{G-v}(A', B') \geq (1/2-\varepsilon_0)n \geq 3K^2\alpha n/2$. If $|B'| = n/2$, then $|A'| = |B'|$ and Proposition 2.2.1(i) implies that $e_G(A', B') \geq |B'| = n/2 \geq 2K^2(\alpha+\varepsilon')n$. Moreover, $|A'| = |B'|$ together with the fact that $\delta(G) \geq n/2$ also implies that $d_{G[A',B']}(v) \geq 1$ for any vertex $v \in V(G)$. Hence $e_{G-v}(A', B') \geq n/2 - 1 \geq 3K^2\alpha n/2$. Thus regardless of the size of B', we always have

$$e_G(A', B') \geq 2K^2(\alpha+\varepsilon')n \tag{2.5.2}$$

and

$$e_{G-v}(A', B') \geq 3K^2\alpha n/2 \geq K^2(\alpha+\varepsilon')n \quad \text{for any } v \in V(G). \tag{2.5.3}$$

Set $G^\diamond := G - G[A] - G[B] - G[A_0] - G[B_0]$. Note that each vertex $v \in V_0$ satisfies

$$d_{G^\diamond}(v) \geq (1/2-\varepsilon_0)n \geq 2K^2(\alpha+\varepsilon')n. \tag{2.5.4}$$

Moreover, both (2.5.2) and (2.5.3) also hold for $G^\diamond$, and since $(G - G[A] - G[B], \mathcal{P})$ is a $(K, m, \varepsilon_0, \varepsilon)$-exceptional scheme, $(G^\diamond, \mathcal{P})$ is also a $(K, m, \varepsilon_0, \varepsilon)$-exceptional scheme. Thus we can apply Lemma 2.5.2 to $G^\diamond$ to obtain edge-disjoint spanning subgraphs $H(i,i')$, $H'(i,i')$ of $G^\diamond$ (for all $i, i' \leq K$) which satisfy (a$_1$)–(a$_5$) of Lemma 2.5.2. Set $G(i,i') := H(i,i') + H'(i,i')$ for all $i, i' \leq K$. We claim that each $G(i,i')$ satisfies the following properties:

(i) All edges of $G(i,i')$ lie in $G^\diamond[A_0 \cup A_i \cup B_0 \cup B_{i'}]$.
(ii) $e_{G(i,i')}(A', B') \geq 2(\alpha+\sqrt{\varepsilon})n$.
(iii) $e_{G(i,i')-v}(A', B') \geq \alpha n$ for all $v \in V_0$.
(iv) $d_{G(i,i')}(v) \geq 2(\alpha+\sqrt{\varepsilon})n$ for all $v \in V_0$.
(v) $d_{G(i,i')}(v, A') \geq d_{G(i,i')}(v, B') - \sqrt{\varepsilon}n$ for all $v \in A_0$ and $d_{G(i,i')}(v, B') \geq d_{G(i,i')}(v, A') - \sqrt{\varepsilon}n$ for all $v \in B_0$.

Indeed, (i) follows from (a$_1$) and (a$_2$). To prove (ii), note that $e_{G(i,i')}(A', B') = e(H'(i,i'))$. Now apply (a$_3$) and (2.5.2). For (iii), note that (a$_4$) and $\Delta(G[A', B']) \leq n/2$ imply that for all $v \in V_0$,

$$d_{G(i,i')[A',B']}(v) = d_{H'(i,i')}(v) \leq (d_{G[A',B']}(v) + 2\varepsilon n)/K^2 \leq (1/2+2\varepsilon)n/K^2.$$

If $e_G(A', B') \geq n$, then (a$_3$) implies that $e_{G(i,i')}(A', B') \geq (1-4\varepsilon)n/K^2 \geq \alpha n + d_{G(i,i')[A',B']}(v)$ and so (iii) follows. If $e_G(A', B') < n$, then for all $v \in V_0$

$$\begin{aligned} e_{G(i,i')-v}(A', B') &= e(H'(i,i')) - d_{H'(i,i')}(v) \\ &\overset{(a_3),(a_4)}{\geq} (e_{G-v}(A', B') - 6\varepsilon n)/K^2 \overset{(2.5.3)}{\geq} \alpha n. \end{aligned}$$

So (iii) follows again. (iv) follows from (a$_5$) and (2.5.4). For (v), note that (a$_1$) and (a$_2$) imply that for $v \in A_0$,

$$\begin{aligned} d_{G(i,i')}(v, A') &= d_{G(i,i')}(v) - d_{H'(i,i')}(v) \overset{(a_4),(a_5)}{\geq} (d_G(v, A') - 6\varepsilon n)/K^2 \\ &\geq (d_G(v, B') - 6\varepsilon n)/K^2 \overset{(a_4)}{\geq} d_{H'(i,i')}(v) - 8\varepsilon n = d_{G(i,i')}(v, B') - 8\varepsilon n. \end{aligned}$$

The second part of (v) follows similarly.

Note that each $(G(i,i'), A, A_0, B, B_0)$ is an (ε_0, K)-framework since this holds for (G, A, A_0, B, B_0). Thus for all $i, i' \leq K$ we can apply Lemma 2.5.1 (with $\sqrt{\varepsilon}$ playing the role of ε) to the (ε_0, K)-framework $(G(i,i'), A, A_0, B, B_0)$ in order to obtain αn edge-disjoint Hamilton exceptional systems with parameter ε_0 in $G(i,i')$. By (i), we may delete any vertices outside $A_0 \cup A_i \cup B_0 \cup B_{i'}$ from these systems without affecting their edges. So each of these Hamilton exceptional systems is in fact an (i,i')-HES. The set $\mathcal{J}$ consisting of all these $K^2\alpha n$ Hamilton exceptional systems is as required in the lemma. □

Given the appropriate set $\mathcal{J}$ of localized Hamilton exceptional systems, the next lemma guarantees a set of $|\mathcal{J}|$ edge-disjoint Hamilton cycles in a graph G such that each of them contains one exceptional system from $\mathcal{J}$, provided that G is sufficiently close to the union of two disjoint copies of $K_{n/2}$. The lemma also allows $\mathcal{J}$ to contain matching exceptional systems (each of these will then be extended into a perfect matching of G). Note that with a suitable $\mathcal{J}$ and an appropriate choice of parameters we can achieve that the 'uncovered' graph has density $2\rho \pm 2/K \ll 1$, i.e. we do have an approximate decomposition. We defer the proof of the lemma until Chapter 5.

LEMMA 2.5.4. *Suppose that* $0 < 1/n \ll \varepsilon_0 \ll 1/K \ll \rho \ll 1$ *and* $0 \leq \mu \ll 1$, *where* $n, K \in \mathbb{N}$ *and* K *is odd. Suppose that* G *is a graph on* n *vertices and* $\mathcal{P}$ *is a* (K, m, ε_0)*-partition of* $V(G)$. *Furthermore, suppose that the following conditions hold:*

(a) $d(v, A_i) = (1-4\mu \pm 4/K)m$ *and* $d(w, B_i) = (1-4\mu \pm 4/K)m$ *for all* $v \in A$, $w \in B$ *and* $1 \leq i \leq K$.
(b) *There is a set* $\mathcal{J}$ *which consists of at most* $(1/4 - \mu - \rho)n$ *edge-disjoint exceptional systems with parameter* ε_0 *in* G.
(c) $\mathcal{J}$ *has a partition into* K^2 *sets* $\mathcal{J}_{i,i'}$ *(one for all* $1 \leq i, i' \leq K$*) such that each* $\mathcal{J}_{i,i'}$ *consists of precisely* $|\mathcal{J}|/K^2$ (i,i')*-ES with respect to* $\mathcal{P}$.
(d) *If* $\mathcal{J}$ *contains matching exceptional systems then* $|A'| = |B'|$ *is even.*

Then G *contains* $|\mathcal{J}|$ *edge-disjoint spanning subgraphs* $H_1, \dots, H_{|\mathcal{J}|}$ *which satisfy the following properties:*

- *For each* H_s *there is some* $J_s \in \mathcal{J}$ *such that* $J_s \subseteq H_s$.
- *If* J_s *is a Hamilton exceptional system, then* H_s *is a Hamilton cycle of* G. *If* J_s *is a matching exceptional system, then* H_s *is the edge-disjoint union of two perfect matchings in* G.

Matching exceptional systems do no play any role in the current application to prove Theorem 1.3.9, but they will occur when we use Lemma 2.5.4 again in the proof of Theorem 1.3.3.

To prove Theorem 1.3.9, we first apply Lemma 2.5.3 to find suitable localized Hamilton exceptional systems and then apply Lemma 2.5.4 to transform these into Hamilton cycles.

Proof of Theorem 1.3.9. Choose new constants $\varepsilon_{\rm ex}$, ε_0, ε_1, ε_2, ϕ and an odd number $K \in \mathbb{N}$ such that

$$1/n_0 \ll \varepsilon_{\rm ex} \ll \varepsilon_0 \ll \varepsilon_1 \ll \varepsilon_2 \ll 1/K \ll \phi \ll \varepsilon.$$

Further, we may assume that $\varepsilon \ll 1$. Let $n \geq n_0$ and let G be any graph on n vertices such that $\delta(G) \geq n/2$ and such that G is $\varepsilon_{\rm ex}$-close to two disjoint copies of $K_{n/2}$. By modifying ϕ slightly, we may assume that $(1/4-\phi)n/K^2 \in \mathbb{N}$.

Apply Proposition 2.2.5 to obtain a partition A, A_0, B, B_0 of $V(G)$ such that such that (G, A, A_0, B, B_0) is an (ε_0, K)-framework, $d(v, A') \geq d(v)/2$ for all $v \in A'$ and $d(v, B') \geq d(v)/2$ for all $v \in B'$. Let $m := |A|/K = |B|/K$. Apply Lemma 2.4.2 with ε_0 playing the role of μ to obtain partitions $A_1, \dots, A_K$ of A and $B_1, \dots, B_K$ of B which satisfy the following properties, where $\mathcal{P} = \{A_0, A_1, \dots, A_K, B_0, B_1, \dots, B_K\}$:

- $(G[A] + G[B], \mathcal{P})$ is a $(K, m, \varepsilon_0, \varepsilon_2)$-scheme.
- $(G - G[A] - G[B], \mathcal{P})$ is a $(K, m, \varepsilon_0, \varepsilon_1)$-exceptional scheme.

Apply Lemma 2.5.3 to obtain a set $\mathcal{J}$ of $(1/4-\phi)n$ edge-disjoint Hamilton exceptional systems with parameter ε_0 in G such that, for each $i, i' \leq K$, $\mathcal{J}$ contains precisely $(1/4-\phi)n/K^2$ (i,i')-HES. Finally, our aim is to apply Lemma 2.5.4 with $\mu := 1/K$ and $\rho := \phi - 1/K$. So let us check that conditions (a)–(c) of Lemma 2.5.4 hold (note that (d) is not relevant). Clearly (b) and (c) hold. To verify (a) note that (Sch3) implies that for all $v \in A$ we have $d(v, A_i) \geq (1-\varepsilon_2)m \geq (1-1/K)m \geq (1-4\mu-4/K)m$. Similarly, for all $w \in B$ we have $d(w, B_i) \geq (1-4\mu-4/K)m$. So we can apply Lemma 2.5.4 to obtain $|\mathcal{J}| \geq (1/4-\varepsilon)n$ edge-disjoint Hamilton cycles. □

2.6. Eliminating the Edges inside A_0 and B_0

This and the remaining sections of the chapter are all devoted to the proof of Theorem 1.3.3. Suppose that G is a D-regular graph and (G, A, A_0, B, B_0) is an (ε_0, K)-framework with $\Delta(G[A', B']) \leq D/2$. The aim of this section is to construct a small number of Hamilton cycles (and perfect matchings if appropriate) which together cover all the edges of $G[A_0]$ and $G[B_0]$. The first step is to construct a small number of exceptional systems containing all the edges of $G[A_0]$ and $G[B_0]$.

LEMMA 2.6.1. *Suppose that $0 < 1/n \ll \varepsilon_0 \leq \lambda \ll 1$ and that $n, \lambda n, D, K \in \mathbb{N}$. Let G be a D-regular graph on n vertices with $D \geq n - 2\lfloor n/4 \rfloor - 1$. Suppose that (G, A, A_0, B, B_0) is an (ε_0, K)-framework with $\Delta(G[A', B']) \leq D/2$. Let*

$$\ell := \left\lfloor \frac{\max\{0, D - e_G(A', B')\}}{2} \right\rfloor \quad \text{and} \quad \phi n := \begin{cases} 2\lambda n + 1 & \text{if } D \text{ is odd,} \\ 2\lambda n & \text{if } D \text{ is even.} \end{cases}$$

Let w_1 and w_2 be vertices of G such that $d_{G[A',B']}(w_1) \geq d_{G[A',B']}(w_2) \geq d_{G[A',B']}(v)$ for all $v \in V(G) \setminus \{w_1, w_2\}$. Then there exist $\lambda n + 1$ edge-disjoint subgraphs $J_0, J_1, \dots, J_{\lambda n}$ of G which cover all the edges in $G[A_0] + G[B_0]$ and satisfy the following properties:

(i) *If D is odd, then J_0 is a perfect matching in G with $e_{J_0}(A', B') \leq 1$. If D is even, then J_0 is empty.*
(ii) *J_s is a matching exceptional system with parameter ε_0 for all $1 \leq s \leq \min\{\ell, \lambda n\}$.*
(iii) *J_s is a Hamilton exceptional system with parameter ε_0 and such that $e_{J_s}(A', B') = 2$ for all $\ell < s \leq \lambda n$.*

(iv) *Let* $\mathcal{J}$ *be the union of all the* J_s *and let* $H^\diamond := G[A', B'] - \mathcal{J}$. *Then* $e_{\mathcal{J}}(A', B') \le \phi n$ *and* $d_{\mathcal{J}}(v) = \phi n$ *for all* $v \in V_0$. *Moreover,* $e(H^\diamond)$ *is even.*

(v) $d_{H^\diamond}(w_1) \le (D - \phi n)/2$. *Furthermore, if* $D = n/2 - 1$ *then* $d_{H^\diamond}(w_2) \le (D - \phi n)/2$.

(vi) *If* $e_G(A', B') < D$*, then* $e(H^\diamond) \le D - \phi n$ *and* $\Delta(H^\diamond) \le e(H^\diamond)/2$.

As indicated in Section 2.1, the main proof of Theorem 1.3.3 splits into three cases: (a) the non-critical case with $e_G(A', B') \ge D$, (b) the critical case with $e_G(A', B') \ge D$ and (c) the case with $e_G(A', B') < D$. The formal definition of 'critical' and a more detailed discussion of the different cases is given in Section 2.7.

The above lemma will be used in all three cases. In these different cases, we will need that the Hamilton cycles or perfect matchings produced by the lemma use appropriate edges between A' and B' (and thus the 'leftover' $H^\diamond$ has suitable properties). In particular, (v) will ensure that we can apply Lemma 2.7.4 in case (b). Similarly, (vi) will ensure that we can apply Lemma 2.7.5 in case (c). (ii) and (vi) will only be relevant in case (c).

Proof of Lemma 2.6.1. Set $H := G[A', B']$ and $W := \{w_1, w_2\}$. First, we construct J_0. If D is even, then (i) is trivial, so we may assume that D is odd (and so n is even). We will construct J_0 such that it satisfies (i) as well as the following additional property:

(i′) If w_1w_2 is an edge in $G[A'] + G[B']$, then w_1w_2 lies in J_0. Moreover, $e_{J_0}(A', B') = 1$ if $|A'|$ is odd and $e_{J_0}(A', B') = 0$ if $|A'|$ is even.

Suppose first that $|A'|$ is even (and so $|B'|$ is even as well). Since our assumptions imply that $\delta(G[A']) \ge \lceil D/2 \rceil \ge 3\varepsilon_0 n$, there exists a matching M'_A in $G[A']$ of size at most $|A_0|+2$ covering all the vertices of $A_0 \cup (A' \cap W)$. Moreover, if w_1w_2 is an edge in $G[A']$, then we can ensure that $w_1w_2 \in M'_A$. Note that $A'' := A' \setminus V(M'_A)$ is a subset of A and $|A''|$ is even. (FR4) implies that $\delta(G[A'']) \ge D - \varepsilon_0 n - 2(|A_0|+2) \ge |A''|/2$. Therefore, there exists a perfect matching M''_A in $G[A'']$ (e.g. by Dirac's theorem). Hence, $M_A := M'_A + M''_A$ is a perfect matching in $G[A']$. Similarly, there is a perfect matching M_B in $G[B']$ such that if w_1w_2 is an edge in $G[B']$, then w_1w_2 is in M_B. Set $J_0 := M_A + M_B$.

Next assume that $|A'|$ is odd. If $D \ge \lfloor n/2 \rfloor$, then Proposition 2.2.3 implies that $e(H - W) > 0$. If $D = n/2 - 1$, then $n = 0 \pmod 4$ and so $|B'| \le n/2 - 1$ since $|A'|$ is odd. Together with Proposition 2.2.1(ii) this implies that $e(H) \ge n/2 - 1$. Since in this case we also have that $\Delta(H) \le \lfloor D/2 \rfloor = n/4 - 1$, it follows that $e(H - W) \ge e(H) - 2\Delta(H) > 0$. Thus in both cases there exists an edge ab in $H - W$ with $a \in A'$ and $b \in B'$. Note that both $|A' \setminus \{a\}|$ and $|B' \setminus \{b\}|$ are even. Moreover, $\delta(G[A' \setminus \{a\}]) \ge \lceil D/2 \rceil - 1 \ge 3\varepsilon_0 n$ and $\delta(G[B' \setminus \{b\}]) \ge \lceil D/2 \rceil - 1 \ge 3\varepsilon_0 n$. Thus we can argue as in the case when $|A'|$ is even to find perfect matchings M_A and M_B in $G[A' \setminus \{a\}]$ and $G[B' \setminus \{b\}]$ respectively such that if w_1w_2 is an edge in $G[A'] + G[B']$ then $w_1w_2 \in M_A + M_B$. Set $J_0 := M_A + M_B + ab$.

This completes the construction of J_0. (If D is even we set $J_0 := \emptyset$.) So (i) and (i′) hold. Let $G' := G - J_0$ and $H' := G'[A', B']$. Since $|A_0| + |B_0| \le \varepsilon_0 n \le \lambda n$, Vizing's theorem implies that we can decompose $G'[A_0] + G'[B_0]$ into λn edge-disjoint (possibly empty) matchings $M_1, \dots, M_{\lambda n}$. By relabeling these matchings if necessary, we may assume that if $w_1w_2 \in E_{G'}(A_0)$ or $w_1w_2 \in E_{G'}(B_0)$, then $w_1w_2 \in M_1$.

Case 1: $e(H) \geq D$.

Note that in this case $\ell = 0$ and $e(H') \geq D - 1$. For each $s = 1, \dots, \lambda n$ in turn we will extend M_s into a Hamilton exceptional system J_s with $e_{J_s}(A', B') = 2$ and such that J_s and $J_{s'}$ are edge-disjoint for all $0 \leq s' < s$. In order to do this, we will first extend M_s into a Hamilton exceptional system candidate F_s by adding two independent $A'B'$-edges f_s and f'_s. We will then use Lemma 2.3.2 to extend F_s into a Hamilton exceptional system J_s. For all s with $1 \leq s \leq \lambda n$, we will choose these edges and sets to satisfy the following:

(α_1) J_s is a Hamilton exceptional system with parameter ε_0 such that $e_{J_s}(A', B') = 2$.

(α_2) Suppose that $d_H(w_1) \geq 2\lambda n$. Then w_1 is an endpoint of f_s.

(α_3) Suppose that $d_H(w_2) \geq 2\lambda n$. Then w_2 is an endpoint of f'_s, unless both $s = 1$ and $w_1 w_2 \in M_1$.

(α_4) J_s contains M_s as well as the edges f_s and f'_s. $J_s - M_s - f_s - f'_s$ only contains A_0A-edges and B_0B-edges of G. J_s is edge-disjoint from $J_0, \dots, J_{s-1}$.

First suppose that $w_1w_2 \in M_1$. We construct J_1 satisfying the above. Our assumption means that w_1w_2 is an edge in $G[A'] + G[B']$, so D is even (or else $w_1w_2 \in J_0$ by (i$'$)). Moreover, $H' = H$ and $D \geq \lfloor n/2 \rfloor$ by (1.3.1) and the fact that D is even. Together with Proposition 2.2.3 this implies that $e(H' - W) = e(H - W) > 0$. Pick an $A'B'$-edge f'_1 in $H' - W$. Let U_1 be the connected component in $M_1 + f'_1$ containing f'_1. So $|U_1| \leq 4$ and $w_1 \notin U_1$. If $d_H(w_1) \geq 2\lambda n$, we can find an $A'B'$-edge f_1 such that w_1 is one endpoint of f_1 and the other endpoint of f_1 does not lie in U_1. If $d_H(w_1) < 2\lambda n$, then the choice of w_1 implies that $\Delta(H) \leq 2\lambda n$. So there exists an $A'B'$-edge f_1 in $H' - V(U_1) = H - V(U_1)$ since $e(H - V(U_1)) \geq e(H) - |U_1|\Delta(H) \geq e(H) - 8\lambda n > 0$. Set $F_1 := M_1 + f_1 + f'_1$. Note that f_1 satisfies (α_2) and that F_1 is a Hamilton exceptional system candidate with $e_{F_1}(A', B') = 2$. By Lemma 2.3.2, we can extend F_1 into a Hamilton exceptional system J_1 with parameter ε_0 in G such that $F_1 \subseteq J_1$ and such that $J_1 - F_1$ only contains A_0A-edges and B_0B-edges of G.

Next, suppose that for some $1 \leq s \leq \lambda n$ we have already constructed $J_0, \dots, J_{s-1}$ satisfying (α_1)–(α_4). So $s \geq 2$ if $w_1w_2 \in M_1$. Let $G_s := G - \sum_{j=s}^{\lambda n} M_j - \sum_{j=0}^{s-1} J_j$ and $H_s := G_s[A', B']$. Note that

$$e(H_s) \geq e(H) - 2(s-1) - 1 \geq D - 2\lambda n. \tag{2.6.1}$$

Moreover, note that $d_{G_s}(v, A) \geq d_G(v, A) - 2(s-1) - 1 \geq \sqrt{\varepsilon_0} n$ for all $v \in A_0$ and $d_{G_s}(v, B) \geq \sqrt{\varepsilon_0} n$ for all $v \in B_0$.

We first pick the edge f'_s as follows. If $d_H(w_2) \geq 2\lambda n$, then $d_{H_s}(w_2) \geq d_H(w_2) - s \geq \lambda n$. So we can pick an $A'B'$-edge f'_s of H_s such that w_2 is an endpoint of f'_s and the connected component U_s of $M_s + f'_s$ containing f'_s does not contain w_1. If $d_H(w_2) < 2\lambda n$, then pick an $A'B'$-edge f'_s of H_s such that the connected component U_s of $M_s + f'_s$ containing f'_s does not contain w_1. To see that such an edge exists, note that in this case the neighbour w'_1 of w_1 in M_s satisfies $d_H(w'_1) \leq d_H(w_2) < 2\lambda n$ (if w'_1 exists) and that (2.6.1) implies that $e(H_s) \geq D - 2\lambda n > D/2 + 2\lambda n \geq d_H(w_1) + 2\lambda n$. Observe that in both cases $|U_s| \leq 4$.

We now pick the edge f_s as follows. If $d_H(w_1) \geq 2\lambda n$, then $d_{H_s}(w_1) \geq d_H(w_1) - s \geq \lambda n$. So we can find an $A'B'$-edge f_s of H_s such that w_1 is one endpoint of f_s and the other endpoint of f_s does not lie in U_s. If $d_H(w_1) < 2\lambda n$, then $\Delta(H) \leq 2\lambda n$

and thus (2.6.1) implies that

$$e(H_s - V(U_s)) \geq D - 2\lambda n - 2\lambda n|U_s| \geq 1.$$

So there exists an $A'B'$-edge f_s in $H_s - V(U_s)$.

In all cases the edges f_s and f'_s satisfy (α_2) and (α_3). Set $F_s := M_s + f_s + f'_s$. Clearly, F_s is a Hamilton exceptional system candidate with $e_{F_s}(A', B') = 2$. Recall that $d_{G_s}(v, A) \geq \sqrt{\varepsilon_0}n$ for all $v \in A_0$ and $d_{G_s}(v, B) \geq \sqrt{\varepsilon_0}n$ for all $v \in B_0$. Thus by Lemma 2.3.2, we can extend F_s into a Hamilton exceptional system J_s with parameter ε_0 such that $F_s \subseteq J_s \subseteq G_s + F_s$ and such that $J_s - F_s$ only contains A_0A-edges and B_0B-edges of G_s. Hence we have constructed $J_1, \dots, J_{\lambda n}$ satisfying (α_1)–(α_4). So (iii) holds. Note (ii) and (vi) are vacuously true.

To verify (iv), recall that $\mathcal{J} := J_0 \cup \dots \cup J_{\lambda n}$ and $H^\diamond = G[A', B'] - \mathcal{J}$. For all $1 \leq s \leq \lambda n$ we have $e_{J_s}(A', B') = 2$ by (iii). Moreover, (i) and (i′) together imply that $e_{J_0}(A', B') = 1$ if and only if both $|A'|$ and D are odd. Therefore, $e_{\mathcal{J}}(A', B') \leq \phi n$. Moreover, since $e(H^\diamond) = e(H) - 2\lambda n - e_{J_0}(A', B')$, Proposition 2.2.2(i) implies that $e(H^\diamond)$ is even. Thus (iv) holds.

To verify (v), note that if $d_H(w_1) \leq 2\lambda n$ then clearly $d_{H^\diamond}(w_1) \leq 2\lambda n \leq (D - \phi n)/2$. If $d_H(w_1) \geq 2\lambda n$ then (α_2) implies that $d_{J_s[A',B']}(w_1) = 1$ for all $1 \leq s \leq \lambda n$. Hence $d_{H^\diamond}(w_1) \leq \lfloor D/2 \rfloor - \lambda n = (D - \phi n)/2$. Now suppose that $D = n/2 - 1$ and so $n = 0 \pmod 4$ by (1.3.1). Thus D is odd and so (i′) implies that if w_1w_2 is an edge in $G[A'] + G[B']$, then $w_1w_2 \in J_0$. In particular $w_1w_2 \notin M_1$. (Note that if $w_1w_2 \in G[A', B']$, then w_1w_2 is not contained in M_1 either since $M_1 \subseteq G[A_0] + G[B_0]$.) Thus in the case when $d_H(w_2) \geq 2\lambda n$, (α_3) implies that $d_{J_s[A',B']}(w_2) = 1$ for all $1 \leq s \leq \lambda n$. Hence $d_{H^\diamond}(w_2) \leq \lfloor D/2 \rfloor - \lambda n = (D - \phi n)/2$. If $d_H(w_2) \leq 2\lambda n$ then clearly $d_{H^\diamond}(w_2) \leq 2\lambda n \leq (D - \phi n)/2$. Therefore (v) holds.

Case 2: $e(H) < D$

Together with Proposition 2.2.1(ii) this implies that $n = 0 \pmod 4$, $D = n/2 - 1$ and $|A'| = n/2 = |B'|$. So D is odd and $|A'|$ is even. In particular, by Proposition 2.2.2(i) $e(H)$ is even and by (i) and (i′) J_0 is a perfect matching with $e_{J_0}(A', B') = 0$. Moreover, Proposition 2.2.4 implies that $\Delta(H) \leq e(H)/2$ in this case (recall that $H := G[A', B']$).

Note that each M_s is a matching exceptional system candidate. By Lemma 2.3.2, for each $1 \leq s \leq \min\{\ell, \lambda n\}$ in turn, we can extend M_s into a matching exceptional system J_s with parameter ε_0 in $G' = G - J_0$ such that $M_s \subseteq J_s$, and such that J_s and $J_{s'}$ are edge-disjoint whenever $1 \leq s' < s \leq \min\{\ell, \lambda n\}$. Thus (ii) holds.

If $\ell \geq \lambda n$, then $e(H) \leq D - 2\lambda n = D - \phi n + 1$. But since $e(H)$ is even and $D - \phi n + 1$ is odd this means that $e(H) \leq D - \phi n$. Thus $\Delta(H) \leq e(H)/2 \leq (D - \phi n)/2$. Moreover, $d_{\mathcal{J}}(v) = 2\lambda n + d_{J_0}(v) = \phi n$ for all $v \in V_0$. Hence (iv)–(vi) hold since $H^\diamond = H$. ((iii) is vacuously true.)

Therefore, we may assume that $\ell < \lambda n$. Using a similar argument as in Case 1, for all $\ell < s \leq \lambda n$ we can extend the matchings M_s into edge-disjoint Hamilton exceptional systems J_s satisfying (α_1)–(α_4) and which are edge-disjoint from $J_0, \dots, J_\ell$. Indeed, suppose that for $\ell < s \leq \lambda n$ we have already constructed $J_{\ell+1}, \dots, J_{s-1}$ satisfying (α_1)–(α_4). (Note that (i′) implies that the exception in (α_3) is not relevant.) The fact that D is odd and $e(H)$ is even implies that $\ell = (D - e(H) - 1)/2$. Then defining H_s analogously to Case 1, we have

$$e(H_s) \geq e(H) - 2(s - 1 - \ell) = D - 2s \geq D - 2\lambda n,$$

where in the first inequality we use that $e_{J_0}(A', B') = 0$ by (i′). So the analogue of (2.6.1) holds. Hence we can proceed exactly as in Case 1 to construct J_s (the remaining calculations go through as before). Thus (iii) holds.

To verify (iv), note that $e_{\mathcal{J}}(A', B') = 2(\lambda n - \ell)$. So

$$e(H^\diamond) = e(H) - 2(\lambda n - \ell) = e(H) - 2\lambda n + (D - e(H) - 1) = D - \phi n. \tag{2.6.2}$$

In particular, $e(H^\diamond)$ is even and $e_{\mathcal{J}}(A', B') = e(H) - e(H^\diamond) < \phi n$. So (iv) holds.

In order to verify (vi), recall that $\Delta(H) \leq e(H)/2$. Moreover, note that (α_2) implies that if $d_H(w_1) \geq 2\lambda n$, then $d_{J_s[A',B']}(w_1) = 1$ for all $\ell < s \leq \lambda n$. Hence

$$\begin{aligned} d_{H^\diamond}(w_1) &= d_H(w_1) - (\lambda n - \ell) = \Delta(H) - \lambda n + \ell \\ &\leq e(H)/2 - \lambda n + (D - e(H) - 1)/2 = (D - \phi n)/2 \overset{(2.6.2)}{=} e(H^\diamond)/2. \end{aligned}$$

Similarly if $d_H(w_2) \geq 2\lambda n$, then $d_{H^\diamond}(w_2) \leq e(H^\diamond)/2$. If $d_H(w_1) \leq 2\lambda n$, then $d_{H^\diamond}(w_1) \leq 2\lambda n \leq e(H^\diamond)/2$ by (2.6.2) and the analogue also holds for w_2. Thus in all cases $d_H(w_1), d_H(w_2) \leq e(H^\diamond)/2$. Our choice of w_1 and w_2 implies that for all $v \in V(G) \setminus W$ we have

$$d_H(v) \leq (e(H) + 3)/3 \leq (D + 3)/3 \overset{(2.6.2)}{<} e(H^\diamond)/2.$$

Therefore, $\Delta(H^\diamond) \leq e(H^\diamond)/2$. Together with (2.6.2) this implies (vi) and thus (v). □

The next lemma implies that each of the exceptional systems J_s guaranteed by Lemma 2.6.1 can be extended into a Hamilton cycle (if J_s is a Hamilton exceptional system) or into two perfect matchings (if J_s is a matching exceptional system and both $|A'|$ and $|B'|$ are even).

LEMMA 2.6.2. *Suppose that* $0 < 1/n \ll \varepsilon_0 \leq \lambda \ll 1$ *and that* $n, \lambda n, K \in \mathbb{N}$. *Suppose that* (G, A, A_0, B, B_0) *is an* (ε_0, K)*-framework such that* $\delta(G[A]) \geq 4|A|/5$ *and* $\delta(G[B]) \geq 4|B|/5$. *Let* $J_1, \dots, J_{\lambda n}$ *be exceptional systems with parameter* ε_0. *Suppose that* G *and* $J_1, \dots, J_{\lambda n}$ *are pairwise edge-disjoint. Then there are edge-disjoint subgraphs* $H_1, \dots, H_{\lambda n}$ *in* $G + \sum_{s=1}^{\lambda n} J_s$ *which satisfy the following properties:*

(i) $J_s \subseteq H_s$ *and* $E(H_s - J_s) \subseteq E(G[A] + G[B])$ *for all* $1 \leq s \leq \lambda n$.

(ii) *If* J_s *is a Hamilton exceptional system, then* H_s *is a Hamilton cycle on* $V(G)$.

(iii) *If* J_s *is a matching exceptional system, then* H_s *is an union of a Hamilton cycle on* $A' = A \cup A_0$ *and a Hamilton cycle on* $B' = B \cup B_0$.

Proof. Recall that, given an exceptional system J, we have defined matchings J^*_A, J^*_B and $J^* = J^*_A + J^*_B$ in Section 2.3. We will write $J^*_{s,A} := (J_s)^*_A$ and $J^*_{s,B} := (J_s)^*_B$. For each $s \leq \lambda n$ in turn, we will find a subgraph H^*_s of $G[A] + G[B] + J^*_s$ containing J^*_s such that H^*_s is edge-disjoint from $H^*_1, \dots, H^*_{s-1}$. Moreover, H^*_s will be the union of two cycles C_A and C_B such that C_A is a Hamilton cycle on A which is consistent with $J^*_{s,A}$ and C_B is a Hamilton cycle on B which is consistent with $J^*_{s,B}$. (Recall from Section 2.3 that we always view different J^*_i as being edge-disjoint from each other. So asking H^*_s to be edge-disjoint from $H^*_1, \dots, H^*_{s-1}$ is the same as asking $H^*_s - J^*_s$ to be edge-disjoint from $H^*_1 - J^*_1, \dots, H^*_{s-1} - J^*_{s-1}$.)

Suppose that for some $1 \leq s \leq \lambda n$ we have already found $H^*_1, \dots, H^*_{s-1}$. For all $i < s$, let $H_i := H^*_i - J^*_i + J_i$. Let $G_s := G - (H_1 \cup \dots \cup H_{s-1})$. First we construct C_A as follows. Recall from (2.3.1) that $J^*_{s,A}$ is a matching of size at most

$2\sqrt{\varepsilon_0}n$. Note that $\delta(G_s[A]) \geq \delta(G[A]) - 2s \geq (4/5 - 5\lambda n)|A|$. So we can greedily find a path P_A of length at most $6\sqrt{\varepsilon_0}n$ in $G_s[A] + J^*_{s,A}$ such that P_A is consistent with $J^*_{s,A}$. Let u and v denote the endpoints of P_A. Let G^A_s be the graph obtained from $G_s[A] - V(P_A)$ by adding a new vertex w whose neighbourhood is precisely $(N_{G_s}(u) \cap N_{G_s}(v)) \setminus V(P_A)$. Note that $\delta(G^A_s) \geq |G^A_s|/2$ (with room to spare). Thus G^A_s contains a Hamilton cycle C'_A by Dirac's theorem. But C'_A corresponds to a Hamilton cycle C_A of $G_s[A] + J^*_{s,A}$ that is consistent with $J^*_{s,A}$. Similarly, we can find a Hamilton cycle C_B of $G_s[B] + J^*_{s,B}$ that is consistent with $J^*_{s,B}$. Let $H^*_s = C_A + C_B$. This completes the construction of $H^*_1, \dots, H^*_{\lambda n}$.

For each $1 \leq s \leq \lambda n$ we take $H_s := H^*_s - J^*_s + J_s$. Then (i) holds. Proposition 2.3.1 implies (ii) and (iii). □

By combining Lemmas 2.6.1 and 2.6.2 we obtain the following result, which guarantees a set of edge-disjoint Hamilton cycles covering all edges of $G[A_0]$ and $G[B_0]$.

LEMMA 2.6.3. *Suppose that* $0 < 1/n \ll \varepsilon_0 \ll \phi \ll 1$ *and that* $D, n, (D - \phi n)/2, K \in \mathbb{N}$. *Let* G *be a* D*-regular graph on* n *vertices with* $D \geq n - 2\lfloor n/4 \rfloor - 1$. *Suppose that* (G, A, A_0, B, B_0) *is an* (ε_0, K)*-framework with* $\Delta(G[A', B']) \leq D/2$. *Let* w_1 *and* w_2 *be (fixed) vertices of* G *such that* $d_{G[A',B']}(w_1) \geq d_{G[A',B']}(w_2) \geq d_{G[A',B']}(v)$ *for all* $v \in V(G) \setminus \{w_1, w_2\}$. *Then there exists a* ϕn*-regular spanning subgraph* G_0 *of* G *which satisfies the following properties:*

(i) $G[A_0] + G[B_0] \subseteq G_0$.
(ii) $e_{G_0}(A', B') \leq \phi n$ *and* $e_{G-G_0}(A', B')$ *is even.*
(iii) G_0 *can be decomposed into* $\lfloor e_{G_0}(A', B')/2 \rfloor$ *Hamilton cycles and* $\phi n - 2\lfloor e_{G_0}(A', B')/2 \rfloor$ *perfect matchings. Moreover, if* $e_G(A', B') \geq D$, *then this decomposition of* G_0 *uses* $\lfloor \phi n/2 \rfloor$ *Hamilton cycles and one perfect matching if* D *is odd.*
(iv) *Let* $H^\diamond := G[A', B'] - G_0$. *Then* $d_{H^\diamond}(w_1) \leq (D - \phi n)/2$. *Furthermore, if* $D = n/2 - 1$ *then* $d_{H^\diamond}(w_2) \leq (D - \phi n)/2$.
(v) *If* $e_G(A', B') < D$, *then* $\Delta(H^\diamond) \leq e(H^\diamond)/2 \leq (D - \phi n)/2$.

Proof. Let

$$\ell := \left\lfloor \frac{\max\{0, D - e_G(A', B')\}}{2} \right\rfloor \quad \text{and} \quad \lambda n := \lfloor \phi n/2 \rfloor = \begin{cases} (\phi n - 1)/2 & \text{if } D \text{ is odd,} \\ \phi n/2 & \text{if } D \text{ is even.} \end{cases}$$

(The last equality holds since our assumption that $(D - \phi n)/2 \in \mathbb{N}$ implies that D is odd if and only if ϕn is odd.) So ℓ, ϕ and λ are as in Lemma 2.6.1. Thus we can apply Lemma 2.6.1 to G in order to obtain $\lambda n + 1$ subgraphs $J_0, \dots, J_{\lambda n}$ as described there. Let G' be the graph obtained from $G[A'] + G[B']$ by removing all the edges in $J_0 \cup \dots \cup J_{\lambda n}$. Recall that J_0 is either a perfect matching in G or empty. Since each of $J_1, \dots, J_{\lambda n}$ is an exceptional system and so by (EC3) we have $e_{J_s}(A) = 0$ for all $1 \leq s \leq \lambda n$, it follows that $\delta(G'[A]) \geq \delta(G[A]) - 1 \geq 4|A|/5$, where the final inequality follows from (FR3) and (FR4). Similarly $\delta(G'[B]) \geq 4|B|/5$. So we can apply Lemma 2.6.2 with G' playing the role of G in order to extend $J_1, \dots, J_{\lambda n}$ into edge-disjoint subgraphs $H_1, \dots, H_{\lambda n}$ of $G' + \sum_{s=1}^{\lambda n} J_s$ such that

(a) H_s is a Hamilton cycle on $V(G)$ which contains precisely two $A'B'$-edges for all $\ell < s \leq \lambda n$;

(b) H_s is the union of a Hamilton cycle on A' and a Hamilton cycle on B' for all $1 \leq s \leq \min\{\ell, \lambda n\}$.

Indeed, the property $e_{H_s}(A', B') = 2$ in (a) follows from Lemma 2.6.1(iii) and 2.6.2(i). Let $G_0 := J_0 + \sum_{s=1}^{\lambda n} H_s$. Then (i) holds since by Lemma 2.6.1 all the $J_0, \dots, J_{\lambda n}$ together cover all edges in $G[A_0]$ and $G[B_0]$. Let $\mathcal{J}_{\rm HC}$ be the union of all J_s with $\ell < s \leq \lambda n$ and let $\mathcal{J}$ be the union of all J_s with $0 \leq s \leq \lambda n$. The definition of G_0, Lemma 2.6.1(ii),(iii) and Lemma 2.6.2(i) together imply that $G_0[A', B'] = \mathcal{J}[A', B'] = J_0[A', B'] + \mathcal{J}_{\rm HC}[A', B']$ and so

$$e_{G_0}(A', B') = e_{\mathcal{J}}(A', B') \tag{2.6.3}$$

$$= e_{J_0}(A', B') + 2(\max\{0, \lambda n - \ell\}). \tag{2.6.4}$$

Together with Lemma 2.6.1(iv), (2.6.3) implies (ii). Moreover, the graph $H^\diamond$ defined in (iv) is the same as the graph $H^\diamond$ defined in Lemma 2.6.1(iv). Thus (iv) and (v) follow from Lemma 2.6.1(v) and (vi).

So it remains to verify (iii). Note that if $\ell > 0$ then $e_G(A', B') < D$ and so $n = 0 \pmod 4$, $D = n/2 - 1$ and $|A'| = n/2 = |B'|$ by Proposition 2.2.1(ii). In particular, both A' and B' are even and so for all $1 \leq s \leq \ell$ the graph H_s can be decomposed into two edge-disjoint perfect matchings. Recall that by Lemma 2.6.1(i) the graph J_0 is a perfect matching if D is odd and empty if D is even. Thus, if $\ell \leq \lambda n$, then G_0 can be decomposed into $\lambda n - \ell$ edge-disjoint Hamilton cycles and $n_{\rm match}$ edge-disjoint perfect matchings, where $n_{\rm match} = 2\ell$ if D is even and $n_{\rm match} = 2\ell + 1$ if D is odd. In particular, this implies the 'moreover part' of (iii) (since $\ell = 0$ if $e_G(A', B') \geq D$). Also, (2.6.4) together with the fact that $e_{J_0}(A', B') \leq 1$ by Lemma 2.6.1(i) implies that $\lambda n - \ell = \lfloor e_{G_0}(A', B')/2 \rfloor$ and so $\phi n - 2\lfloor e_{G_0}(A', B')/2 \rfloor = n_{\rm match}$. Thus (iii) holds in this case. If $\ell > \lambda n$, then (a) implies that there are no Hamilton cycles at all in the decomposition. Also (2.6.4) implies that $\lfloor e_{G_0}(A', B')/2 \rfloor = 0$, as required in (iii). Similarly, (b) implies that $n_{\rm match} = 2\lambda n$ if D is even and $n_{\rm match} = 2\lambda n + 1$ if D is odd, which also agrees with (iii). □

2.7. Constructing Localized Exceptional Systems

Suppose that (G, A, A_0, B, B_0) is an (ε_0, K)-framework and that G_0 is the spanning subgraph of our given D-regular graph G obtained by Lemma 2.6.3. Set $G' := G - G_0$. (So G' has no edges inside A_0 or B_0.) Roughly speaking, the aim of this section is to decompose $G' - G'[A] - G'[B]$ into edge-disjoint exceptional systems. Each of these exceptional systems J will then be extended into a Hamilton cycle (in the case when J is a Hamilton exceptional system) or into two perfect matchings (in the case when J is a matching exceptional system). We will ensure that all but a small number of these exceptional systems are localized (with respect to some (K, m, ε_0)-partition $\mathcal{P}$ of $V(G)$ refining the partition A, A_0, B, B_0). Moreover, for all $1 \leq i, i' \leq K$, the number of (i, i')-localized exceptional systems in our decomposition will be the same. (Recall that (i, i')-localized exceptional systems were defined in Section 2.3.)

However, rather than decomposing the above 'leftover' $G' - G'[A] - G'[B]$ in a single step, we actually need to proceed in two steps: initially, we find a small number of exceptional systems J which have some additional useful properties (e.g. the number of $A'B'$-edges of J is either zero or two). These exceptional systems will be used to construct the robustly decomposable graph $G^{\rm rob}$. (Recall that the role of $G^{\rm rob}$ was discussed in Section 2.1.) Let $G'' := G - G_0 - G^{\rm rob}$. Some of the additional properties of the exceptional systems contained in $G^{\rm rob}$ then allow us to find the desired decomposition of $G^\diamond := G'' - G''[A] - G''[B]$. (We need to proceed in two steps rather than one as we have little control over the structure of $G^{\rm rob}$.)

Recall that in order to construct the required (localized) exceptional systems, we will distinguish three cases:

(a) the case when G is 'non-critical' and contains at least D $A'B'$-edges (see Lemma 2.7.3);
(b) the case when G is 'critical' and contains at least D $A'B'$-edges (see Lemma 2.7.4);
(c) the case when G contains less than D $A'B'$-edges (see Lemma 2.7.5).

Each of the three lemmas above is formulated in such a way that we can apply it twice: firstly to obtain the small number of exceptional systems needed for the robustly decomposable graph $G^{\rm rob}$ and secondly for the decomposition of the graph $G^\diamond$ into exceptional systems. The proofs of all the results in this section are deferred until Chapter 3.

2.7.1. Critical Graphs. Roughly speaking, G is critical if most of its $A'B'$-edges are incident to only a few vertices. More precisely, given a partition A', B' of $V(G)$ and $D \in \mathbb{N}$, we say that G is *critical* (with respect to A', B' and D) if both of the following hold:

- $\Delta(G[A', B']) \geq 11D/40$;
- $e(H) \leq 41D/40$ for all subgraphs H of $G[A', B']$ with $\Delta(H) \leq 11D/40$.

Note that the property of G being critical depends only on D and the partition $A' = A \cup A_0$ and $B' = B \cup B_0$ of $V(G)$, which is fixed after we have applied Proposition 2.2.5 to obtain a framework (G, A, A_0, B, B_0). In particular, it does not depend on the choice of the (K, m, ε_0)-partition $\mathcal{P}$ of $V(G)$ refining A, A_0, B, B_0. (In the proof of Theorem 1.3.3 we will fix a framework (G, A, A_0, B, B_0), but will then choose two different partitions refining A, A_0, B, B_0.)

One example of a critical graph is the following: $G_{\rm crit}$ consists of two disjoint cliques on $(n-1)/2$ vertices with vertex set A and B respectively, where $n = 1 \pmod 4$. In addition, there is a vertex a which is adjacent to exactly half of the vertices in each of A and B. Also, add a perfect matching M between those vertices of A and those vertices in B not adjacent to a. Let $A' := A \cup \{a\}$, $B' := B$ and $D := (n-1)/2$. Then $G_{\rm crit}$ is critical, and D-regular with $e(A', B') = D$. Note that $e(M) = D/2$. To obtain a Hamilton decomposition of $G_{\rm crit}$, we will need to decompose $G_{\rm crit}[A', B']$ into $D/2$ Hamilton exceptional system candidates J_s (which need to be matchings of size exactly two in this case). In this example, this decomposition is essentially unique: every J_s has to consist of exactly one edge in M and one edge incident to a. Note that in this way, every edge between a and B yields a 'connection' (i.e. a maximal path) between A' and B' required in (ESC4).

The following lemma (proved in Section 3.1) collects some properties of critical graphs. In particular, there is a set W consisting of between one and three vertices with many neighbours in both A and B. We will need to use $A'B'$-edges incident to one or two vertices of W to provide 'connections' between A' and B' when constructing the Hamilton exceptional system candidates in the critical case (b).

LEMMA 2.7.1. *Suppose that $0 < 1/n \ll 1$ and that $D, n \in \mathbb{N}$ with $D \geq n - 2\lfloor n/4 \rfloor - 1$. Let G be a D-regular graph on n vertices and let A', B' be a partition of $V(G)$ with $|A'|, |B'| \geq D/2$ and $\Delta(G[A', B']) \leq D/2$. Suppose that G is critical. Let W be the set of vertices $w \in V(G)$ such that $d_{G[A',B']}(w) \geq 11D/40$. Then the following properties are satisfied:*

(i) $1 \leq |W| \leq 3$.
(ii) *Either $D = (n-1)/2$ and $n = 1 \pmod 4$, or $D = n/2 - 1$ and $n = 0 \pmod 4$. Furthermore, if $n = 1 \pmod 4$, then $|W| = 1$.*
(iii) $e_G(A', B') \leq 17D/10 + 5 < n$.

Recall from Proposition 2.2.1(ii) that we have $e_G(A', B') \geq D$ unless $D = n/2 - 1$, $n = 0 \pmod 4$ and $|A| = |B| = n/2$. Together with Lemma 2.7.1(ii) this shows that in order to find the decomposition into exceptional systems, we can distinguish the following three cases.

COROLLARY 2.7.2. *Suppose that $0 < 1/n \ll 1$ and that $D, n \in \mathbb{N}$ with $D \geq n - 2\lfloor n/4 \rfloor - 1$. Let G be a D-regular graph on n vertices and let A', B' be a partition of $V(G)$ with $|A'|, |B'| \geq D/2$ and $\Delta(G[A', B']) \leq D/2$. Then exactly one of the following holds:*

(a) *$e_G(A', B') \geq D$ and G is not critical.*
(b) *$e_G(A', B') \geq D$ and G is critical. In particular, $e_G(A', B') < n$ and either $D = (n-1)/2$ and $n = 1 \pmod 4$, or $D = n/2 - 1$ and $n = 0 \pmod 4$.*
(c) *$e_G(A', B') < D$. In particular, $D = n/2 - 1$, $n = 0 \pmod 4$ and $|A| = |B| = n/2$.*

2.7.2. Decomposition into Exceptional Systems. Recall from the beginning of Section 2.7 that our aim is to find a decomposition of $G - G_0 - G[A] - G[B]$ into suitable exceptional systems (in particular, most of these exceptional systems have to be localized). The following lemma (proved in Section 3.2) states that this can be done if we are in Case (a) of Corollary 2.7.2, i.e. if G is not critical and $e_G(A', B') \geq D$.

LEMMA 2.7.3. *Suppose that $0 < 1/n \ll \varepsilon_0 \ll \varepsilon \ll \lambda, 1/K \ll 1$, that $D \geq n/3$, that $0 \leq \phi \ll 1$ and that $D, n, K, m, \lambda n/K^2, (D - \phi n)/(2K^2) \in \mathbb{N}$. Suppose that the following conditions hold:*

(i) *G is a D-regular graph on n vertices.*
(ii) *$\mathcal{P}$ is a (K, m, ε_0)-partition of $V(G)$ such that $D \leq e_G(A', B') \leq \varepsilon_0 n^2$ and $\Delta(G[A', B']) \leq D/2$. Furthermore, G is not critical.*
(iii) *G_0 is a subgraph of G such that $G[A_0] + G[B_0] \subseteq G_0$, $e_{G_0}(A', B') \leq \phi n$ and $d_{G_0}(v) = \phi n$ for all $v \in V_0$.*
(iv) *Let $G^\diamond := G - G[A] - G[B] - G_0$. $e_{G^\diamond}(A', B')$ is even and $(G^\diamond, \mathcal{P})$ is a $(K, m, \varepsilon_0, \varepsilon)$-exceptional scheme.*

Then there exists a set $\mathcal{J}$ consisting of $(D - \phi n)/2$ edge-disjoint Hamilton exceptional systems with parameter ε_0 in $G^\diamond$ which satisfies the following properties:

(a) *Together all the Hamilton exceptional systems in $\mathcal{J}$ cover all edges of $G^\diamond$.*
(b) *For all $1 \leq i, i' \leq K$, the set $\mathcal{J}$ contains $(D - (\phi + 2\lambda)n)/(2K^2)$ (i, i')-HES. Moreover, $\lambda n/K^2$ of these (i, i')-HES J are such that $e_J(A', B') = 2$.*

Note that (b) implies that $\mathcal{J}$ contains λn Hamilton exceptional systems which might not be localized. This will make them less useful for our purposes and we extend them into Hamilton cycles in a separate step. On the other hand, the lemma is 'robust' in the sense that we can remove a sparse subgraph G_0 before we find the decomposition $\mathcal{J}$ into Hamilton exceptional systems. In our first application of Lemma 2.7.3 (i.e. to construct the exceptional systems for the robustly decomposable graph G^{rob}), we will let G_0 be the graph obtained from Lemma 2.6.3. In the second application, G_0 also includes G^{rob}. In our first application of Lemma 2.7.3, we will only use the (i,i')-HES J with $e_J(A',B')=2$.

The next lemma is an analogue of Lemma 2.7.3 for the case when G is critical and $e_G(A',B') \geq D$. By Corollary 2.7.2(b) we know that in this case $D=(n-1)/2$ or $D=n/2-1$. (Again we defer the proof to Section 3.3.)

LEMMA 2.7.4. *Suppose that* $0 < 1/n \ll \varepsilon_0 \ll \varepsilon \ll \lambda, 1/K \ll 1$, *that* $D \geq n - 2\lfloor n/4 \rfloor - 1$, *that* $0 \leq \phi \ll 1$ *and that* $n, K, m, \lambda n/K^2, (D-\phi n)/(400K^2) \in \mathbb{N}$. *Suppose that the following conditions hold:*

(i) *G is a D-regular graph on n vertices.*
(ii) *$\mathcal{P}$ is a (K,m,ε_0)-partition of $V(G)$ such that $e_G(A',B') \geq D$ and $\Delta(G[A',B']) \leq D/2$. Furthermore, G is critical. In particular, $e_G(A',B') < n$ and $D=(n-1)/2$ or $D=n/2-1$ by Lemma 3.1.1(ii) and (iii).*
(iii) *G_0 is a subgraph of G such that $G[A_0]+G[B_0] \subseteq G_0$, $e_{G_0}(A',B') \leq \phi n$ and $d_{G_0}(v)=\phi n$ for all $v \in V_0$.*
(iv) *Let $G^\diamond := G - G[A] - G[B] - G_0$. $e_{G^\diamond}(A',B')$ is even and $(G^\diamond, \mathcal{P})$ is a $(K,m,\varepsilon_0,\varepsilon)$-exceptional scheme.*
(v) *Let w_1 and w_2 be (fixed) vertices such that $d_{G[A',B']}(w_1) \geq d_{G[A',B']}(w_2) \geq d_{G[A',B']}(v)$ for all $v \in V(G) \setminus \{w_1,w_2\}$. Suppose that*

$$d_{G^\diamond[A',B']}(w_1), d_{G^\diamond[A',B']}(w_2) \leq (D-\phi n)/2. \tag{2.7.1}$$

Then there exists a set $\mathcal{J}$ consisting of $(D-\phi n)/2$ edge-disjoint Hamilton exceptional systems with parameter ε_0 in $G^\diamond$ which satisfies the following properties:

(a) *Together the Hamilton exceptional systems in $\mathcal{J}$ cover all edges of $G^\diamond$.*
(b) *For each $1 \leq i,i' \leq K$, the set $\mathcal{J}$ contains $(D-(\phi+2\lambda)n)/(2K^2)$ (i,i')-HES. Moreover, $\lambda n/K^2$ of these (i,i')-HES are such that*
    (b_1) *$e_J(A',B')=2$ and*
    (b_2) *$d_{J[A',B']}(w)=1$ for all $w \in \{w_1,w_2\}$ with $d_{G[A',B']}(w) \geq 11D/40$.*

Similarly as for Lemma 2.7.3, (b) implies that $\mathcal{J}$ contains λn Hamilton exceptional systems which might not be localized. Another similarity is that when constructing the robustly decomposable graph G^{rob}, we only use those Hamilton exceptional systems J which have some additional useful properties, namely (b_1) and (b_2) in this case. This guarantees that (2.7.1) will be satisfied in the second application of Lemma 2.7.4 (i.e. after the removal of G^{rob}), by 'tracking' the degrees of the high degree vertices w_1 and w_2. Indeed, if $d_{G[A',B']}(w_2) \geq 11D/40$, then ($b_2$) will imply that $d_{G^{\mathrm{rob}}[A',B']}(w_i)$ is large for $i=1,2$. This in turn means that after removing G^{rob}, in the leftover graph $G^\diamond$, $d_{G^\diamond[A',B']}(w_i)$ is comparatively small, i.e. condition (2.7.1) will hold in the second application of Lemma 2.7.4.

Condition (2.7.1) itself is natural for the following reason: suppose for example that it is violated for w_1 and that $w_1 \in A_0$. Then for some Hamilton exceptional system J returned by the lemma, both edges of J incident to w_1 will have their other endpoint in B'. So (the edges at) w_1 cannot be used as a 'connection' between A' and B' in the Hamilton cycle which will extend J, and it may be impossible to find these connections elsewhere.

The next lemma is an analogue of Lemma 2.7.3 for the case when $e_G(A', B') < D$. (Again we defer the proof to Section 3.4.) Recall that Proposition 2.2.1(ii) (or Corollary 2.7.2) implies that in this case we have $n = 0 \pmod 4$, $D = n/2 - 1$ and $|A'| = |B'| = n/2$. In particular, $|A'|$ and $|B'|$ are both even. This agrees with the fact that the decomposition may also involve matching exceptional systems in the current case: we will later extend each such system to a cycle spanning A' and one spanning B'. As $|A'|$ and $|B'|$ are both even, these cycles correspond to two edge-disjoint perfect matchings in G.

LEMMA 2.7.5. *Suppose that* $0 < 1/n \ll \varepsilon_0 \ll \varepsilon \ll \lambda, 1/K \ll 1$, *that* $0 \le \phi \ll 1$ *and that* $n/4, K, m, \lambda n/K^2, (n/2 - 1 - \phi n)/(2K^2) \in \mathbb{N}$. *Suppose that the following conditions hold:*

(i) *G is an $(n/2 - 1)$-regular graph on n vertices.*
(ii) *$\mathcal{P}$ is a (K, m, ε_0)-partition of $V(G)$ such that $\Delta(G[A', B']) \le n/4$ and $|A'| = |B'| = n/2$.*
(iii) *G_0 is a subgraph of G such that $G[A_0] + G[B_0] \subseteq G_0$ and $d_{G_0}(v) = \phi n$ for all $v \in V_0$.*
(iv) *Let $G^\diamond := G - G[A] - G[B] - G_0$. $e_{G^\diamond}(A', B')$ is even and $(G^\diamond, \mathcal{P})$ is a $(K, m, \varepsilon_0, \varepsilon)$-exceptional scheme.*
(v) $\Delta(G^\diamond[A', B']) \le e_{G^\diamond}(A', B')/2 \le (n/2 - 1 - \phi n)/2$.

Then there exists a set $\mathcal{J}$ consisting of $(n/2 - 1 - \phi n)/2$ edge-disjoint exceptional systems in $G^\diamond$ which satisfies the following properties:

(a) *Together the exceptional systems in $\mathcal{J}$ cover all edges of $G^\diamond$. Each J in $\mathcal{J}$ is either a Hamilton exceptional system with $e_J(A', B') = 2$ or a matching exceptional system.*
(b) *For all $1 \le i, i' \le K$, the set $\mathcal{J}$ contains $(n/2 - 1 - (\phi n + 2\lambda))/(2K^2)$ (i, i')-ES.*

As in the other two cases, we will use the exceptional systems in (b) to construct the robustly decomposable graph $G^{\rm rob}$. Unlike the critical case with $e_G(A', B') \ge D$, there is no need to 'track' the degrees of the vertices w_i of high degree in $G[A', B']$ this time. Indeed, let $G'' := G - G_0 - G^{\rm rob}$, where G_0 is the graph defined by Lemma 2.6.3. Then $G''[A', B']$ is the union of all those J in $\mathcal{J}$ (from the first application of Lemma 2.7.5) not used in the construction of $G^{\rm rob}$. So (a) implies that $G''[A', B']$ is a union of matchings of size two. So (v) will be trivially satisfied when we apply Lemma 2.7.5 for the second time (i.e. with $G_0 + G^{\rm rob}$ playing the role of G_0).

2.8. Special Factors and Exceptional Factors

As discussed in the proof sketch, the main proof proceeds as follows. First we remove a sparse 'robustly decomposable' graph $G^{\rm rob}$ from the original graph G. Then we find an approximate decomposition of $G - G^{\rm rob}$. Finally we find a decomposition of $G^{\rm rob} + G'$, where G' is the (very sparse) leftover from the approximate decomposition.

Both the approximate decomposition as well as the actual decomposition step assume that we work with a graph with two components, one on A and the other on B. So in both steps, we would need $A_0 \cup B_0$ to be empty, which we clearly cannot assume. We build on the ideas of Section 2.3 to deal with this problem. In both steps, one can choose 'exceptional path systems' in G with the following crucial property: one can replace each such exceptional path system EPS with a path system EPS^* so that

(α_1) EPS^* can be partitioned into EPS^*_A and EPS^*_B with the vertex sets of EPS^*_A and EPS^*_B being contained in A and B respectively;
(α_2) the union of any Hamilton cycle C^*_A in $G^*_A := G[A] - EPS + EPS^*_A$ containing EPS^*_A and any Hamilton cycle C^*_B in $G^*_B := G[B] - EPS + EPS^*_B$ containing EPS^*_B corresponds to either a Hamilton cycle of G containing EPS or to the union of two edge-disjoint perfect matchings in G containing EPS.

Each exceptional path system EPS will contain one of the exceptional systems J constructed in Section 2.7. EPS^* will then be obtained from EPS by replacing J by J^*. (Recall that J^* was defined in Section 2.3 and that we view the edges of J^* as 'fictive edges' which are different from the edges of G.) So G^*_A is obtained from $G[A]$ by adding $J^*_A = J^*[A]$. Furthermore, J determines which of the cases in (α_2) holds: If J is a Hamilton exceptional system, then (α_2) will give a Hamilton cycle of G, while in the case when J is a matching exceptional system, (α_2) will give the union of two edge-disjoint perfect matchings in G.

So, roughly speaking, this allows us to work with G^*_A and G^*_B rather than G in the two steps. A convenient way of handling these exceptional path systems is to combine many of them into an 'exceptional factor' EF (see Section 2.8.2 for the definition).

One complication is that the 'robust decomposition lemma' (Lemma 2.9.4) we use from [**21**] deals with digraphs rather than undirected graphs. So in order to be able to apply it, we need to suitably orient the edges of G and so we will actually consider a directed path system $EPS^*_{\rm dir}$ instead of the EPS^* above (the exceptional path system EPS itself will still be undirected). Moreover, we have to apply the robust decomposition lemma twice, once to G^*_A and once to G^*_B.

The formulation of the robust decomposition lemma is quite general and rather than guaranteeing (α_2) directly, it assumes the existence of certain directed 'special paths systems' SPS which are combined into 'special factors' SF. These are introduced in Section 2.8.1. Each of the Hamilton cycles produced by the lemma then contains exactly one of these special path systems. So to apply the lemma, it suffices to check that each of our exceptional path systems EPS corresponds to two path systems $EPS^*_{A,\rm dir}$ and $EPS^*_{B,\rm dir}$ which both satisfy the conditions required of a special path system.

2.8.1. Special Path Systems and Special Factors. As mentioned above, the robust decomposition lemma requires 'special path systems' and 'special factors' as an input when constructing the robustly decomposable graph. These are defined in this subsection.

Let $K, m \in \mathbb{N}$. A *(K,m)-equipartition* $\mathcal{Q}$ of a set V of vertices is a partition of V into sets $V_1, \dots, V_K$ such that $|V_i| = m$ for all $i \leq K$. The V_i are called *clusters* of $\mathcal{Q}$. Suppose that $\mathcal{Q} = \{V_1, \dots, V_K\}$ is a (K,m)-equipartition of V and $L, m/L \in \mathbb{N}$. We say that $(\mathcal{Q}, \mathcal{Q}')$ is a *(K,L,m)-equipartition of V* if $\mathcal{Q}'$ is obtained from $\mathcal{Q}$ by partitioning each cluster V_i of $\mathcal{Q}$ into L sets $V_{i,1}, \dots, V_{i,L}$ of size m/L. So $\mathcal{Q}'$ consists of the KL clusters $V_{i,j}$.

Let $(\mathcal{Q}, \mathcal{Q}')$ be a (K,L,m)-equipartition of V. Consider a spanning cycle $C = V_1 \dots V_K$ on the clusters of $\mathcal{Q}$. Given an integer f dividing K, the *canonical interval partition* $\mathcal{I}$ of C into f intervals consists of the intervals

$$V_{(i-1)K/f+1}V_{(i-1)K/f+2}\dots V_{iK/f+1}$$

for all $i \le f$ (with addition modulo K).

Suppose that G is a digraph on V and $h \le L$. Let $I = V_jV_{j+1}\dots V_{j'}$ be an interval in $\mathcal{I}$. A *special path system SPS of style h in G spanning the interval I* consists of m/L vertex-disjoint directed paths $P_1, \dots, P_{m/L}$ such that the following conditions hold:

(SPS1) Every P_s has its initial vertex in $V_{j,h}$ and its final vertex in $V_{j',h}$.
(SPS2) SPS contains a matching $\mathrm{Fict}(SPS)$ such that all the edges in $\mathrm{Fict}(SPS)$ avoid the endclusters V_j and $V_{j'}$ of I and such that $E(P_s) \setminus \mathrm{Fict}(SPS) \subseteq E(G)$.
(SPS3) The vertex set of SPS is $V_{j,h} \cup V_{j+1,h} \cup \dots \cup V_{j',h}$.

The edges in $\mathrm{Fict}(SPS)$ are called *fictive edges of SPS*.

Let $\mathcal{I} = \{I_1, \dots, I_f\}$. A *special factor SF with parameters* (L, f) *in G (with respect to C,* $\mathcal{Q}'$*)* is a 1-regular digraph on V which is the union of Lf digraphs $SPS_{j,h}$ (one for all $j \le f$ and $h \le L$) such that each $SPS_{j,h}$ is a special path system of style h in G which spans I_j. We write $\mathrm{Fict}(SF)$ for the union of the sets $\mathrm{Fict}(SPS_{j,h})$ over all $j \le f$ and $h \le L$ and call the edges in $\mathrm{Fict}(SF)$ *fictive edges of SF*.

We will always view fictive edges as being distinct from each other and from the edges in other digraphs. So if we say that special factors $SF_1, \dots, SF_r$ are pairwise edge-disjoint from each other and from some digraph Q on V, then this means that Q and all the $SF_i - \mathrm{Fict}(SF_i)$ are pairwise edge-disjoint, but for example there could be an edge from x to y in Q as well as in $\mathrm{Fict}(SF_i)$ for several indices $i \le r$. But these are the only instances of multiedges that we allow, i.e. if there is more than one edge from x to y, then all but at most one of these edges are fictive edges.

2.8.2. Exceptional Path Systems and Exceptional Factors. We now introduce 'exceptional path systems' which will be combined into 'exceptional factors'. These will satisfy the requirements of special path systems and special factors respectively. So they can be used as an 'input' for the robust decomposition lemma. Moreover, they will satisfy the properties (α_1) and (α_2) described at the beginning of Section 2.8 (see Proposition 2.8.1). More precisely, suppose that

$$\mathcal{P} = \{A_0, A_1, \dots, A_K, B_0, B_1, \dots, B_K\}$$

is a (K, m, ε_0)-partition of a vertex set V and $L, m/L \in \mathbb{N}$. We say that $(\mathcal{P}, \mathcal{P}')$ is a (K, L, m, ε_0)*-partition of* V if $\mathcal{P}'$ is obtained from $\mathcal{P}$ by partitioning each cluster A_i of $\mathcal{P}$ into L sets $A_{i,1}, \dots, A_{i,L}$ of size m/L and partitioning each cluster B_i of $\mathcal{P}$ into L sets $B_{i,1}, \dots, B_{i,L}$ of size m/L. (So $\mathcal{P}'$ consists of the exceptional sets A_0, B_0, the KL clusters $A_{i,j}$ and the KL clusters $B_{i,j}$.) Set

$$\mathcal{Q}_A := \{A_1, \dots, A_K\}, \qquad \mathcal{Q}'_A := \{A_{1,1}, \dots, A_{K,L}\},$$
$$\mathcal{Q}_B := \{B_1, \dots, B_K\}, \qquad \mathcal{Q}'_B := \{B_{1,1}, \dots, B_{K,L}\}. \tag{2.8.1}$$

Note that $(\mathcal{Q}_A, \mathcal{Q}'_A)$ and $(\mathcal{Q}_B, \mathcal{Q}'_B)$ are (K, L, m)-equipartitions of A and B respectively (where we recall that $A = \bigcup_{i=1}^K A_i$ and $B = \bigcup_{i=1}^K B_i$).

Suppose that J is a Hamilton exceptional system (for the partition A, A_0, B, B_0) with $e_J(A', B') = 2$. Thus J contains precisely two AB-paths. Let $P_1 = a_1 \dots b_1$ and $P_2 = a_2 \dots b_2$ be these two paths, where $a_1, a_2 \in A$ and $b_1, b_2 \in B$. Recall from

Section 2.3 that J_A^* is the matching consisting of the edge a_1a_2 and an edge between any two vertices $a, a' \in A$ for which J contains a path $P_{aa'}$ whose endvertices are a and a'. We also defined a matching J_B^* in a similar way and set $J^* := J_A^* \cup J_B^*$. We say that an *orientation of J is good* if every path in J is oriented consistently and one of the paths P_1, P_2 is oriented towards B while the other is oriented towards A. Given a good orientation $J_{\rm dir}$ of J, the *orientation $J_{\rm dir}^*$ of J^* induced by $J_{\rm dir}$* is defined as follows:

- For every path $P_{aa'}$ in J whose endvertices a, a' both belong to A, we orient the edge aa' of J^* towards its endpoint of the (oriented) path $P_{aa'}$ in $J_{\rm dir}$.
- If in $J_{\rm dir}$ the path P_1 is oriented towards b_1 (and thus P_2 is oriented towards a_2), then we orient the edge a_1a_2 of J^* towards a_2 and the edge b_1b_2 of J^* towards b_1. The analogue holds if P_1 is oriented towards a_1 (and thus P_2 is oriented towards b_2).

If J is a matching exceptional system, we define good orientations of J and the corresponding induced orientations of J^* in a similar way.

We now define exceptional path systems. As mentioned at the beginning of Section 2.8, each such exceptional path system EPS will correspond to two directed path systems $EPS_{A,\rm dir}^*$ and $EPS_{B,\rm dir}^*$ satisfying the conditions of a special path system (for $(\mathcal{Q}_A, \mathcal{Q}'_A)$ and $(\mathcal{Q}_B, \mathcal{Q}'_B)$ respectively).

Let $(\mathcal{P}, \mathcal{P}')$ be a (K, L, m, ε_0)-partition of a vertex set V. Suppose that $K/f \in \mathbb{N}$. The *canonical interval partition* $\mathcal{I}(f, K)$ of $[K] := \{1, \dots, K\}$ into f intervals consists of the intervals

$$\{(i-1)K/f + 1, (i-1)K/f + 2, \dots, iK/f + 1\}$$

for all $i \leq f$ (with addition modulo K).

Suppose that G is an oriented graph on $A \cup B$ such that $G = G[A] + G[B]$. Let $h \leq L$ and suppose that $I \in \mathcal{I}(f, K)$ is an interval with $I = \{j, j+1, \dots, j'\}$. An *exceptional path system EPS of style h for G spanning I* consists of $2m/L$ vertex-disjoint undirected paths $P_0, P_0', P_1^A, \dots, P_{m/L-1}^A, P_1^B, \dots, P_{m/L-1}^B$, such that the following conditions hold:

(EPS1) $V(P_s^A) \subseteq A$ and P_s^A has one endvertex in $A_{j,h}$ and its other endvertex in $A_{j',h}$ (for all $1 \leq s < m/L$). The analogue holds for every P_s^B.

(EPS2) Each of P_0 and P_0' has one endvertex in $A_{j,h} \cup B_{j,h}$ and its other endvertex in $A_{j',h} \cup B_{j',h}$.

(EPS3) $J := EPS - EPS[A] - EPS[B]$ is either a Hamilton exceptional system with $e_J(A', B') = 2$ or a matching exceptional system (with respect to the partition A, A_0, B, B_0). Moreover $E(J) \subseteq E(P_0) \cup E(P_0')$ and no edge of J has an endvertex in $A_{j,h} \cup A_{j',h} \cup B_{j,h} \cup B_{j',h}$.

(EPS4) Let $P_{0,\rm dir}$ and $P'_{0,\rm dir}$ be the paths obtained by orienting P_0 and P_0' towards their endvertices in $A_{j',h} \cup B_{j',h}$. Then the orientation $J_{\rm dir}$ of J obtained in this way is good. Let $J_{\rm dir}^*$ be the orientation of J^* induced by $J_{\rm dir}$. Then $(P_{0,\rm dir} + P'_{0,\rm dir}) - J_{\rm dir} + J_{\rm dir}^*$ consists of two vertex-disjoint paths $P_{0,\rm dir}^A$ and $P_{0,\rm dir}^B$ such that $V(P_{0,\rm dir}^A) \subseteq A$, $P_{0,\rm dir}^A$ has one endvertex in $A_{j,h}$ and its other endvertex in $A_{j',h}$ and such that the analogue holds for $P_{0,\rm dir}^B$.

(EPS5) The vertex set of EPS is $V_0 \cup A_{j,h} \cup A_{j+1,h} \cdots \cup A_{j',h} \cup B_{j,h} \cup B_{j+1,h} \cdots \cup B_{j',h}$.

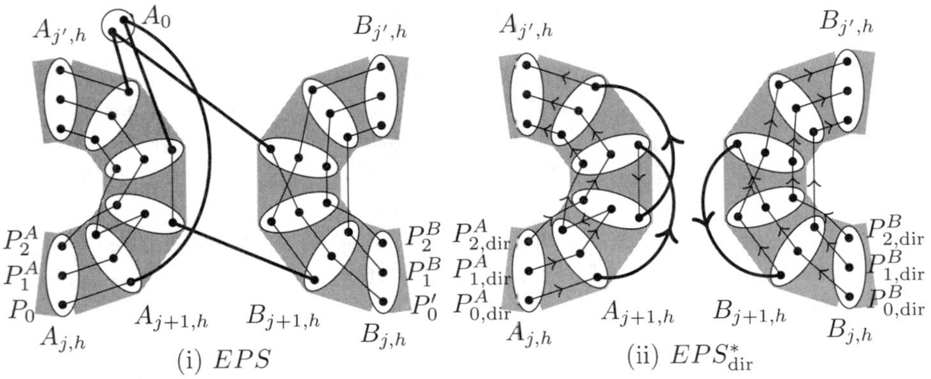

FIGURE 2.8.1. An example of an exceptional path system EPS and the corresponding directed version EPS^*_{dir} in the case when $|A_0| = 2$, $B_0 = \emptyset$, $m/L = 3$ and $|I| = 6$. The thick edges indicate J and J^*_{dir} respectively.

(EPS6) For each $1 \leq s < m/L$, let $P^A_{s,\mathrm{dir}}$ be the path obtained by orienting P^A_s towards its endvertex in $A_{j',h}$. Define $P^B_{s,\mathrm{dir}}$ in a similar way. Then $E(P^A_{0,\mathrm{dir}}) \setminus E(J_{\mathrm{dir}}), E(P^B_{0,\mathrm{dir}}) \setminus E(J_{\mathrm{dir}}) \subseteq E(G)$ and $E(P^A_{s,\mathrm{dir}}), E(P^B_{s,\mathrm{dir}}) \subseteq E(G)$ for every $1 \leq s < m/L$.

We call EPS *a Hamilton exceptional path system* if J (as defined in (EPS3)) is a Hamilton exceptional system, and a *matching exceptional path system* otherwise. Let $EPS^*_{A,\mathrm{dir}}$ be the (directed) path system consisting of $P^A_{0,\mathrm{dir}}, P^A_{1,\mathrm{dir}}, \dots, P^A_{m/L-1,\mathrm{dir}}$. Then $EPS^*_{A,\mathrm{dir}}$ is a special path system of style h in $G[A]$ which spans the interval $A_j A_{j+1} \dots A_{j'}$ of the cycle $A_1 \dots A_K$ and satisfies $\mathrm{Fict}(EPS^*_{A,\mathrm{dir}}) = J^*_{\mathrm{dir}}[A]$. Define $EPS^*_{B,\mathrm{dir}}$ similarly and let $EPS^*_{\mathrm{dir}} := EPS^*_{A,\mathrm{dir}} + EPS^*_{B,\mathrm{dir}}$ and $\mathrm{Fict}(EPS^*_{\mathrm{dir}}) := \mathrm{Fict}(EPS^*_{A,\mathrm{dir}}) \cup \mathrm{Fict}(EPS^*_{B,\mathrm{dir}})$ (see Figure 2.8.1).

Let $\mathcal{I}(f, K) = \{I_1, \dots, I_f\}$. An *exceptional factor EF with parameters (L, f) for G (with respect to $(\mathcal{P}, \mathcal{P}')$)* is the union of Lf edge-disjoint undirected graphs $EPS_{j,h}$ (one for all $j \leq f$ and $h \leq L$) such that each $EPS_{j,h}$ is an exceptional path system of style h for G which spans I_j. We write $EF^*_{A,\mathrm{dir}}$ for the union of $EPS^*_{j,h,A,\mathrm{dir}}$ over all $j \leq f$ and $h \leq L$. Note that $EF^*_{A,\mathrm{dir}}$ is a special factor with parameters (L, f) in $G[A]$ (with respect to $C = A_1 \dots A_K$, $\mathcal{Q}'_A$) such that $\mathrm{Fict}(EF^*_{A,\mathrm{dir}})$ is the union of $J^*_{j,h,\mathrm{dir}}[A]$ over all $j \leq f$ and $h \leq L$, where $J_{j,h}$ is the exceptional system contained in $EPS_{j,h}$ (see condition (EPS3)). Define $EF^*_{B,\mathrm{dir}}$ similarly and let $EF^*_{\mathrm{dir}} := EF^*_{A,\mathrm{dir}} + EF^*_{B,\mathrm{dir}}$ and $\mathrm{Fict}(EF^*_{\mathrm{dir}}) := \mathrm{Fict}(EF^*_{A,\mathrm{dir}}) \cup \mathrm{Fict}(EF^*_{B,\mathrm{dir}})$. Note that EF^*_{dir} is a 1-regular directed graph on $A \cup B$ while in EF is an undirected graph on V with

$$d_{EF}(v) = 2 \text{ for all } v \in V \setminus V_0 \quad \text{and} \quad d_{EF}(v) = 2Lf \text{ for all } v \in V_0. \tag{2.8.2}$$

Given an exceptional path system EPS, let J be as in (EPS3) and let

$$EPS^* := EPS - J + J^*, \quad EPS^*_A := EPS^*[A] \text{ and } EPS^*_B := EPS^*[B].$$

(Hence EPS^*, EPS^*_A and EPS^*_B are the undirected graphs obtained from $EPS^*_{\rm dir}$, $EPS^*_{A,{\rm dir}}$ and $EPS^*_{B,{\rm dir}}$ by ignoring the orientations of all edges.) The following result is an immediate consequence of (EPS3), (EPS4) and Proposition 2.3.1. Roughly speaking, it implies that to find a Hamilton cycle in the 'original' graph with vertex set V, it suffices to find a Hamilton cycle on A and one on B, containing (the edges corresponding to) an exceptional path system.

PROPOSITION 2.8.1. *Let $(\mathcal{P}, \mathcal{P}')$ be a (K, L, m, ε_0)-partition of a vertex set V. Suppose that G is a graph on $V \setminus V_0$, that $G_{\rm dir}$ is an orientation of $G[A] + G[B]$ and that EPS is an exceptional path system for $G_{\rm dir}$. Let J be as in (EPS3) and J^*_A as defined in Section 2.3. Let C_A and C_B be two cycles such that*

- *C_A is a Hamilton cycle on A which contains EPS^*_A;*
- *C_B is a Hamilton cycle on B which contains EPS^*_B.*

Then the following assertions hold.

(i) *If EPS is a Hamilton exceptional path system, then $C_A + C_B - EPS^* + EPS$ is a Hamilton cycle on V.*
(ii) *If EPS is a matching exceptional path system, then $C_A + C_B - EPS^* + EPS$ is the union of a Hamilton cycle on A' and a Hamilton cycle on B'. In particular, if both $|A'|$ and $|B'|$ are even, then $C_A + C_B - EPS^* + EPS$ is the union of two edge-disjoint perfect matchings on V.*

Proof. Note that $C_A + C_B - EPS^* + EPS = C_A + C_B - J^* + J$. Recall that J^*_{AB} was defined in Section 2.3. (EPS3) implies that $|E(J^*_A) \setminus E(J^*_{AB})| \leq 1$. Recall from Section 2.3 that a path P is said to consistent with J^*_A if P contains J^*_A and (there is an orientation of P which) visits the endvertices of the edges in $E(J^*_A) \setminus E(J^*_{AB})$ in a prescribed order. Since $E(J^*_A) \setminus E(J^*_{AB})$ contains at most one edge, any path containing J^*_A is also consistent with J^*_A. Therefore, C_A is consistent with J^*_A and, by a similar argument, C_B is consistent with J^*_B. So the proposition follows immediately from Proposition 2.3.1. □

2.8.3. Finding Exceptional Factors in a Scheme. The next lemma (Lemma 2.8.2) will allow us to extend a suitable exceptional system J into an exceptional path system. In particular, we assume that J is 'localized'. This allows us to choose the path system in such a way that it spans only a few clusters. The structure within which we find the path system is called a 'scheme'. Roughly speaking, this is the structure we obtain from $G[A]+G[B]$ (i.e. the union of two almost complete graphs) by considering a random equipartition of A and B and a random orientation of its edges.

We now define this 'oriented' version of the (undirected) schemes which were introduced in Section 2.4. Given an oriented graph G and partitions $\mathcal{P}$ and $\mathcal{P}'$ of a vertex set V, we call $(G, \mathcal{P}, \mathcal{P}')$ a $[K, L, m, \varepsilon_0, \varepsilon]$-*scheme* if the following conditions hold:

(Sch1′) $(\mathcal{P}, \mathcal{P}')$ is a (K, L, m, ε_0)-partition of V.
(Sch2′) $V(G) = A \cup B$ and $e_G(A, B) = 0$.
(Sch3′) $G[A_{i,j}, A_{i',j'}]$ and $G[B_{i,j}, B_{i',j'}]$ are $[\varepsilon, 1/2]$-superregular for all $i, i' \leq K$ and all $j, j' \leq L$ such that $(i,j) \neq (i',j')$. Moreover, $G[A_i, A_{i'}]$ and $G[B_i, B_{i'}]$ are $[\varepsilon, 1/2]$-superregular for all $i \neq i' \leq K$.

(Sch4′) $|N_G^+(x) \cap N_G^-(y) \cap A_{i,j}| \geq (1/5 - \varepsilon)m/L$ for all $x, y \in A$, all $i \leq K$ and all $j \leq L$. Similarly, $|N_G^+(x) \cap N_G^-(y) \cap B_{i,j}| \geq (1/5 - \varepsilon)m/L$ for all $x, y \in B$, all $i \leq K$ and all $j \leq L$.

Note that if $L = 1$ (and so $\mathcal{P} = \mathcal{P}'$), then (Sch1′) just says that $\mathcal{P}$ is a (K, m, ε_0)-partition of V.

Suppose that J is an (i, i')-ES with respect to $\mathcal{P}$. Given $h \leq L$, we say that J has *style* h *(with respect to the* (K, L, m, ε_0)*-partition* $(\mathcal{P}, \mathcal{P}')$*)* if all the edges of J have their endvertices in $V_0 \cup A_{i,h} \cup B_{i',h}$.

LEMMA 2.8.2. *Suppose that* $K, L, n, m/L \in \mathbb{N}$, *that* $0 < 1/n \ll \varepsilon, \varepsilon_0 \ll 1$ *and* $\varepsilon_0 \ll 1/K, 1/L$. *Let* $(G, \mathcal{P}, \mathcal{P}')$ *be a* $[K, L, m, \varepsilon_0, \varepsilon]$*-scheme with* $|V(G) \cup V_0| = n$. *Let* $I = \{j, j+1, \dots, j'\} \subseteq [K]$ *be an integer interval with* $|I| \geq 4$. *Let* J *be either an* (i_1, i_2)*-HES of style* $h \leq L$ *with* $e_J(A', B') = 2$ *or an* (i_1, i_2)*-MES of style* $h \leq L$ *(with respect to* $(\mathcal{P}, \mathcal{P}')$*), for some* $i_1, i_2 \in \{j+1, \dots, j'-1\}$. *Then there exists an exceptional path system of style* h *for* G *which spans the interval* I *and contains all edges of* J.

Proof. Let $J_{\rm dir}$ be a good orientation of J and let $J^*_{\rm dir}$ be the induced orientation of J^*. Let $x_1x_2, \dots, x_{2s'-1}x_{2s'}$ be the edges of $J^*_{A,{\rm dir}} := J^*_{\rm dir}[A]$. Since J is an (i_1, i_2)-ES of style h with $e_J(A', B') \leq 2$ it follows that $s' = e(J^*_A) \leq |V_0| + 1 \leq 2\varepsilon_0 n$ and $x_i \in A_{i_1,h}$ for all $i \leq 2s'$. Since $|I| \geq 4$ we have $i_1 + 1 \in \{j+1, \dots, j'-1\}$ or $i_1 - 1 \in \{j+1, \dots, j'-1\}$. We will only consider the case when $i_1 + 1 \in \{j+1, \dots, j'-1\}$. (The argument for the other case is similar.)

Our assumption that $\varepsilon_0 \ll 1/K, 1/L$ implies that $\varepsilon_0 n \leq m/100L$ (say). Together with (Sch4′) this ensures that for every $1 \leq r < s'$, we can pick a vertex $w_r \in A_{i_1+1,h}$ such that $x_{2r}w_r$ and $w_r x_{2r+1}$ are (directed) edges in G and such that $w_1, \dots, w_{s'-1}$ are distinct from each other. We also pick a vertex $w_{s'} \in A_{i_1+1,h} \setminus \{w_1, \dots, w_{s'-1}\}$ such that $x_{2s'}w_{s'}$ is a (directed) edge in G. Let Q_0 be the path $x_1x_2w_1x_3x_4w_2 \dots x_{2s'-1}x_{2s'}w_{s'}$. Thus Q_0 is a directed path from $A_{i_1,h}$ to $A_{i_1+1,h}$ in $G + J^*_{\rm dir}$ which contains all edges of $J^*_{A,{\rm dir}}$. Note that $|V(Q_0) \cap A_{i_1,h}| = 2s'$ and $|V(Q_0) \cap A_{i_1+1,h}| = s'$. Moreover, $V(Q_0) \cap A_i = \emptyset$ for all $i \notin \{i_1, i_1+1\}$ and $V(Q_0) \cap B = \emptyset$.

Pick a vertex $w_0 \in A_{j,h}$ so that w_0x_1 is an edge of G. Find a path Q'_0 from $w_{s'}$ to $A_{j',h}$ in G such that the vertex set of Q'_0 consists of $w_{s'}$ and precisely one vertex in each $A_{i,h}$ for all $i \in \{j+1, \dots, j'\} \setminus \{i_1, i_1+1\}$ and no other vertices. (Sch4′) ensures that this can be done greedily. Define $P^A_{0,{\rm dir}}$ to be the concatenation of w_0x_1, Q_0 and Q'_0. Note that $P^A_{0,{\rm dir}}$ is a directed path from $A_{j,h}$ to $A_{j',h}$ in $G + J^*_{\rm dir}$ which contains $J^*_{A,{\rm dir}}$. Moreover,

$$|V(P^A_{0,{\rm dir}}) \cap A_{i,h}| = \begin{cases} 1 & \text{for } i \in \{j, \dots, j'\} \setminus \{i_1, i_1+1\}, \\ 2s' & \text{for } i = i_1, \\ s' & \text{for } i = i_1 + 1, \\ 0 & \text{otherwise,} \end{cases}$$

while $V(P^A_{0,{\rm dir}}) \cap B = \emptyset$ and $V(P^A_{0,{\rm dir}}) \cap A_{i,h'} = \emptyset$ for all $i \leq K$ and all $h' \neq h$. (Sch4′) ensures that we can also choose $2s' - 1$ (directed) paths $P^A_{1,{\rm dir}}, \dots, P^A_{2s'-1,{\rm dir}}$ in G such that the following conditions hold:

- For all $1 \leq r < 2s'$, $P^A_{r,{\rm dir}}$ is a path from $A_{j,h}$ to $A_{j',h}$.

- For all $1 \le r \le s'$, $P^A_{r,\mathrm{dir}}$ contains precisely one vertex in $A_{i,h}$ for each $i \in \{j, \dots, j'\} \setminus \{i_1\}$ and no other vertices.
- For all $s' < r < 2s'$, $P^A_{r,\mathrm{dir}}$ contains precisely one vertex in $A_{i,h}$ for each $i \in \{j, \dots, j'\} \setminus \{i_1, i_1 + 1\}$ and no other vertices.
- $P^A_{0,\mathrm{dir}}, \dots, P^A_{2s'-1,\mathrm{dir}}$ are pairwise vertex-disjoint.

Let Q be the union of $P^A_{0,\mathrm{dir}}, \dots, P^A_{2s'-1,\mathrm{dir}}$. Thus Q is a path system consisting of $2s'$ vertex-disjoint directed paths from $A_{j,h}$ to $A_{j',h}$. Moreover, $V(Q)$ consists of precisely $2s'$ vertices in $A_{i,h}$ for every $j \le i \le j'$ and no other vertices. Set $A'_{i,h} := A_{i,h} \setminus V(Q)$ for all $i \le K$. Note that

$$|A'_{i,h}| = \frac{m}{L} - 2s' \ge \frac{m}{L} - 4\varepsilon_0 n \ge \frac{m}{L} - 10\varepsilon_0 mK \ge (1 - \sqrt{\varepsilon_0})\frac{m}{L} \tag{2.8.3}$$

since $\varepsilon_0 \ll 1/K, 1/L$. Pick a new constant ε' such that $\varepsilon, \varepsilon_0 \ll \varepsilon' \ll 1$. Then Proposition 1.4.1, (Sch3′) and (2.8.3) together imply that $G[A'_{i,h}, A'_{i+1,h}]$ is still $[\varepsilon', 1/2]$-superregular and so by Proposition 1.4.2 we can find a perfect matching in $G[A'_{i,h}, A'_{i+1,h}]$ for all $j \le i < j'$. The union Q' of all these matchings forms $m/L - 2s'$ vertex-disjoint directed paths $P^A_{2s',\mathrm{dir}}, \dots, P^A_{m/L-1,\mathrm{dir}}$. Note that $P^A_{0,\mathrm{dir}}, P^A_{1,\mathrm{dir}}, \dots, P^A_{m/L-1,\mathrm{dir}}$ are pairwise vertex-disjoint and together cover precisely the vertices in $\bigcup_{i=j}^{j'} A_{i,h}$. Moreover, $P^A_{0,\mathrm{dir}}$ contains $J^*_{A,\mathrm{dir}}$.

Similarly, we find m/L vertex-disjoint directed paths $P^B_{0,\mathrm{dir}}, P^B_{1,\mathrm{dir}}, \dots, P^B_{m/L-1,\mathrm{dir}}$ from $B_{j,h}$ to $B_{j',h}$ such that $P^B_{0,\mathrm{dir}}$ contains $J^*_{B,\mathrm{dir}}$ and together the paths cover precisely the vertices in $\bigcup_{i=j}^{j'} B_{i,h}$. For each $1 \le r < m/L$, let P^A_r and P^B_r be the undirected paths obtained from $P^A_{r,\mathrm{dir}}$ and $P^B_{r,\mathrm{dir}}$ by ignoring the directions of all the edges.

Since $J^*_{A,\mathrm{dir}} \subseteq P^A_{0,\mathrm{dir}}$ and $J^*_{B,\mathrm{dir}} \subseteq P^B_{0,\mathrm{dir}}$ and since J^*_{dir} is the orientation of J^* induced by J_{dir}, it follows that $P^A_{0,\mathrm{dir}} + P^B_{0,\mathrm{dir}} - J^*_{\mathrm{dir}} + J_{\mathrm{dir}}$ consists of two vertex-disjoint paths $P_{0,\mathrm{dir}}$ and $P'_{0,\mathrm{dir}}$ from $A_{j,h} \cup B_{j,h}$ to $A_{j',h} \cup B_{j',h}$ with $V(P_{0,\mathrm{dir}}) \cup V(P'_{0,\mathrm{dir}}) = V_0 \cup V(P^A_{0,\mathrm{dir}}) \cup V(P^B_{0,\mathrm{dir}})$. Let P_0 and P'_0 be the undirected paths obtained from $P_{0,\mathrm{dir}}$ and $P'_{0,\mathrm{dir}}$ by ignoring the directions of all the edges. Let EPS be the union of $P_0, P'_0, P^A_1, \dots, P^A_{m/L-1}, P^B_1, \dots, P^B_{m/L-1}$. Then EPS is an exceptional path system for G, as required. To see this, note that $J = EPS - EPS[A] - EPS[B]$ since $e_J(A), e_J(B) = 0$ by the definition of an exceptional system (see (EC3) in Section 2.3). □

The next lemma uses the previous one to show that we can obtain many edge-disjoint exceptional factors by extending exceptional systems with suitable properties.

LEMMA 2.8.3. *Suppose that $L, f, q, n, m/L, K/f \in \mathbb{N}$, that $K/f \ge 3$, that $0 < 1/n \ll \varepsilon, \varepsilon_0 \ll 1$, that $\varepsilon_0 \ll 1/K, 1/L$ and $Lq/m \ll 1$. Let $(G, \mathcal{P}, \mathcal{P}')$ be a $[K, L, m, \varepsilon_0, \varepsilon]$-scheme with $|V(G) \cup V_0| = n$. Suppose that there exists a set $\mathcal{J}$ of Lfq edge-disjoint exceptional systems satisfying the following conditions:*

(i) *Each $J \in \mathcal{J}$ is either a Hamilton exceptional system with $e_J(A', B') = 2$ or a matching exceptional system.*
(ii) *For all $i \le f$ and all $h \le L$, $\mathcal{J}$ contains precisely q (i_1, i_2)-ES of style h (with respect to $(\mathcal{P}, \mathcal{P}')$) for which $i_1, i_2 \in \{(i-1)K/f + 2, \dots, iK/f\}$.*

Then there exist q edge-disjoint exceptional factors with parameters (L, f) for G (with respect to $(\mathcal{P}, \mathcal{P}')$) covering all edges in $\bigcup \mathcal{J}$.

Recall that the canonical interval partition $\mathcal{I}(f, K)$ of $[K]$ into f intervals consists of the intervals $\{(i-1)K/f + 1, \dots, iK/f + 1\}$ for all $i \leq f$. So (ii) ensures that for each interval $I \in \mathcal{I}(f, K)$ and each $h \leq L$, the set $\mathcal{J}$ contains precisely q exceptional systems of style h whose edges are only incident to vertices in V_0 and vertices belonging to clusters A_{i_1} and B_{i_2} for which both i_1 and i_2 lie in the interior of I. We will use Lemma 2.8.2 to extend each such exceptional system into an exceptional path system of style h spanning I.

Proof of Lemma 2.8.3. Choose a new constant ε' with $\varepsilon, Lq/m \ll \varepsilon' \ll 1$. Let $\mathcal{J}_1, \dots, \mathcal{J}_q$ be a partition of $\mathcal{J}$ such that for all $j \leq q$, $h \leq L$ and $i \leq f$, the set $\mathcal{J}_j$ contains precisely one (i_1, i_2)-ES of style h with $i_1, i_2 \in \{(i-1)K/f + 2, \dots, iK/f\}$. Thus each $\mathcal{J}_j$ consists of Lf exceptional systems. For each $j \leq q$ in turn, we will choose an exceptional factor EF_j with parameters (L, f) for G (with respect to $(\mathcal{P}, \mathcal{P}')$) such that EF_j and $EF_{j'}$ are edge-disjoint for all $j' < j$ and EF_j contains all edges of the exceptional systems in $\mathcal{J}_j$. Assume that for some $1 \leq j \leq q$ we have already constructed $EF_1, \dots, EF_{j-1}$. In order to construct EF_j, we will choose the Lf exceptional path systems forming EF_j one by one, such that each of these exceptional path systems is edge-disjoint from $EF_1, \dots, EF_{j-1}$ and contains precisely one of the exceptional systems in $\mathcal{J}_j$. Suppose that we have already chosen some of these exceptional path systems and that next we wish to choose an exceptional path system of style h which spans the interval I of the canonical interval partition $\mathcal{I}(f, K)$ and contains $J \in \mathcal{J}_j$. Let G' be the oriented graph obtained from G by deleting all the edges in the path systems already chosen for EF_j as well as deleting all the edges in $EF_1, \dots, EF_{j-1}$. Recall that $V(G) = A \cup B$. Thus $\Delta(G - G') \leq 2j < 3q$ by (2.8.2). Together with Proposition 1.4.1 this implies that $(G', \mathcal{P}, \mathcal{P}')$ is still a $[K, L, m, \varepsilon_0, \varepsilon']$-scheme. (Here we use that $\Delta(G - G') < 3q = 3Lq/m \cdot m/L$ and $\varepsilon, Lq/m \ll \varepsilon' \ll 1$.) So we can apply Lemma 2.8.2 with ε' playing the role of ε to obtain an exceptional path system of style h for G' (and thus for G) which spans I and contains all edges of J. This completes the proof of the lemma. □

2.9. The Robust Decomposition Lemma

The aim of this section is to state the robust decomposition lemma (Lemma 2.9.4). This is the key lemma proved in [**21**] and guarantees the existence of a 'robustly decomposable' digraph $G^{\rm rob}_{\rm dir}$ within a 'setup'. For our purposes, we will then derive an undirected version in Corollary 2.9.5 to construct a robustly decomposable graph $G^{\rm rob}$. Then $G^{\rm rob} + H$ will have a Hamilton decomposition for any sparse regular graph H which is edge-disjoint from $G^{\rm rob}$. The crucial ingredient of a setup is a 'universal walk', which we introduce in the next subsection. The (proof of the) robust decomposition lemma then uses edges guaranteed by this universal walk to 'balance out' edges of the graph H when constructing the Hamilton decomposition of $G^{\rm rob} + H$.

2.9.1. Chord Sequences and Universal Walks.

Let R be a digraph whose vertices are $V_1, \dots, V_k$ and suppose that $C = V_1 \dots V_k$ is a Hamilton cycle of R. (Later on the vertices of R will be clusters. So we denote them by capital letters.)

A *chord sequence* $CS(V_i, V_j)$ from V_i to V_j in R is an ordered sequence of edges of the form

$$CS(V_i, V_j) = (V_{i_1-1}V_{i_2}, V_{i_2-1}V_{i_3}, \dots, V_{i_t-1}V_{i_{t+1}}),$$

where $V_{i_1} = V_i$, $V_{i_{t+1}} = V_j$ and the edge $V_{i_s-1}V_{i_{s+1}}$ belongs to R for each $s \leq t$. If $i = j$ then we consider the empty set to be a chord sequence from V_i to V_j. Without loss of generality, we may assume that $CS(V_i, V_j)$ does not contain any edges of C. (Indeed, suppose that $V_{i_s-1}V_{i_{s+1}}$ is an edge of C. Then $i_s = i_{s+1}$ and so we can obtain a chord sequence from V_i to V_j with fewer edges.) For example, if $V_{i-1}V_{i+1} \in E(R)$, then the edge $V_{i-1}V_{i+1}$ is a chord sequence from V_i to V_{i+1}.

The crucial property of chord sequences is that they satisfy a 'local balance' condition. Suppose that CS is obtained by concatenating several chord sequences

$$CS(V_{i_1}, V_{i_2}), CS(V_{i_2}, V_{i_3}), \dots, CS(V_{i_{k-1}}, V_{i_k})$$

so that $V_{i_1} = V_{i_k}$. Then for every cluster V_i, the number of edges of CS leaving V_{i-1} equals the number of edges entering V_i. We will not use this property explicitly, but it underlies the proof of the robust decomposition lemma (Lemma 2.9.4) that we apply and appears implicitly e.g. in (U3).

A closed walk U in R is a *universal walk for* C *with parameter* ℓ' if the following conditions hold:

(U1) For every $i \leq k$ there is a chord sequence $ECS(V_i, V_{i+1})$ from V_i to V_{i+1} such that U contains all edges of all these chord sequences (counted with multiplicities) and all remaining edges of U lie on C.

(U2) Each $ECS(V_i, V_{i+1})$ consists of at most $\sqrt{\ell'}/2$ edges.

(U3) U enters each V_i exactly ℓ' times and leaves each V_i exactly ℓ' times.

Note that condition (U1) means that if an edge $V_iV_j \in E(R) \setminus E(C)$ occurs in total 5 times (say) in $ECS(V_1, V_2), \dots, ECS(V_k, V_1)$ then it occurs precisely 5 times in U. We will identify each occurrence of V_iV_j in $ECS(V_1, V_2), \dots, ECS(V_k, V_1)$ with a (different) occurrence of V_iV_j in U. Note that the edges of $ECS(V_i, V_{i+1})$ are allowed to appear in a different order within $ECS(V_i, V_{i+1})$ and within U.

LEMMA 2.9.1. *Let R be a digraph with vertices $V_1, \dots, V_k$. Suppose that $C = V_1 \dots V_k$ is a Hamilton cycle of R and that $V_iV_{i+2} \in E(R)$ for every $1 \leq i \leq k$. Let $\ell' \geq 4$ be an integer. Let $U_{\ell'}$ the multiset obtained from $\ell' - 1$ copies of $E(C)$ by adding $V_iV_{i+2} \in E(R)$ for every $1 \leq i \leq k$. Then the edges in $U_{\ell'}$ can be ordered so that the resulting sequence forms a universal walk for C with parameter ℓ'.*

In the remainder of this section, we will also write $U_{\ell'}$ for the universal walk guaranteed by Lemma 2.9.1.

Proof. Let us first show that the edges in $U_{\ell'}$ can be ordered so that the resulting sequence forms a closed walk in R. To see this, consider the multidigraph U obtained from $U_{\ell'}$ by deleting one copy of $E(C)$. Then U is $(\ell' - 1)$-regular and thus has a decomposition into 1-factors. We order the edges of $U_{\ell'}$ as follows: We first traverse all cycles of the 1-factor decomposition of U which contain the cluster V_1. Next, we traverse the edge V_1V_2 of C. Next we traverse all those cycles of the 1-factor decomposition which contain V_2 and which have not been traversed so far. Next we traverse the edge V_2V_3 of C and so on until we reach V_1 again.

Recall that, for each $1 \leq i \leq k$, the edge $V_{i-1}V_{i+1}$ is a chord sequence from V_i to V_{i+1}. Thus we can take $ECS(V_i, V_{i+1}) := V_{i-1}V_{i+1}$. Then $U_{\ell'}$ satisfies (U1)–(U3). □

2.9.2. Setups and the Robust Decomposition Lemma. The aim of this subsection is to state the robust decomposition lemma (Lemma 2.9.4, proved in [**21**]) and derive Corollary 2.9.5, which we shall use later on in order to prove Theorem 1.3.3. The robust decomposition lemma guarantees the existence of a 'robustly decomposable' digraph $G^{\rm rob}_{\rm dir}$ within a 'setup'. Roughly speaking, a setup is a digraph G together with its 'reduced digraph' R, which contains a Hamilton cycle C and a universal walk U. In our application, we will have two setups: $G[A]$ and $G[B]$ will play the role of G, and R will be the complete digraph in both cases. To define a setup formally, we first need to define certain 'refinements' of partitions.

Given a digraph G and a partition $\mathcal{P}$ of $V(G)$ into k clusters $V_1, \dots, V_k$ of equal size, we say that a partition $\mathcal{P}'$ of V is an *ℓ'-refinement of $\mathcal{P}$* if $\mathcal{P}'$ is obtained by splitting each V_i into ℓ' subclusters of equal size. (So $\mathcal{P}'$ consists of $\ell' k$ clusters.) $\mathcal{P}'$ is an *ε-uniform ℓ-refinement* of $\mathcal{P}$ if it is an ℓ-refinement of $\mathcal{P}$ which satisfies the following condition: Whenever x is a vertex of G, V is a cluster in $\mathcal{P}$ and $|N^+_G(x) \cap V| \geq \varepsilon |V|$ then $|N^+_G(x) \cap V'| = (1 \pm \varepsilon)|N^+_G(x) \cap V|/\ell$ for each cluster $V' \in \mathcal{P}'$ with $V' \subseteq V$. The inneighbourhoods of the vertices of G satisfy an analogous condition. We need the following simple observation from [**21**]. The proof proceeds by considering a random partition to obtain a uniform refinement.

LEMMA 2.9.2. *Suppose that $0 < 1/m \ll 1/k, \varepsilon \ll \varepsilon', d, 1/\ell \leq 1$ and that $n, k, \ell, m/\ell \in \mathbb{N}$. Suppose that G is a digraph on $n = km$ vertices and that $\mathcal{P}$ is a partition of $V(G)$ into k clusters of size m. Then there exists an ε-uniform ℓ-refinement of $\mathcal{P}$. Moreover, any ε-uniform ℓ-refinement $\mathcal{P}'$ of $\mathcal{P}$ automatically satisfies the following condition:*

- *Suppose that V, W are clusters in $\mathcal{P}$ and V', W' are clusters in $\mathcal{P}'$ with $V' \subseteq V$ and $W' \subseteq W$. If $G[V, W]$ is $[\varepsilon, d']$-superregular for some $d' \geq d$ then $G[V', W']$ is $[\varepsilon', d']$-superregular.*

We will also need the following definition from [**21**]. $(G, \mathcal{P}, \mathcal{P}', R, C, U, U')$ is called an *$(\ell', k, m, \varepsilon, d)$-setup* if the following properties are satisfied:

(ST1) G and R are digraphs. $\mathcal{P}$ is a partition of $V(G)$ into k clusters of size m. The vertex set of R consists of these clusters.

(ST2) For every edge VW of R the corresponding pair $G[V, W]$ is $(\varepsilon, \geq d)$-regular.

(ST3) C is a Hamilton cycle of R and for every edge VW of C the corresponding pair $G[V, W]$ is $[\varepsilon, \geq d]$-superregular.

(ST4) U is a universal walk for C with parameter ℓ' and $\mathcal{P}'$ is an ε-uniform ℓ'-refinement of $\mathcal{P}$.

(ST5) Suppose that $C = V_1 \dots V_k$ and let $V_j^1, \dots, V_j^{\ell'}$ denote the clusters in $\mathcal{P}'$ which are contained in V_j (for each $1 \leq j \leq k$). Then U' is a closed walk on the clusters in $\mathcal{P}'$ which is obtained from U as follows: When U visits V_j for the ath time, we let U' visit the subcluster V_j^a (for all $1 \leq a \leq \ell'$).

(ST6) Each edge of U' corresponds to an $[\varepsilon, \geq d]$-superregular pair in G.

In [**21**], in a setup, the digraph G could also contain an exceptional set, but since we are only using the definition in the case when there is no such exceptional set, we have only stated it in this special case.

Suppose that $(G, \mathcal{P}, \mathcal{P}')$ is a $[K, L, m, \varepsilon_0, \varepsilon]$-scheme. Recall that $A_1, \dots, A_K$ and $B_1, \dots, B_K$ denote the clusters of $\mathcal{P}$. Let $\mathcal{Q}_A := \{A_1, \dots, A_K\}$, $\mathcal{Q}_B := \{B_1, \dots, B_K\}$ and let $C_A = A_1 \dots A_K$ and $C_B = B_1 \dots B_K$ be (directed) cycles.

Suppose that $\ell', m/\ell' \in \mathbb{N}$ with $\ell' \geq 4$. Let $\mathcal{Q}'_A$ be an ε-uniform ℓ'-refinement of $\mathcal{Q}_A$. Let R_A be the complete digraph whose vertices are the clusters in $\mathcal{Q}_A$. Let $U_{A,\ell'}$ be a universal walk for C_A with parameter ℓ' as defined in Lemma 2.9.1. Let $U'_{A,\ell'}$ be the closed walk obtained from $U_{A,\ell'}$ as described in (ST5). We will call

$$(G[A], \mathcal{Q}_A, \mathcal{Q}'_A, R_A, C_A, U_{A,\ell'}, U'_{A,\ell'})$$

the *A-setup associated to* $(G, \mathcal{P}, \mathcal{P}')$. Define $\mathcal{Q}'_B$, R_B, $U_{B,\ell'}$ and $U'_{B,\ell'}$ similarly. We will call

$$(G[B], \mathcal{Q}_B, \mathcal{Q}'_B, R_B, C_B, U_{B,\ell'}, U'_{B,\ell'})$$

the *B-setup associated to* $(G, \mathcal{P}, \mathcal{P}')$. The following lemma shows that both the A-setup and the B-setup indeed satisfy all the conditions in the definition of a setup.

LEMMA 2.9.3. *Suppose that $1/m \ll 1/K, \varepsilon_0, \varepsilon \ll \varepsilon', 1/\ell'$ and $K, L, m/L, \ell'$, $m/\ell' \in \mathbb{N}$ with $\ell' \geq 4$. Suppose that $(G, \mathcal{P}, \mathcal{P}')$ is a $[K, L, m, \varepsilon_0, \varepsilon]$-scheme. Then each of*

$$(G[A], \mathcal{Q}_A, \mathcal{Q}'_A, R_A, C_A, U_{A,\ell'}, U'_{A,\ell'}) \quad \text{and} \quad (G[B], \mathcal{Q}_B, \mathcal{Q}'_B, R_B, C_B, U_{B,\ell'}, U'_{B,\ell'})$$

is an $(\ell', K, m, \varepsilon', 1/2)$-setup.

Proof. It suffices to show that $(G[A], \mathcal{Q}_A, \mathcal{Q}'_A, R_A, C_A, U_{A,\ell'}, U'_{A,\ell'})$ is an $(\ell', K, m, \varepsilon', 1/2)$-setup. Clearly, (ST1) holds. (Sch3$'$) implies that (ST2) and (ST3) hold. Lemma 2.9.1 implies (ST4). (ST5) follows from the definition of $U'_{A,\ell'}$. (ST6) follows from Lemma 2.9.2 since $\mathcal{Q}'_A$ is an ε-uniform ℓ'-refinement of $\mathcal{Q}_A$. □

We now state the robust decomposition lemma from [**21**]. Recall that this guarantees the existence of a 'robustly decomposable' digraph $G^{\rm rob}_{\rm dir}$, whose crucial property is that $H + G^{\rm rob}_{\rm dir}$ has a Hamilton decomposition for any sparse regular digraph H which is edge-disjoint from $G^{\rm rob}_{\rm dir}$.

$G^{\rm rob}_{\rm dir}$ consists of digraphs $CA_{\rm dir}(r)$ (the 'chord absorber') and $PCA_{\rm dir}(r)$ (the 'parity extended cycle switcher') together with some special factors. $G^{\rm rob}_{\rm dir}$ is constructed in two steps: given a suitable set $\mathcal{SF}$ of special factors, the lemma first 'constructs' $CA_{\rm dir}(r)$ and then, given another suitable set $\mathcal{SF}'$ of special factors, the lemma 'constructs' $PCA_{\rm dir}(r)$. The reason for having two separate steps is that in [**21**], it is not clear how to construct $CA_{\rm dir}(r)$ after constructing $\mathcal{SF}'$ (rather than before), as the removal of $\mathcal{SF}'$ from the digraph under consideration affects its properties considerably.

LEMMA 2.9.4. *Suppose that $0 < 1/m \ll 1/k \ll \varepsilon \ll 1/q \ll 1/f \ll r_1/m \ll d \ll 1/\ell', 1/g \ll 1$ and that $rk^2 \leq m$. Let*

$$r_2 := 96\ell' g^2 kr, \quad r_3 := rfk/q, \quad r^\diamond := r_1 + r_2 + r - (q-1)r_3, \quad s' := rfk + 7r^\diamond$$

and suppose that $k/14, k/f, k/g, q/f, m/4\ell', fm/q, 2fk/3g(g-1) \in \mathbb{N}$. Suppose that $(G, \mathcal{P}, \mathcal{P}', R, C, U, U')$ is an $(\ell', k, m, \varepsilon, d)$-setup and $C = V_1 \dots V_k$. Suppose that $\mathcal{P}^*$ is a (q/f)-refinement of $\mathcal{P}$ and that $SF_1, \dots, SF_{r_3}$ are edge-disjoint special factors with parameters $(q/f, f)$ with respect to C, $\mathcal{P}^*$ in G. Let $\mathcal{SF} := SF_1 + \dots + SF_{r_3}$. Then there exists a digraph $CA_{\rm dir}(r)$ for which the following holds:

(i) *$CA_{\rm dir}(r)$ is an $(r_1 + r_2)$-regular spanning subdigraph of G which is edge-disjoint from $\mathcal{SF}$.*

(ii) *Suppose that* $SF'_1, \dots, SF'_{r^\diamond}$ *are special factors with parameters* $(1,7)$ *with respect to* C, $\mathcal{P}$ *in* G *which are edge-disjoint from each other and from* $CA_{\rm dir}(r) + \mathcal{SF}$. *Let* $\mathcal{SF}' := SF'_1 + \dots + SF'_{r^\diamond}$. *Then there exists a digraph* $PCA_{\rm dir}(r)$ *for which the following holds:*
   (a) $PCA_{\rm dir}(r)$ *is a* $5r^\diamond$*-regular spanning subdigraph of* G *which is edge-disjoint from* $CA_{\rm dir}(r) + \mathcal{SF} + \mathcal{SF}'$.
   (b) *Let* $\mathcal{SPS}$ *be the set consisting of all the* s' *special path systems contained in* $\mathcal{SF} + \mathcal{SF}'$. *Suppose that* H *is an* r*-regular digraph on* $V(G)$ *which is edge-disjoint from* $G^{\rm rob}_{\rm dir} := CA_{\rm dir}(r) + PCA_{\rm dir}(r) + \mathcal{SF} + \mathcal{SF}'$. *Then* $H + G^{\rm rob}_{\rm dir}$ *has a decomposition into* s' *edge-disjoint Hamilton cycles* $C_1, \dots, C_{s'}$. *Moreover,* C_i *contains one of the special path systems from* $\mathcal{SPS}$, *for each* $i \le s'$.

Recall from Section 2.8.1 that we always view fictive edges in special factors as being distinct from each other and from the edges in other graphs. So for example, saying that $CA_{\rm dir}(r)$ and $\mathcal{SF}$ are edge-disjoint in Lemma 2.9.4 still allows for a fictive edge xy in $\mathcal{SF}$ to occur in $CA_{\rm dir}(r)$ as well (but $CA_{\rm dir}(r)$ will avoid all non-fictive edges in $\mathcal{SF}$).

In the proof of Theorem 1.3.3 we will use the following 'undirected' consequence of Lemma 2.9.4.

COROLLARY 2.9.5. *Suppose that* $0 < 1/m \ll \varepsilon_0, 1/K \ll \varepsilon \ll 1/L \ll 1/f \ll r_1/m \ll 1/\ell', 1/g \ll 1$ *and that* $rK^2 \le m$. *Let*

$$r_2 := 96\ell' g^2 K r, \quad r_3 := rK/L, \quad r^\diamond := r_1 + r_2 + r - (Lf - 1)r_3, \quad s' := rfK + 7r^\diamond$$

and suppose that $K/14, K/f, K/g, m/4\ell', m/L, 2fK/3g(g-1) \in \mathbb{N}$. *Suppose that* $(G_{\rm dir}, \mathcal{P}, \mathcal{P}')$ *is a* $[K, L, m, \varepsilon_0, \varepsilon]$*-scheme and let* G' *denote the underlying undirected graph of* $G_{\rm dir}$. *Suppose that* $EF_1, \dots, EF_{r_3}$ *are edge-disjoint exceptional factors with parameters* (L, f) *for* $G_{\rm dir}$ *(with respect to* $(\mathcal{P}, \mathcal{P}')$*). Let* $\mathcal{EF} := EF_1 + \dots + EF_{r_3}$. *Then there exists a graph* $CA(r)$ *for which the following holds:*

(i) $CA(r)$ *is a* $2(r_1 + r_2)$*-regular spanning subgraph of* G' *which is edge-disjoint from* $\mathcal{EF}$.
(ii) *Suppose that* $EF'_1, \dots, EF'_{r^\diamond}$ *are exceptional factors with parameters* $(1,7)$ *for* $G_{\rm dir}$ *(with respect to* $(\mathcal{P}, \mathcal{P})$*) which are edge-disjoint from each other and from* $CA(r) + \mathcal{EF}$. *Let* $\mathcal{EF}' := EF'_1 + \dots + EF'_{r^\diamond}$. *Then there exists a graph* $PCA(r)$ *for which the following holds:*
   (a) $PCA(r)$ *is a* $10r^\diamond$*-regular spanning subgraph of* G' *which is edge-disjoint from* $CA(r) + \mathcal{EF} + \mathcal{EF}'$.
   (b) *Let* $\mathcal{EPS}$ *be the set consisting of all the* s' *exceptional path systems contained in* $\mathcal{EF} + \mathcal{EF}'$. *Suppose that* H_A *is a* $2r$*-regular graph on* $A = \bigcup_{i=1}^K A_i$ *and* H_B *is a* $2r$*-regular graph on* $B = \bigcup_{i=1}^K B_i$. *Suppose that* $H := H_A + H_B$ *is edge-disjoint from* $G^{\rm rob} := CA(r) + PCA(r) + \mathcal{EF} + \mathcal{EF}'$. *Then* $H + G^{\rm rob}$ *has a decomposition into* s' *edge-disjoint* 2*-factors* $H_1, \dots, H_{s'}$ *such that each* H_i *contains one of the exceptional path systems from* $\mathcal{EPS}$. *Moreover, for each* $i \le s'$, *the following assertions hold:*
      (b_1) *If the exceptional path system contained in* H_i *is a Hamilton exceptional path system, then* H_i *is a Hamilton cycle on* $V(G_{\rm dir}) \cup V_0$.

(b$_2$) *If the exceptional path system contained in* H_i *is a matching exceptional path system, then* H_i *is the union of a Hamilton cycle on* $A' = A \cup A_0$ *and a Hamilton cycle on* $B' = B \cup B_0$. *In particular, if both* $|A'|$ *and* $|B'|$ *are even, then* H_i *is the union of two edge-disjoint perfect matchings on* $V(G_{\rm dir}) \cup V_0$.

We remark that, as usual, in Corollary 2.9.5 we write A_0 and B_0 for the exceptional sets of $\mathcal{P}$, V_0 for $A_0 \cup B_0$, and $A_1, \dots, A_K, B_1, \dots, B_K$ for the clusters in $\mathcal{P}$. Note that the vertex set of each of $\mathcal{EF}$, $\mathcal{EF}'$, $G^{\rm rob}$ includes V_0 while that of $G_{\rm dir}$, $CA(r)$, $PCA(r)$, H does not.

Moreover, note that matching exceptional systems are only constructed if both $|A'|$ and $|B'|$ are even. Indeed, we only construct matching exceptional systems in the case when $e_G(A', B') < D$. But by Proposition 2.2.1(ii), in this case we have that $n = 0 \pmod 4$ and $|A'| = |B'| = n/2$. Therefore, Corollary 2.9.5(ii)(b) implies that $H + G^{\rm rob}$ has a decomposition into Hamilton cycles and perfect matchings. The proportion of Hamilton cycles (and perfect matchings) in this decomposition is determined by $\mathcal{EF} + \mathcal{EF}'$, and does not depend on H.

Proof of Corollary 2.9.5. Choose new constants ε', d such that $\varepsilon \ll \varepsilon' \ll 1/L$ and $r_1/m \ll d \ll 1/\ell', 1/g$. Consider the A-setup $(G_{\rm dir}[A], \mathcal{Q}_A, \mathcal{Q}'_A, R_A, C_A, U_{A,\ell'}, U'_{A,\ell'})$ associated to $(G_{\rm dir}, \mathcal{P}, \mathcal{P}')$. By Lemma 2.9.3, this is an $(\ell', K, m, \varepsilon', 1/2)$-setup and thus also an $(\ell', K, m, \varepsilon', d)$-setup.

Recall that $\mathcal{P}'$ is obtained from $\mathcal{P}$ by partitioning each cluster A_i of $\mathcal{P}$ into L sets $A_{i,1}, \dots, A_{i,L}$ of equal size and partitioning each cluster B_i of $\mathcal{P}$ into L sets $B_{i,1}, \dots, B_{i,L}$ of equal size. Let $\mathcal{Q}''_A := \{A_{1,1}, \dots, A_{K,L}\}$. (So $\mathcal{Q}''_A$ plays the role of $\mathcal{Q}'_A$ in (2.8.1).) Let $EF^*_{i,A,{\rm dir}}$ be as defined in Section 2.8.2. Recall from there that, for each $i \le r_3$, $EF^*_{i,A,{\rm dir}}$ is a special factor with parameters (L, f) with respect to $C_A = A_1 \dots A_K$, $\mathcal{Q}''_A$ in $G_{\rm dir}[A]$ such that ${\rm Fict}(EF^*_{i,A,{\rm dir}})$ is the union of $J^*[A]$ over all the Lf exceptional systems J contained in EF_i. Thus we can apply Lemma 2.9.4 to $(G_{\rm dir}[A], \mathcal{Q}_A, \mathcal{Q}'_A, R_A, C_A, U_{A,\ell'}, U'_{A,\ell'})$ with K, Lf, ε' playing the roles of k, q, ε in order to obtain a spanning subdigraph $CA_{A,{\rm dir}}(r)$ of $G_{\rm dir}[A]$ which satisfies Lemma 2.9.4(i). Similarly, we obtain a spanning subdigraph $CA_{B,{\rm dir}}(r)$ of $G_{\rm dir}[B]$ which satisfies Lemma 2.9.4(i) (with $G_{\rm dir}[B]$ playing the role of G). Thus the underlying undirected graph $CA(r)$ of $CA_{A,{\rm dir}}(r) + CA_{B,{\rm dir}}(r)$ satisfies Corollary 2.9.5(i).

Now let $EF'_1, \dots, EF'_{r^\diamond}$ be exceptional factors as described in Corollary 2.9.5(ii). Similarly as before, for each $i \le r^\diamond$, $(EF'_i)^*_{A,{\rm dir}}$ is a special factor with parameters $(1, 7)$ with respect to C_A, $\mathcal{Q}_A$ in $G_{\rm dir}[A]$ such that ${\rm Fict}((EF'_i)^*_{A,{\rm dir}})$ is the union of $J^*[A]$ over all the 7 exceptional systems J contained in EF'_i. Thus we can apply Lemma 2.9.4 (with $G_{\rm dir}[A]$ playing the role of G) to obtain a spanning subdigraph $PCA_{A,{\rm dir}}(r)$ of $G_{\rm dir}[A]$ which satisfies Lemma 2.9.4(ii)(a) and (ii)(b). Similarly, we obtain a spanning subdigraph $PCA_{B,{\rm dir}}(r)$ of $G_{\rm dir}[B]$ which satisfies Lemma 2.9.4(ii)(a) and (ii)(b) (with $G_{\rm dir}[B]$ playing the role of G). Thus the underlying undirected graph $PCA(r)$ of $PCA_{A,{\rm dir}}(r) + PCA_{B,{\rm dir}}(r)$ satisfies Corollary 2.9.5(ii)(a).

It remains to check that Corollary 2.9.5(ii)(b) holds too. Thus let $H = H_A + H_B$ be as described in Corollary 2.9.5(ii)(b). Let $H_{A,{\rm dir}}$ be an r-regular orientation of H_A. (To see that such an orientation exists, apply Petersen's theorem, i.e. Theorem 1.3.10, to obtain a decomposition of H_A into 2-factors and then orient each 2-factor to obtain a (directed) 1-factor.) Let $\mathcal{EF}^*_{A,{\rm dir}} := EF^*_{1,A,{\rm dir}} + \dots + EF^*_{r_3,A,{\rm dir}}$

and let $(\mathcal{EF}')^*_{A,\mathrm{dir}} := (EF'_1)^*_{A,\mathrm{dir}} + \cdots + (EF'_{r^\diamond})^*_{A,\mathrm{dir}}$. Then Lemma 2.9.4(ii)(b) implies that $H_{A,\mathrm{dir}} + CA_{A,\mathrm{dir}}(r) + PCA_{A,\mathrm{dir}}(r) + \mathcal{EF}^*_{A,\mathrm{dir}} + (\mathcal{EF}')^*_{A,\mathrm{dir}}$ has a decomposition into s' edge-disjoint (directed) Hamilton cycles $C'_{1,A}, \dots, C'_{s',A}$ such that each $C'_{i,A}$ contains $EPS^*_{i',A,\mathrm{dir}}$ for some exceptional path system $EPS_{i'} \in \mathcal{EPS}$. Similarly, let $H_{B,\mathrm{dir}}$ be an r-regular orientation of H_B. Then $H_{B,\mathrm{dir}} + CA_{B,\mathrm{dir}}(r) + PCA_{B,\mathrm{dir}}(r) + \mathcal{EF}^*_{B,\mathrm{dir}} + (\mathcal{EF}')^*_{B,\mathrm{dir}}$ has a decomposition into s' edge-disjoint (directed) Hamilton cycles $C'_{1,B}, \dots, C'_{s',B}$ such that each $C'_{i,B}$ contains $EPS^*_{i'',B,\mathrm{dir}}$ for some exceptional path system $EPS_{i''} \in \mathcal{EPS}$. By relabeling the $C'_{i,A}$ and $C'_{i,B}$ if necessary, we may assume that $C'_{i,A}$ contains $EPS^*_{i,A,\mathrm{dir}}$ and $C'_{i,B}$ contains $EPS^*_{i,B,\mathrm{dir}}$. Let $C_{i,A}$ and $C_{i,B}$ be the undirected cycles obtained from $C'_{i,A}$ and $C'_{i,B}$ by ignoring the directions of all the edges. So $C_{i,A}$ contains $EPS^*_{i,A}$ and $C_{i,B}$ contains $EPS^*_{i,B}$. Let $H_i := C_{i,A} + C_{i,B} - EPS^*_i + EPS_i$. Then Proposition 2.8.1 (applied with G' playing the role of G) implies that $H_1, \dots, H_{s'}$ is a decomposition of $H + G^{\mathrm{rob}} = H + CA(r) + PCA(r) + \mathcal{EF} + \mathcal{EF}'$ into edge-disjoint 2-factors satisfying Corollary 2.9.5(ii)(b_1) and (b_2). □

2.10. Proof of Theorem 1.3.3

Before we can prove Theorem 1.3.3, we need the following two observations. Recall that a $(K, m, \varepsilon_0, \varepsilon)$-scheme was defined in Section 2.4 and that a $[K, L, m, \varepsilon_0, \varepsilon']$-scheme was defined in Section 2.8.3.

PROPOSITION 2.10.1. *Suppose that* $0 < 1/m \ll \varepsilon, \varepsilon_0 \ll \varepsilon' \ll 1/K, 1/L \ll 1$ *and that* $K, L, m/L \in \mathbb{N}$. *Suppose that* $(G, \mathcal{P}')$ *is a* $(KL, m/L, \varepsilon_0, \varepsilon)$*-scheme. Suppose that* $\mathcal{P}$ *is a* (K, m, ε_0)*-partition such that* $\mathcal{P}'$ *is an* L*-refinement of* $\mathcal{P}$. *Then there exists an orientation* G_{dir} *of* G *such that* $(G_{\mathrm{dir}}, \mathcal{P}, \mathcal{P}')$ *is a* $[K, L, m, \varepsilon_0, \varepsilon']$*-scheme.*

Proof. Randomly orient every edge in G to obtain an oriented graph G_{dir}. (So given any edge xy in G with probability 1/2, $xy \in E(G_{\mathrm{dir}})$ and with probability 1/2, $yx \in E(G_{\mathrm{dir}})$.) (Sch1′) and (Sch2′) follow immediately from (Sch1) and (Sch2).

Note that (Sch3) imply that $G[A_{i,j}, B_{i',j'}]$ is $[1, \sqrt{\varepsilon}]$-superregular with density at least $1 - \varepsilon$, for all $i, i' \leq K$ and $j, j' \leq L$. Using this, (Sch3′) follows easily from the large deviation bound in Proposition 1.4.4. (Sch4′) follows from Proposition 1.4.4 in a similar way. □

PROPOSITION 2.10.2. *Suppose that* G *is a* D*-regular graph on* n *vertices which is* ε*-close to the union of two disjoint copies of* $K_{n/2}$. *Then* $D \leq (1/2 + 4\varepsilon)n$.

Proof. Let $B \subseteq V(G)$ with $|B| = \lfloor n/2 \rfloor$ be such that $e(B, V(G) \setminus B) \leq \varepsilon n^2$. Note that B exists since G is ε-close to the union of two disjoint copies of $K_{n/2}$. Let $A = V(G) \setminus B$. If $D > (1/2 + 4\varepsilon)n$, then Proposition 2.2.1(i) implies that $e(A, B) > \varepsilon n^2$, a contradiction. □

We can now put everything together and prove Theorem 1.3.3 in the following steps. We choose the (localized) exceptional systems needed as an 'input' for Corollary 2.9.5 to construct the robustly decomposable graph G^{rob} in Step 3. For this, we first choose appropriate constants and a suitable vertex partition in Steps 1 and 2 respectively (in Step 1, we also find some Hamilton cycles covering 'bad' edges). In Step 4, we then apply Corollary 2.9.5 to find G^{rob}. Similarly, we then choose the (localized) exceptional systems needed as an 'input' for the 'approximate decomposition lemma' (Lemma 2.5.4) in Step 6 (in this step, we also find some Hamilton cycles which extend those exceptional systems which are not localized). For Step 6, we first choose a suitable vertex partition in Step 5. In Step 7, we find an approximate decomposition using Lemma 2.5.4 and in Step 8, we decompose the union of the 'leftover' and $G^{\rm rob}$ via Corollary 2.9.5.

Proof of Theorem 1.3.3.
Step 1: Choosing the constants and a framework. Choose $n_0 \in \mathbb{N}$ to be sufficiently large compared to $1/\varepsilon_{\rm ex}$. Let G and D be as in Theorem 1.3.3. By Proposition 2.10.2

$$n/2 - 1 \leq D \leq (1/2 + 4\varepsilon_{\rm ex})n. \tag{2.10.1}$$

Define new constants such that

$$0 < 1/n_0 \ll \varepsilon_{\rm ex} \ll \varepsilon_0 \ll \phi_0 \ll \varepsilon_* \ll \varepsilon'_* \ll \varepsilon'_1 \ll \lambda_{K_2} \ll 1/K_2 \ll \gamma \ll 1/K_1$$
$$\ll \varepsilon''_* \ll 1/L \ll 1/f \ll \gamma_1 \ll 1/g \ll \varepsilon'_2, \lambda_{K_1L} \ll \varepsilon \ll 1,$$

where $K_1, K_2, L, f, g \in \mathbb{N}$ and K_2 is odd. Note that we can choose the constants such that

$$\frac{D - \phi_0 n}{400(K_1LK_2)^2}, \phi_0 n, \frac{\lambda_{K_1L}n}{(K_1L)^2}, \frac{\lambda_{K_2}n}{K_2^2}, \frac{K_1}{14fg}, \frac{2fK_1}{3g(g-1)} \in \mathbb{N}. \tag{2.10.2}$$

Apply Proposition 2.2.5 to obtain a partition A, A_0, B, B_0 of $V(G)$ such that (G, A, A_0, B, B_0) is an $(\varepsilon_0, 4gK_1LK_2)$-framework with $\Delta(G[A', B']) \leq D/2$ (where $A' := A \cup A_0$ and $B' := B \cup B_0$). Let w_1 and w_2 be two vertices of G such that $d_{G[A',B']}(w_1) \geq d_{G[A',B']}(w_2) \geq d_{G[A',B']}(v)$ for all $v \in V(G) \setminus \{w_1, w_2\}$. Note that the partition A, A_0, B, B_0 of $V(G)$ and the two vertices w_1 and w_2 are fixed throughout the proof. Moreover, in the remainder of the proof, given a graph H on $V(G)$, we will always write $H^\diamond$ for $H - H[A] - H[B]$.

Next we apply Lemma 2.6.3 with ϕ_0 and $4gK_1LK_2$ playing the roles of ϕ and K to find a spanning subgraph $\mathcal{H}'_1$ of G. Let $G_1 := G - \mathcal{H}'_1$. Thus the following properties are satisfied:

- (α_1) $G[A_0] + G[B_0] \subseteq \mathcal{H}'_1$ and $\mathcal{H}'_1$ is a $\phi_0 n$-regular spanning graph of G.
- (α_2) $e_{\mathcal{H}'_1}(A', B') \leq \phi_0 n$ and $e_{G_1}(A', B')$ is even.
- (α_3) The edges of $\mathcal{H}'_1$ can be decomposed into $\lfloor e_{\mathcal{H}'_1}(A', B')/2 \rfloor$ Hamilton cycles and $\phi_0 n - 2\lfloor e_{\mathcal{H}'_1}(A', B')/2 \rfloor$ perfect matchings. Moreover, if $e_G(A', B') \geq D$, then this decomposition consists of $\lfloor \phi_0 n/2 \rfloor$ Hamilton cycles and one perfect matching if D is odd.
- (α_4) $d_{G_1[A',B']}(w_1) \leq (D - \phi_0 n)/2$. Furthermore, if $D = n/2 - 1$ then $d_{G_1[A',B']}(w_2) \leq (D - \phi_0 n)/2$.
- (α_5) If $e_G(A', B') < D$, then $\Delta(G_1[A', B']) \leq e(G_1[A', B'])/2 \leq (D - \phi_0 n)/2$.

Let $\mathcal{H}_1$ be the collection of Hamilton cycles and perfect matchings guaranteed by (α_3). (So $\mathcal{H}'_1 = \bigcup \mathcal{H}_1$.) Note that

$$D_1 := D - \phi_0 n \tag{2.10.3}$$

is even (since (2.10.2) implies that D and $\phi_0 n$ have the same parity) and that G_1 is D_1-regular. Moreover, (G_1, A, A_0, B, B_0) is an $(\varepsilon_0, 4gK_1LK_2)$-framework with

$\Delta(G_1[A',B']) \le D/2$. Let

$$m_1 := \frac{|A|}{K_1} = \frac{|B|}{K_1}, \qquad r := \gamma m_1, \qquad r_1 := \gamma_1 m_1, \qquad r_2 := 96 g^3 K_1 r,$$
$$r_3 := \frac{rK_1}{L}, \qquad r^\diamond := r_1 + r_2 + r - (Lf-1)r_3,$$
$$m_2 := \frac{|A|}{K_2} = \frac{|B|}{K_2}, \qquad D_4 := D_1 - 2(Lfr_3 + 7r^\diamond). \tag{2.10.4}$$

Note that (FR3) implies $m_1/L \in \mathbb{N}$. Moreover,

$$r_2, r_3 \le \gamma^{1/2} m_1 \le \gamma^{1/3} r_1, \qquad r_1/2 \le r^\diamond \le 2r_1. \tag{2.10.5}$$

Furthermore, by changing γ, γ_1 slightly, we may assume that $r/400LK_2^2, r_1/400K_2^2 \in \mathbb{N}$. This implies that $r_2/400K_2^2, r_3/400K_2^2, r^\diamond/400K_2^2 \in \mathbb{N}$. Together with the fact that $D_1/400K_2^2 = (D-\phi_0 n)/400K_2^2 \in \mathbb{N}$ by (2.10.2), this in turn implies that

$$D_4/400K_2^2 \in \mathbb{N}. \tag{2.10.6}$$

Step 2: Choosing a $(K_1, L, m_1, \varepsilon_0)$-partition $(\mathcal{P}_1, \mathcal{P}_1')$. We now prepare the ground for the construction of the robustly decomposable graph $G^{\rm rob}$, which we will obtain via the robust decomposition lemma (Corollary 2.9.5) in Step 4.

Since (G_1, A, A_0, B, B_0) is an $(\varepsilon_0, 4gK_1LK_2)$-framework, it is also an (ε_0, K_1L)-framework. Recall that G_1 is D_1-regular and $D_1 = D - \phi_0 n \ge (1-3\phi_0)n/2$ (as $D \ge n/2 - 1$). Apply Lemma 2.4.2 with G_1, m_1/L, $3\phi_0$, K_1L, ε_*, ε_* playing the roles of G, m, μ, K, ε_1, ε_2 to obtain partitions $A_1', \dots, A_{K_1L}'$ of A and $B_1', \dots, B_{K_1L}'$ of B into sets of size m_1/L such that the following properties are satisfied:

- (S_1a) Together with A_0 and B_0 all these sets A_i' and B_i' form a $(K_1L, m_1/L, \varepsilon_0)$-partition $\mathcal{P}_1'$ of $V(G_1)$.
- (S_1b) $(G_1[A] + G_1[B], \mathcal{P}_1')$ is a $(K_1L, m_1/L, \varepsilon_0, \varepsilon_*)$-scheme.
- (S_1c) $(G_1^\diamond, \mathcal{P}_1')$ is a $(K_1L, m_1/L, \varepsilon_0, \varepsilon_*)$-exceptional scheme (where $G_1^\diamond := G_1 - G_1[A] - G_2[B]$).

Note that $(1-\varepsilon_0)n \le n - |A_0 \cup B_0| = 2K_1m_1 \le n$ by (FR3). For all $i \le K_1$ and all $h \le L$, let $A_{i,h} := A'_{(i-1)L+h}$. (So this is just a relabeling of the sets A_i'.) Define $B_{i,h}$ similarly and let $A_i := \bigcup_{h \le L} A_{i,h}$ and $B_i := \bigcup_{h \le L} B_{i,h}$. Let $\mathcal{P}_1 := \{A_0, B_0, A_1, \dots, A_{K_1}, B_1, \dots, B_{K_1}\}$ denote the corresponding $(K_1, m_1, \varepsilon_0)$-partition of $V(G)$. Thus $(\mathcal{P}_1, \mathcal{P}_1')$ is a $(K_1, L, m_1, \varepsilon_0)$-partition of $V(G)$, as defined in Section 2.8.2.

Step 3: Exceptional systems for the robustly decomposable graph. In order to be able to apply Corollary 2.9.5 to obtain the robustly decomposable graph $G^{\rm rob}$, we first need to construct suitable exceptional systems with parameter ε_0. The construction of these exceptional systems depends on whether G is critical and whether $e_G(A', B') \ge D$. First we show that in each case, for all $1 \le i_1', i_2' \le K_1L$, we can always find sets $\mathcal{J}_{i_1', i_2'}$ of $\lambda_{K_1L} n/(K_1L)^2$ (i_1', i_2')-ES with respect to $\mathcal{P}_1'$.

Case 1: $e_G(A', B') \ge D$ and G is not critical. Our aim is to apply Lemma 2.7.3 to G with $\mathcal{H}_1'$, m_1/L, K_1L, $\mathcal{P}_1'$, ε_*, ϕ_0, λ_{K_1L} playing the roles of G_0, m, K, $\mathcal{P}$, ε, ϕ, λ. First we verify that Lemma 2.7.3(i)–(iv) are satisfied. Lemma 2.7.3(i) holds trivially. (FR2) implies that $e_G(A', B') \le \varepsilon_0 n^2$. Moreover, recall from (S_1a) that $\mathcal{P}_1'$ is a $(K_1L, m_1/L, \varepsilon_0)$-partition of $V(G)$ and that A' and B' were chosen (by Proposition 2.2.5) such that $\Delta(G[A', B']) \le D/2$. Altogether this shows that Lemma 2.7.3(ii) holds. Lemma 2.7.3(iii) follows from (α_1) and (α_2). To verify

Lemma 2.7.3(iv), note that $G_1^\diamond$ plays the role of $G^\diamond$ in Lemma 2.7.3 and $G_1^\diamond[A', B'] = G_1[A', B']$. So $e_{G_1^\diamond}(A', B')$ is even by (α_2). Together with the fact that $(G_1^\diamond, \mathcal{P}_1')$ is a $(K_1L, m_1/L, \varepsilon_0, \varepsilon_*)$-exceptional scheme by ($S_1$c), this implies Lemma 2.7.3(iv).

By Lemma 2.7.3, we obtain a set $\mathcal{J}$ of $\lambda_{K_1L}n$ edge-disjoint Hamilton exceptional systems J in $G_1^\diamond$ such that $e_J(A', B') = 2$ for each $J \in \mathcal{J}$ and such that for all $1 \le i_1', i_2' \le K_1L$ the set $\mathcal{J}$ contains precisely $\lambda_{K_1L}n/(K_1L)^2$ (i_1', i_2')-HES with respect to the partition $\mathcal{P}_1'$. For all $1 \le i_1', i_2' \le K_1L$, let $\mathcal{J}_{i_1', i_2'}$ be the set of these $\lambda_{K_1L}n/(K_1L)^2$ (i_1', i_2')-HES in $\mathcal{J}$. So $\mathcal{J}$ is the union of all the sets $\mathcal{J}_{i_1', i_2'}$. (Note that the set $\mathcal{J}$ here is a subset of the set $\mathcal{J}$ in Lemma 2.7.3, i.e. we do not use all the Hamilton exceptional systems constructed by Lemma 2.7.3. So we do not need the full strength of Lemma 2.7.3 at this point.)

Case 2: $e_G(A', B') \ge D$ and G is critical. Recall from Lemma 2.7.1(ii) that in this case we have $D = (n-1)/2$ or $D = n/2 - 1$. Our aim is to apply Lemma 2.7.4 to G with $\mathcal{H}_1'$, m_1/L, K_1L, $\mathcal{P}_1'$, ε_*, ϕ_0, λ_{K_1L} playing the roles of G_0, m, K, $\mathcal{P}$, ε, ϕ, λ. Similar arguments as in Case 1 show that Lemma 2.7.4(i)–(iv) hold. Recall that w_1 and w_2 are (fixed) vertices in $V(G)$ such that $d_{G[A',B']}(w_1) \ge d_{G[A',B']}(w_2) \ge d_{G[A',B']}(v)$ for all $v \in V(G) \setminus \{w_1, w_2\}$. Since $G_1^\diamond[A', B'] = G_1[A', B']$, ($\alpha_4$) implies that $d_{G_1^\diamond[A',B']}(w_1) \le (D - \phi_0 n)/2$. Moreover, if $D = n/2 - 1$, then $d_{G_1^\diamond[A',B']}(w_2) \le (D - \phi_0 n)/2$. Let W be the set of vertices $w \in V(G)$ such that $d_{G[A',B']}(w) \ge 11D/40$, as defined in Lemma 2.7.1. If $D = (n-1)/2$, then $|W| = 1$ by Lemma 2.7.1(ii). This means that $w_2 \notin W$ and so $d_{G_1^\diamond[A',B']}(w_2) \le d_{G[A',B']}(w_2) \le 11D/40$. Thus in both cases we have that

$$d_{G_1^\diamond[A',B']}(w_1), d_{G_1^\diamond[A',B']}(w_2) \le (D - \phi_0 n)/2. \tag{2.10.7}$$

Therefore, Lemma 2.7.4(v) holds.

By Lemma 2.7.4, we obtain a set $\mathcal{J}$ of $\lambda_{K_1L}n$ edge-disjoint Hamilton exceptional systems J in $G_1^\diamond$ such that, for all $1 \le i_1', i_2' \le K_1L$, the set $\mathcal{J}$ contains precisely $\lambda_{K_1L}n/(K_1L)^2$ (i_1', i_2')-HES with respect to the partition $\mathcal{P}_1'$. Moreover, each $J \in \mathcal{J}$ satisfies $e_J(A', B') = 2$ and $d_{J[A',B']}(w) = 1$ for all $w \in \{w_1, w_2\}$ with $d_{G[A',B']}(w) \ge 11D/40$. For all $1 \le i_1', i_2' \le K_1L$, let $\mathcal{J}_{i_1', i_2'}$ be the set of these $\lambda_{K_1L}n/(K_1L)^2$ (i_1', i_2')-HES. So $\mathcal{J}$ is the union of all the sets $\mathcal{J}_{i_1', i_2'}$. (So similarly as in Case 1, we do not use all the Hamilton exceptional systems constructed by Lemma 2.7.4 at this point.)

Case 3: $e_G(A', B') < D$. Recall from Proposition 2.2.1(ii) that in this case we have $D = n/2 - 1$, $n = 0 \pmod 4$ and $|A'| = |B'| = n/2$. Our aim is to apply Lemma 2.7.5 to G with $\mathcal{H}_1'$, m_1/L, K_1L, $\mathcal{P}_1'$, ε_*, ϕ_0, λ_{K_1L} playing the roles of G_0, m, K, $\mathcal{P}$, ε, ϕ, λ. Similar arguments as in Case 1 show that Lemma 2.7.5(i)–(iv) hold. Since $G_1^\diamond[A', B'] = G_1[A', B']$ and $D = n/2 - 1$, Lemma 2.7.5(v) follows from (α_5).

By Lemma 2.7.5, $G_1^\diamond$ can be decomposed into a set $\mathcal{J}'$ of $D_1/2$ edge-disjoint exceptional systems such that each of these exceptional systems J is either a Hamilton exceptional system with $e_J(A', B') = 2$ or a matching exceptional system. (So $\mathcal{J}'$ plays the role of the set $\mathcal{J}$ in Lemma 2.7.5.) Lemma 2.7.5(b) guarantees that we can choose a subset $\mathcal{J}$ of $\mathcal{J}'$ such that $\mathcal{J}$ consists of $\lambda_{K_1L}n$ edge-disjoint exceptional systems J in $G_1^\diamond$ such that for all $1 \le i_1', i_2' \le K_1L$ the set $\mathcal{J}$ contains precisely $\lambda_{K_1L}n/(K_1L)^2$ (i_1', i_2')-ES with respect to the partition $\mathcal{P}_1'$. For all $1 \le i_1', i_2' \le K_1L$, let $\mathcal{J}_{i_1', i_2'}$ be the set of these $\lambda_{K_1L}n/(K_1L)^2$ (i_1', i_2')-ES. So $\mathcal{J}$ is the union of all the sets $\mathcal{J}_{i_1', i_2'}$. (Note that to construct the robustly decomposable graph we will only use the exceptional systems in $\mathcal{J}$. However, in order to prove condition (β_5) below, we will also use the fact that $G_1^\diamond$ has a decomposition into edge-disjoint exceptional systems.)

Thus in each of the three cases, $\mathcal{J}$ is the union of all the sets $\mathcal{J}_{i'_1,i'_2}$, where for all $1 \le i'_1, i'_2 \le K_1L$, the set $\mathcal{J}$ consists of precisely $\lambda_{K_1L}n/(K_1L)^2$ (i'_1,i'_2)-ES with respect to the partition $\mathcal{P}'_1$. Moreover, all the $\lambda_{K_1L}n$ exceptional systems in $\mathcal{J}$ are edge-disjoint.

Our next aim is to choose two disjoint subsets $\mathcal{J}_{\rm CA}$ and $\mathcal{J}_{\rm PCA}$ of $\mathcal{J}$ with the following properties:

(a) In total $\mathcal{J}_{\rm CA}$ contains Lfr_3 exceptional systems. For each $i \le f$ and each $h \le L$, $\mathcal{J}_{\rm CA}$ contains precisely r_3 (i_1,i_2)-ES of style h (with respect to the $(K_1, L, m_1, \varepsilon_0)$-partition $(\mathcal{P}_1, \mathcal{P}'_1)$) such that $i_1, i_2 \in \{(i-1)K_1/f + 2, \dots, iK_1/f\}$.
(b) In total $\mathcal{J}_{\rm PCA}$ contains $7r^\diamond$ exceptional systems. For each $i \le 7$, $\mathcal{J}_{\rm PCA}$ contains precisely $r^\diamond$ (i_1,i_2)-ES (with respect to the partition $\mathcal{P}_1$) with $i_1, i_2 \in \{(i-1)K_1/7 + 2, \dots, iK_1/7\}$.
(c) Each exceptional system $J \in \mathcal{J}_{\rm CA} \cup \mathcal{J}_{\rm PCA}$ is either a Hamilton exceptional system with $e_J(A', B') = 2$ or a matching exceptional system.

(Recall that we defined in Section 2.8.3 when an (i_1,i_2)-ES has style h with respect to a $(K_1, L, m_1, \varepsilon_0)$-partition $(\mathcal{P}_1, \mathcal{P}'_1)$.) To see that it is possible to choose $\mathcal{J}_{\rm CA}$ and $\mathcal{J}_{\rm PCA}$, split $\mathcal{J}$ into two sets $\mathcal{J}_1$ and $\mathcal{J}_2$ such that both $\mathcal{J}_1$ and $\mathcal{J}_2$ contain at least $\lambda_{K_1L}n/3(K_1L)^2$ (i'_1,i'_2)-ES with respect to $\mathcal{P}'_1$, for all $1 \le i'_1, i'_2 \le K_1L$. Note that, for each $i \le f$, there are $(K_1/f-1)^2$ choices of pairs (i_1,i_2) with $i_1, i_2 \in \{(i-1)K_1/f+2, \dots, iK_1/f\}$. Moreover, for each such pair (i_1,i_2) and each $h \le L$ there is precisely one pair (i'_1,i'_2) with $1 \le i'_1, i'_2 \le K_1L$ and such that any (i'_1,i'_2)-ES with respect to $\mathcal{P}'_1$ is an (i_1,i_2)-ES of style h with respect to $(\mathcal{P}_1, \mathcal{P}'_1)$. Together with the fact that $\gamma \ll \lambda_{K_1L}, 1/L, 1/f$ and

$$\frac{(K_1/f-1)^2\lambda_{K_1L}n}{3(K_1L)^2} \ge \frac{\gamma n}{L} \ge \frac{\gamma K_1 m_1}{L} = \frac{rK_1}{L} = r_3,$$

this implies that we can choose a set $\mathcal{J}_{\rm CA} \subseteq \mathcal{J}_1$ satisfying (a).

Similarly, for each $i \le 7$, there are $(K_1/7-1)^2$ choices of pairs (i_1,i_2) with $i_1, i_2 \in \{(i-1)K_1/7+2, \dots, iK_1/7\}$. Moreover, for each such pair (i_1,i_2) there are L^2 distinct pairs (i'_1,i'_2) with $1 \le i'_1, i'_2 \le K_1L$ and such that any (i'_1,i'_2)-ES with respect to $\mathcal{P}'_1$ is an (i_1,i_2)-ES with respect to $\mathcal{P}_1$. Together with the fact that $\gamma_1 \ll \lambda_{K_1L}$ and

$$\frac{(K_1/7-1)^2L^2\lambda_{K_1L}n}{3(K_1L)^2} \ge \gamma_1 n \ge 2\gamma_1 m_1 = 2r_1 \overset{(2.10.5)}{\ge} r^\diamond,$$

this implies that we can choose a set $\mathcal{J}_{\rm PCA} \subseteq \mathcal{J}_2$ satisfying (b). Our choice of $\mathcal{J} \supseteq \mathcal{J}_{\rm CA} \cup \mathcal{J}_{\rm PCA}$ guarantees that (c) holds too. Let

$$\mathcal{J}^{\rm rob} := \mathcal{J}_{\rm CA} \cup \mathcal{J}_{\rm PCA}, \quad \phi_0^{\rm rob} := (Lfr_3 + 7r^\diamond)/n \quad\text{and}\quad G_4^\diamond := G_1^\diamond - \bigcup \mathcal{J}^{\rm rob}. \tag{2.10.8}$$

(In Step 5 below we will define a graph G_4 which will satisfy $G_4^\diamond = G_4 - G_4[A] - G_4[B]$. So this will fit with our definition of the operator $^\diamond$.) Note that

$$\phi_0^{\rm rob} \ge \frac{7r^\diamond}{n} \overset{(2.10.5)}{\ge} \frac{3r_1}{n} = \frac{3\gamma_1 m_1}{n} \ge \frac{\gamma_1}{K_1} \ge 2\phi_0 \quad\text{and}\quad 2\phi_0^{\rm rob} n \overset{(2.10.4)}{=} D_1 - D_4. \tag{2.10.9}$$

Moreover, we claim that $\bigcup \mathcal{J}^{\rm rob}$ is a subgraph of $G_1^\diamond \subseteq G$ satisfying the following properties:

(β_1) $d_{\bigcup \mathcal{J}^{\rm rob}}(v) = 2(Lfr_3 + 7r^\diamond) = 2\phi_0^{\rm rob}n$ for each $v \in V_0$.
(β_2) $e_{\bigcup \mathcal{J}^{\rm rob}}(A', B') \leq 2\phi_0^{\rm rob}n$ is even.
(β_3) $\mathcal{J}^{\rm rob}$ contains exactly $\phi_0^{\rm rob}n$ exceptional systems, of which precisely $e_{\bigcup \mathcal{J}^{\rm rob}}(A', B')/2$ are Hamilton exceptional systems. If $e_G(A', B') \geq D$, then $\mathcal{J}^{\rm rob}$ consists entirely of Hamilton exceptional systems. If $\mathcal{J}^{\rm rob}$ contains a matching exceptional system, then $|A'| = |B'| = n/2$ is even.
(β_4) If $e_G(A', B') \geq D$ and G is critical, then $d_{\bigcup \mathcal{J}^{\rm rob}[A',B']}(w) = \phi_0^{\rm rob}n$ for all $w \in \{w_1, w_2\}$ with $d_{G[A',B']}(w) \geq 11D/40$. Moreover, $d_{G_4^\diamond[A',B']}(w_1)$, $d_{G_4^\diamond[A',B']}(w_2) \leq (D - (\phi_0 + 2\phi_0^{\rm rob})n)/2$.
(β_5) If $e_G(A', B') < D$, then $\Delta(G_4^\diamond[A', B']) \leq e(G_4^\diamond[A', B'])/2 \leq D_4/2 = (D - (\phi_0 + 2\phi_0^{\rm rob})n)/2$.

To verify the above, note that $\mathcal{J}^{\rm rob}$ consists of precisely $\phi_0^{\rm rob}n$ exceptional systems J (each of which is an exceptional cover). So (β_1) follows from (EC2). Moreover, each such J is either a Hamilton exceptional system with $e_J(A', B') = 2$ or a matching exceptional system (with $e_J(A', B') = 0$ by (MES)), which implies (β_2) and the first part of (β_3). If $e_G(A', B') \geq D$, then we are in Case 1 or 2 and so the second part of (β_3) follows from our construction of $\mathcal{J} \supseteq \mathcal{J}^{\rm rob}$. The first part of ($\beta_4$) follows from our construction of $\mathcal{J} \supseteq \mathcal{J}^{\rm rob}$ in Case 2. Since $11D/40 < (D - (\phi_0 + 2\phi_0^{\rm rob})n)/2$, we can combine the first part of (β_4) with (2.10.7) to obtain the 'moreover part' of (β_4). Thus it remains to verify (β_5). So suppose that $e_G(A', B') < D$. Recall from Case 3 that $G_1^\diamond$ has a decomposition into a set $\mathcal{J}'$ of $D_1/2$ edge-disjoint exceptional systems J, each of which is either a Hamilton exceptional system with $e_J(A', B') = 2$ or a matching exceptional system. This means that $J[A', B']$ is either empty or a matching of size 2. Note that $G_4^\diamond[A', B']$ is precisely the union of $J[A', B']$ over all those $D_1/2 - \phi_0^{\rm rob}n = D_4/2$ exceptional systems $J \in \mathcal{J}' \setminus \mathcal{J}^{\rm rob}$. So ($\beta_5$) holds.

Step 4: Finding the robustly decomposable graph. Let $G_2 := G_1[A] + G_1[B]$. Recall from (S_1b) that $(G_2, \mathcal{P}_1')$ is a $(K_1L, m_1/L, \varepsilon_0, \varepsilon_*)$-scheme. Apply Proposition 2.10.1 with G_2, $\mathcal{P}_1$, $\mathcal{P}_1'$, K_1, m_1, ε_*, ε_*' playing the roles of G, $\mathcal{P}$, $\mathcal{P}'$, K, m, ε, ε' to obtain an orientation $G_{2,\rm dir}$ of G_2 such that $(G_{2,\rm dir}, \mathcal{P}_1, \mathcal{P}_1')$ is a $[K_1, L, m_1, \varepsilon_0, \varepsilon_*']$-scheme.

Our next aim is to use Lemma 2.8.3 in order to extend the exceptional systems in $\mathcal{J}_{\rm CA}$ into r_3 edge-disjoint exceptional factors with parameters (L, f) for $G_{2,\rm dir}$ (with respect to $(\mathcal{P}_1, \mathcal{P}_1')$). For this, note that (a) and (c) guarantee that $\mathcal{J}_{CA}$ satisfies Lemma 2.8.3(i),(ii) with r_3 playing the role of q. Moreover, $Lr_3/m_1 = rK_1/m_1 = \gamma K_1 \ll 1$. Thus we can indeed apply Lemma 2.8.3 to $(G_{2,\rm dir}, \mathcal{P}_1, \mathcal{P}_1')$ with $\mathcal{J}_{CA}$, m_1, ε_*', K_1, r_3 playing the roles of $\mathcal{J}$, m, ε, K, q in order to obtain r_3 edge-disjoint exceptional factors $EF_1, \dots, EF_{r_3}$ with parameters (L, f) for $G_{2,\rm dir}$ (with respect to $(\mathcal{P}_1, \mathcal{P}_1')$) such that together these exceptional factors cover all edges in $\bigcup \mathcal{J}_{\rm CA}$. Let $\mathcal{EF}_{\rm CA} := EF_1 + \dots + EF_{r_3}$. Since $G_2 = G_1[A] + G_1[B]$, we have $(\mathcal{EF}_{\rm CA})^\diamond = \mathcal{J}_{\rm CA}$. Moreover, each exceptional path system in $\mathcal{EF}_{\rm CA}$ contains a unique exceptional system in $\mathcal{J}_{\rm CA}$ (in particular, their numbers are equal).

Note that $m_1/4g, m_1/L \in \mathbb{N}$ since $m_1 = |A|/K_1$ and $|A|$ is divisible by $4gK_1L$ as (G, A, A_0, B, B_0) is an $(\varepsilon_0, 4gK_1LK_2)$-framework. Furthermore, $rK_1^2 = \gamma m_1 K_1^2 \leq \gamma^{1/2} m_1 \leq m_1$. Thus we can apply Corollary 2.9.5 to the $[K_1, L, m_1, \varepsilon_0, \varepsilon_*'']$-scheme $(G_{2,\rm dir}, \mathcal{P}_1, \mathcal{P}_1')$ with K_1, m_1, ε_*'', g playing the roles of K, m, ε, ℓ' to obtain a spanning subgraph $CA(r)$ of G_2 as described there. (Note that G_2 equals the graph G' defined in Corollary 2.9.5.) In particular, $CA(r)$ is $2(r_1+r_2)$-regular and edge-disjoint from $\mathcal{EF}_{\rm CA}$.

Let G_3 be the graph obtained from G_2 by deleting all the edges of $CA(r)+\mathcal{EF}_{\rm CA}$. Thus G_3 is obtained from G_2 by deleting at most $2(r_1+r_2+r_3)\le 6r_1 = 6\gamma_1 m_1$ edges at every vertex in $A\cup B$. Let $G_{3,\rm dir}$ be the orientation of G_3 in which every edge is oriented in the same way as in $G_{2,\rm dir}$. Since $(G_{2,\rm dir},\mathcal{P}_1,\mathcal{P}'_1)$ is a $[K_1,L,m_1,\varepsilon_0,\varepsilon'_*]$-scheme, Proposition 1.4.1 and the fact that $\varepsilon''_*,\gamma_1\ll\varepsilon$ imply that $(G_{3,\rm dir},\mathcal{P}_1,\mathcal{P}_1)$ is a $[K_1,1,m_1,\varepsilon_0,\varepsilon]$-scheme. Moreover,

$$\frac{r^\diamond}{m_1}\overset{(2.10.5)}{\le}\frac{2r_1}{m_1}=2\gamma_1\ll 1.$$

Together with (b) and (c) this ensures that we can apply Lemma 2.8.3 to $(G_{3,\rm dir},\mathcal{P}_1,\mathcal{P}_1)$ with $\mathcal{J}_{PCA}$, m_1, K_1, 1, 7, $r^\diamond$ playing the roles of $\mathcal{J}$, m, K, L, f, q in order to obtain $r^\diamond$ edge-disjoint exceptional factors $EF'_1,\dots,EF'_{r^\diamond}$ with parameters $(1,7)$ for $G_{3,\rm dir}$ (with respect to $(\mathcal{P}_1,\mathcal{P}_1)$) such that together these exceptional factors cover all edges in $\bigcup\mathcal{J}_{\rm PCA}$. Let $\mathcal{EF}_{\rm PCA}:=EF'_1+\dots+EF'_{r^\diamond}$. Since $G_3\subseteq G_1[A]+G_1[B]$ we have $(\mathcal{EF}_{\rm PCA})^\diamond=\bigcup\mathcal{J}_{\rm PCA}$. Moreover, each exceptional path system in $\mathcal{EF}_{\rm PCA}$ contains a unique exceptional system in $\mathcal{J}_{\rm PCA}$.

Apply Corollary 2.9.5 to obtain a spanning subgraph $PCA(r)$ of G_2 as described there. In particular, $PCA(r)$ is $10r^\diamond$-regular and edge-disjoint from $CA(r)+\mathcal{EF}_{\rm CA}+\mathcal{EF}_{\rm PCA}$.

Let $G^{\rm rob}:=CA(r)+PCA(r)+\mathcal{EF}_{\rm CA}+\mathcal{EF}_{\rm PCA}$. Note that by (2.8.2) all the vertices in $V_0:=A_0\cup B_0$ have the same degree $r_0^{\rm rob}:=2(Lfr_3+7r^\diamond)=2\phi_0^{\rm rob}n$ in $G^{\rm rob}$. So

$$7r_1\overset{(2.10.5)}{\le} r_0^{\rm rob}\overset{(2.10.5)}{\le}30r_1. \tag{2.10.10}$$

Moreover, (2.8.2) also implies that all the vertices in $A\cup B$ have the same degree $r^{\rm rob}$ in $G^{\rm rob}$, where $r^{\rm rob}:=2(r_1+r_2)+10r^\diamond+2r_3+2r^\diamond=2(r_1+r_2+r_3+6r^\diamond)$. So

$$r_0^{\rm rob}-r^{\rm rob}=2\left(Lfr_3+r^\diamond-(r_1+r_2+r_3)\right)=2(Lfr_3+r-(Lf-1)r_3-r_3)=2r.$$

Note that $(G^{\rm rob})^\diamond=\bigcup(\mathcal{J}_{\rm CA}\cup\mathcal{J}_{\rm PCA})=\bigcup\mathcal{J}^{\rm rob}$. Recall that the number of Hamilton exceptional path systems in $\mathcal{EF}_{\rm CA}$ equals the number of Hamilton exceptional systems in $\mathcal{J}_{\rm CA}$, and that the analogue holds for $\mathcal{EF}_{\rm PCA}$. Hence, (β_1), (β_2) and (β_3) imply the follow statements:

- (β'_1) $d_{G^{\rm rob}}(v)=r_0^{\rm rob}=2\phi_0^{\rm rob}n$ for all $v\in V_0$.
- (β'_2) $e_{G^{\rm rob}}(A',B')=e_{\bigcup\mathcal{J}^{\rm rob}}(A',B')\le r_0^{\rm rob}=2\phi_0^{\rm rob}n$ is even.
- (β'_3) $\mathcal{EF}_{\rm CA}+\mathcal{EF}_{\rm PCA}$ contains exactly $\phi_0^{\rm rob}n$ exceptional path systems (and each such path system contains a unique exceptional system in $\mathcal{J}^{\rm rob}$, where $|\mathcal{J}^{\rm rob}|=\phi_0^{\rm rob}n$). Precisely $e_{\bigcup\mathcal{J}^{\rm rob}}(A',B')/2$ of these are Hamilton exceptional path systems. If $e_G(A',B')\ge D$, then every exceptional path system in $\mathcal{EF}_{\rm CA}+\mathcal{EF}_{\rm PCA}$ is a Hamilton exceptional path system. If $\mathcal{EF}_{\rm CA}+\mathcal{EF}_{\rm PCA}$ contains a matching exceptional path system, then $|A'|=|B'|=n/2$ is even.

Step 5: Choosing a (K_2,m_2,ε_0)-partition $\mathcal{P}_2$. We now prepare the ground for the approximate decomposition step (i.e. to apply Lemma 2.5.4). For this, we need to work with a finer partition of $A\cup B$ than the previous one (this will ensure that the leftover from the approximate decomposition step is sufficiently sparse compared to G^{rob}).

So let $G_4 := G_1 - G^{\mathrm{rob}}$ (where G_1 was defined in Step 1) and note that

$$D_4 \overset{(2.10.4)}{=} D_1 - r_0^{\mathrm{rob}} = D_1 - r^{\mathrm{rob}} - 2r. \tag{2.10.11}$$

So

(2.10.12)

$$d_{G_4}(v) = D_4 + 2r \text{ for all } v \in A \cup B \qquad \text{and} \qquad d_{G_4}(v) = D_4 \text{ for all } v \in V_0.$$

Hence

$$\delta(G_4) \geq D_4 \overset{(2.10.9)}{=} D_1 - 2\phi_0^{\mathrm{rob}} n \overset{(2.10.3)}{=} D - (\phi_0 + 2\phi_0^{\mathrm{rob}})n \geq (1 - 6\phi_0^{\mathrm{rob}})n/2$$

as $\phi_0^{\mathrm{rob}} \geq 2\phi_0$ by (2.10.9). Moreover, note that

$$2\phi_0^{\mathrm{rob}} n = r_0^{\mathrm{rob}} \overset{(2.10.10)}{\leq} 30r_1 = 30\gamma_1 m_1 \leq 30\gamma_1 n/K_1,$$

so $\phi_0^{\mathrm{rob}} \ll \varepsilon_2'$. Since (G, A, A_0, B, B_0) is an $(\varepsilon_0, 4gK_1LK_2)$-framework, (G_4, A, A_0, B, B_0) is an (ε_0, K_2)-framework. Now apply Lemma 2.4.2 to (G_4, A, A_0, B, B_0) with $K_2, m_2, \varepsilon_1', \varepsilon_2', 6\phi_0^{\mathrm{rob}}$ playing the roles of $K, m, \varepsilon_1, \varepsilon_2, \mu$ in order to obtain partitions $A_1, \dots, A_{K_2}$ and $B_1, \dots, B_{K_2}$ of A and B satisfying the following conditions:

(S_2a) The vertex partition $\mathcal{P}_2 := \{A_0, B_0, A_1, \dots A_{K_2}, B_1, \dots, B_{K_2}\}$ is a $(K_2, m_2, \varepsilon_0)$-partition of $V(G)$.

(S_2b) $(G_4[A] + G_4[B], \mathcal{P}_2)$ is a $(K_2, m_2, \varepsilon_0, \varepsilon_2')$-scheme.

(S_2c) $(G_4^\diamond, \mathcal{P}_2)$ is a $(K_2, m_2, \varepsilon_0, \varepsilon_1')$-exceptional scheme.

(Recall that $G_4^\diamond := G_1^\diamond - \bigcup \mathcal{J}^{\mathrm{rob}}$ was defined towards the end of Step 3. Since $G_4 = G_1 - G^{\mathrm{rob}}$, we have $(G_4)^\diamond = G_1^\diamond - (G^{\mathrm{rob}})^\diamond = G_1^\diamond - \bigcup \mathcal{J}^{\mathrm{rob}}$, so $(G_4)^\diamond$ is indeed the same as $G_4^\diamond$.) Moreover, by Lemma 2.4.2(iv) we have

(2.10.13)

$$d_{G_4}(v, A_i) = (d_{G_4}(v, A) \pm \varepsilon_0 n)/K_2 \qquad \text{and} \qquad d_{G_4}(v, B_i) = (d_{G_4}(v, B) \pm \varepsilon_0 n)/K_2$$

for all $v \in V(G)$ and $1 \leq i \leq K_2$. (Note that the previous partition of A and B plays no role in the subsequent argument, so denoting the clusters in $\mathcal{P}_2$ by A_i and B_i again will cause no notational conflicts.)

Since (G_4, A, A_0, B, B_0) is an (ε_0, K_2)-framework, (FR3) and (FR4) together imply that each $v \in A$ satisfies $d_{G_4}(v, A_0) \leq |V_0| \leq \varepsilon_0 n$ and $d_{G_4}(v, B') \leq \varepsilon_0 n$. So $d_{G_4}(v, A) = d_{G_4}(v) \pm 2\varepsilon_0 n$. Therefore, for all $v \in A$ and all $1 \leq i \leq K_2$ we have

(2.10.14)

$$d_{G_4}(v, A_i) \overset{(2.10.13)}{=} \frac{d_{G_4}(v, A) \pm \varepsilon_0 n}{K_2} = \frac{d_{G_4}(v) \pm 3\varepsilon_0 n}{K_2} = \frac{d_{G_4}(v) \pm 7\varepsilon_0 K_2 m_2}{K_2}.$$

The analogue holds for $d_{G_4}(v, B_i)$ (where $v \in B$ and $1 \leq i \leq K_2$).

Step 6: Exceptional systems for the approximate decomposition. In order to apply Lemma 2.5.4, we first need to construct suitable exceptional systems. We will show that $G_4^\diamond$ can be decomposed completely into $D_4/2$ exceptional systems with parameter ε_0. Moreover, these exceptional systems can be partitioned into sets $\mathcal{J}_0'$ and $\mathcal{J}_{i_1,i_2}'$ (one set for each pair $1 \leq i_1, i_2 \leq K_2$) such that the following conditions hold, where $\mathcal{J}''$ denotes the union of $\mathcal{J}_{i_1,i_2}'$ over all $1 \leq i_1, i_2 \leq K_2$:

(γ_1) Each $\mathcal{J}_{i_1,i_2}'$ consists of precisely $(D_4 - 2\lambda_{K_2} n)/2K_2^2$ (i_1, i_2)-ES with parameter ε_0 with respect to the partition $\mathcal{P}_2$.

(γ_2) $\mathcal{J}_0'$ contains precisely $\lambda_{K_2}n$ exceptional systems with parameter ε_0.

(γ_3) If $e_G(A',B') \geq D$, then all exceptional systems in $\mathcal{J}_0' \cup \mathcal{J}''$ are Hamilton exceptional systems.

(γ_4) If $e_G(A',B') < D$, then each exceptional system $J \in \mathcal{J}_0' \cup \mathcal{J}''$ is a Hamilton exceptional system with $e_J(A',B') = 2$ or a matching exceptional system. In particular, $\mathcal{J}_0'$ contains precisely $e_{\bigcup \mathcal{J}_0'}(A',B')/2$ Hamilton exceptional systems and $\mathcal{J}''$ contains precisely $e_{\bigcup \mathcal{J}''}(A',B')/2$ Hamilton exceptional systems.

As in Step 3, the construction of $\mathcal{J}_0'$ and the $\mathcal{J}_{i_1,i_2}'$ will depend on whether G is critical and whether $e_G(A',B') \geq D$. Recall that $G_4 = G_1 - G^{\rm rob}$ and note that

$$\frac{D - \phi_0 n - 2\phi_0^{\rm rob} n}{400K_2^2} = \frac{D_4}{400K_2^2} \in \mathbb{N} \tag{2.10.15}$$

by (2.10.6).

Case 1: $e_G(A',B') \geq D$ and G is not critical. Our aim is to apply Lemma 2.7.3 to G with $G - G_4$, m_2, K_2, $\mathcal{P}_2$, ε_1', $\phi_0 + 2\phi_0^{\rm rob}$, λ_{K_2} playing the roles of G_0, m, K, $\mathcal{P}$, ε, ϕ, λ. (So $G_4^\diamond$ will play the role of $G^\diamond$.) First we verify that the conditions in Lemma 2.7.3(i)–(iv) are satisfied. Clearly, Lemma 2.7.3(i) and (ii) hold. Note that $G - G_4 = \mathcal{H}_1' + G^{\rm rob}$, so ($\alpha_1$), ($\alpha_2$), ($\beta_1'$) and ($\beta_2'$) imply Lemma 2.7.3(iii). By (α_2) and (β_2'), $e_{G_4^\diamond}(A',B')$ is even. Together with the fact (S$_2$r) that $(G_4^\diamond, \mathcal{P}_2)$ is a $(K_2, m_2, \varepsilon_0, \varepsilon_1')$-exceptional scheme, this shows that Lemma 2.7.3(iv) holds. Together with (2.10.15) this ensures that we can indeed apply Lemma 2.7.3 to obtain a set of $(D - (\phi_0 + 2\phi_0^{\rm rob})n)/2 = D_4/2$ edge-disjoint Hamilton exceptional systems with parameter ε_0 in G_4. Moreover, these Hamilton exceptional systems can be partitioned into sets $\mathcal{J}_0'$ and $\mathcal{J}_{i_1,i_2}'$ (for all $1 \leq i_1, i_2 \leq K_2$) such that (γ_1)–(γ_3) hold.

Case 2: $e_G(A',B') \geq D$ and G is critical. Our aim is to apply Lemma 2.7.4 to G with $G - G_4$, m_2, K_2, $\mathcal{P}_2$, ε_1', $\phi_0 + 2\phi_0^{\rm rob}$, λ_{K_2} playing the roles of G_0, m, K, $\mathcal{P}$, ε, ϕ, λ. (So as before, $G_4^\diamond$ will play the role of $G^\diamond$.) Similar arguments as in Case 1 show that Lemma 2.7.4(i)–(iv) hold. (β_4) implies Lemma 2.7.4(v). Together with (2.10.15) this ensures that we can indeed apply Lemma 2.7.4 to obtain a set of $D_4/2$ edge-disjoint Hamilton exceptional systems with parameter ε_0 in G_4. Moreover, these Hamilton exceptional systems can be partitioned into sets $\mathcal{J}_0'$ and $\mathcal{J}_{i_1,i_2}'$ (for $1 \leq i_1, i_2 \leq K_2$) such that (γ_1)–(γ_3) hold.

Case 3: $e_G(A',B') < D$. Recall from Proposition 2.2.1(ii) that in this case we have $D = n/2 - 1$, $n = 0 \pmod 4$ and $|A'| = |B'| = n/2$. Our aim is to apply Lemma 2.7.5 to G with $G - G_4$, m_2, K_2, $\mathcal{P}_2$, ε_1', $\phi_0 + 2\phi^{\rm rob}$, λ_{K_2} playing the roles of G_0, m, K, $\mathcal{P}$, ε, ϕ, λ. (So as before, $G_4^\diamond$ will play the role of $G^\diamond$.) Similar arguments as in Case 1 show that Lemma 2.7.5(i)–(iv) hold. (β_5) implies Lemma 2.7.4(v). Together with (2.10.15) this ensures that we can indeed apply Lemma 2.7.5 to obtain a set of $D_4/2$ edge-disjoint exceptional systems in G_4. Moreover, these exceptional systems can be partitioned into sets $\mathcal{J}_0'$ and $\mathcal{J}_{i_1,i_2}'$ (for all $1 \leq i_1, i_2 \leq K_2$) such that (γ_1), (γ_2) and (γ_4) hold. (In particular, (γ_4) implies that each exceptional system in these sets has parameter ε_0.)

Therefore, in each of the three cases we have constructed sets $\mathcal{J}_0'$ and $\mathcal{J}_{i_1,i_2}'$ (for all $1 \leq i_1, i_2 \leq K_2$) satisfying (γ_1)–(γ_4).

We now find Hamilton cycles and perfect matchings covering the 'non-localized' exceptional systems (i.e. the ones in $\mathcal{J}_0'$). Let $G_4' = G_4 - G_4^\diamond$. So G_4' is obtained from G_4 by keeping all edges inside A as well as all edges inside B, and deleting all other edges. Note that (G_4', A, A_0, B, B_0) is an (ε_0, K_2)-framework since (G_4, A, A_0, B, B_0) is an (ε_0, K_2)-framework. Apply Lemma 2.6.2 to (G_4', A, A_0, B, B_0) with K_2, λ_{K_2}, $\mathcal{J}_0'$ playing the roles of K, λ, $\{J_1, \dots, J_{\lambda n}\}$. (Recall from (S$_2$b) that $(G_4[A] + G_4[B], \mathcal{P}_2)$ is a $(K_2, m_2, \varepsilon_0, \varepsilon_2')$-scheme, so $\delta(G_4'[A]) = \delta(G_4[A]) \geq 4|A|/5$ and $\delta(G_4'[B]) = \delta(G_4[B]) \geq 4|B|/5$ by (Sch3).) We obtain edge-disjoint subgraphs $H_1, \dots, H_{|\mathcal{J}_0'|}$ of $G_4' + \bigcup \mathcal{J}_0'$ such that, writing $\mathcal{H}_2 := \{H_1, \dots, H_{|\mathcal{J}_0'|}\}$, the following conditions hold:

(δ_1) For each $H_s \in \mathcal{H}_2$ there is some $J_s \in \mathcal{J}_0'$ such that $J_s \subseteq H_s$.

(δ_2) If J_s is a Hamilton exceptional system, then H_s is a Hamilton cycle on $V(G)$. If J_s is a matching exceptional system, then H_s is the edge-disjoint union of two perfect matchings on $V(G)$.

(δ_3) Let $\mathcal{H}_2' := H_1 + \dots + H_{|\mathcal{J}_0'|}$. If $e_G(A', B') < D$, then $\mathcal{H}_2$ contains precisely $e_{\mathcal{H}_2'}(A', B')/2$ Hamilton cycles on $V(G)$.

Indeed, (δ_1) follows from Lemma 2.6.2(i). (δ_2) follows from Lemma 2.6.2(ii),(iii). (For the second part, note that (γ_3) and (γ_4) imply that $\mathcal{J}_0'$ contains matching exceptional systems only in the case when $e_G(A', B') < D$. But in this case, Proposition 2.2.1(ii) implies that $n = 0 \pmod 4$ and $|A'| = |B'| = n/2$, i.e. $|A'|$ and $|B'|$ are even.) For (δ_3), note that G_4' has no $A'B'$-edges and so $e_{\bigcup \mathcal{J}_0'}(A', B') = e_{\mathcal{H}_2'}(A', B')$. Together with ($\delta_2$) and ($\gamma_4$), this now implies ($\delta_3$).

Recall that $\mathcal{J}''$ is the union of $\mathcal{J}_{i_1,i_2}'$ over all $1 \leq i_1, i_2 \leq K_2$. Let $G_5 := G_4 - \mathcal{H}_2'$ and $D_5 := D_4 - 2|\mathcal{H}_2| = D_4 - 2\lambda_{K_2} n$. So (2.10.12) implies that

(2.10.16)
$$d_{G_5}(v) = D_5 + 2r \text{ for all } v \in A \cup B \qquad \text{and} \qquad d_{G_5}(v) = D_5 \text{ for all } v \in V_0.$$

Note that

$$G_5^\diamond := G_5 - G_5[A] - G_5[B] = G_4^\diamond - \mathcal{H}_2' = G_4^\diamond - \bigcup \mathcal{J}_0' = \bigcup \mathcal{J}''. \tag{2.10.17}$$

Since $d_J(v) = 2$ for all $v \in V_0$ and all $J \in \mathcal{J}''$, it follows that

$$D_5 = 2|\mathcal{J}''|. \tag{2.10.18}$$

Moreover, since $(G_4[A] + G_4[B], \mathcal{P}_2)$ is a $(K_2, m_2, \varepsilon_0, \varepsilon_2')$-scheme and $\varepsilon_2' + 2\lambda_{K_2} \leq \varepsilon$, Proposition 2.4.1 implies that $(G_5[A] + G_5[B], \mathcal{P}_2)$ is a $(K_2, m_2, \varepsilon_0, \varepsilon)$-scheme.

Step 7: Approximate Hamilton cycle decomposition. Our next aim is to apply Lemma 2.5.4 to obtain an approximate decomposition of G_5. Let

$$\mu := (r_0^{\mathrm{rob}} - 2r)/(4K_2 m_2) \qquad \text{and} \qquad \rho := \gamma/(4K_1).$$

We will apply the lemma with G_5, $\mathcal{P}_2$, K_2, m_2, $\mathcal{J}''$, ε playing the roles of G, $\mathcal{P}$, K, m, $\mathcal{J}$, ε. Clearly, conditions (c) and (d) of Lemma 2.5.4 hold.

In order to see that condition (a) is satisfied, recall that $m_1 K_1 = |A| = m_2 K_2$. So

$$0 \leq \frac{7r_1 - 2r}{4K_2 m_2} \overset{(2.10.10)}{\leq} \mu \overset{(2.10.10)}{\leq} \frac{30 r_1}{4K_2 m_2} = \frac{30\gamma_1}{4K_1} \ll 1.$$

Therefore, every vertex $v \in A \cup B$ satisfies

$$\begin{aligned} d_{G_4}(v) &\overset{(2.10.12)}{=} D_4 + 2r \overset{(2.10.11)}{=} D_1 - r_0^{\mathrm{rob}} + 2r \overset{(2.10.3)}{=} D - \phi_0 n - 4K_2 m_2 \mu \\ &\overset{(2.10.1)}{=} (1/2 \pm 4\varepsilon_{\mathrm{ex}})n - \phi_0 n - 4K_2 m_2 \mu \\ &= (1 - 4\mu \pm 3\phi_0)\, K_2 m_2, \end{aligned} \tag{2.10.19}$$

where in the last equality we recall that $(1-\varepsilon_0)n/2 \le |A| = K_2m_2 \le n/2$ and $\varepsilon_0, \varepsilon_{\rm ex} \ll \phi_0$. Recall that $G_5 = G_4 - \mathcal{H}_2'$ and note that

$$\Delta(\mathcal{H}_2') = 2|\mathcal{H}_2| = 2\lambda_{K_2}n \le 5\lambda_{K_2}K_2m_2.$$

Altogether this implies that for each $v \in A$ and for all $1 \le i \le K_2$ we have

$$\begin{aligned} d_{G_5}(v, A_i) &= d_{G_4}(v, A_i) - d_{\mathcal{H}_2'}(v, A_i) = d_{G_4}(v, A_i) \pm 5\lambda_{K_2}K_2m_2 \\ &\overset{(2.10.14)}{=} (d_{G_4}(v) \pm 7\varepsilon_0K_2m_2)/K_2 \pm 5\lambda_{K_2}K_2m_2 \\ &\overset{(2.10.19)}{=} (1 - 4\mu \pm (3\phi_0 + 7\varepsilon_0 + 5\lambda_{K_2}K_2))\, m_2. \end{aligned}$$

Since $\phi_0, \varepsilon_0, \lambda_{K_2} \ll 1/K_2$, it follows that $d_{G_5}(v, A_i) = (1-4\mu \pm 4/K_2)m_2$. Similarly one can show that $d_{G_5}(w, B_j) = (1-4\mu\pm 4/K_2)m_2$ for all $w \in B$. So Lemma 2.5.4(a) holds.

To check condition (b), note that $r = \gamma|A|/K_1 \ge \gamma n/3K_1$. So

$$\begin{aligned} |\mathcal{J}''| &\overset{(2.10.18)}{=} \frac{D_5}{2} \le \frac{D_4}{2} \overset{(2.10.11)}{=} \frac{D - r_0^{\rm rob}}{2} \overset{(2.10.1)}{\le} \frac{n}{4} + 2\varepsilon_{\rm ex}n - \frac{r_0^{\rm rob}}{2} \\ &= \frac{n}{4} + 2\varepsilon_{\rm ex}n - 2K_2m_2\mu - r \le \left(\frac{1}{4} + 2\varepsilon_{\rm ex} - (1-\varepsilon_0)\mu - \frac{\gamma}{3K_1}\right) n \\ &\le \left(\frac{1}{4} - \mu - \frac{\gamma}{4K_1}\right) n = \left(\frac{1}{4} - \mu - \rho\right) n. \end{aligned}$$

Thus Lemma 2.5.4(b) holds.

So we can indeed apply Lemma 2.5.4 to obtain a collection $\mathcal{H}_3$ of $|\mathcal{J}''|$ edge-disjoint spanning subgraphs $H_1', \dots, H_{|\mathcal{J}''|}'$ of G_5 which satisfy the following properties:

(ε_1) For each $H_s' \in \mathcal{H}_3$ there is some $J_s' \in \mathcal{J}''$ such that $J_s' \subseteq H_s'$.

(ε_2) If J_s' is a Hamilton exceptional system then H_s' is a Hamilton cycle on $V(G)$. If J_s' is a matching exceptional system then H_s' is the edge-disjoint union of two perfect matchings on $V(G)$.

(ε_3) Let $\mathcal{H}_3' := H_1' + \dots + H_{|\mathcal{J}''|}'$. If $e_G(A', B') < D$, then $\mathcal{H}_3$ contains precisely $e_{\mathcal{H}_3'}(A', B')/2$ Hamilton cycles on $V(G)$.

For (ε_3), note that (2.10.17) implies $G_5^\diamond = \bigcup \mathcal{J}''$ and thus we have $e_{\bigcup \mathcal{J}''}(A', B') = e_{\mathcal{H}_3'}(A', B')$. Together with ($\varepsilon_2$) and ($\gamma_4$), this now implies ($\varepsilon_3$).

Step 8: Decomposing the leftover and the robustly decomposable graph. Finally, we can apply the 'robust decomposition property' of $G^{\rm rob}$ guaranteed by Corollary 2.9.5 to obtain a decomposition of the leftover from the previous step together with $G^{\rm rob}$ into Hamilton cycles (and perfect matchings if applicable).

To achieve this, let $H' := G_5 - \mathcal{H}_3'$. Thus (2.10.16) and (2.10.18) imply that every vertex in V_0 is isolated in H' while every vertex $v \in A \cup B$ has degree $d_{G_5}(v) - 2|\mathcal{J}''| = D_5 + 2r - 2|\mathcal{J}''| = 2r$ in H' (the last equality follows from (2.10.18)). Moreover, $(H')^\diamond$ contains no edges. (This holds since $\bigcup \mathcal{J}'' \subseteq \mathcal{H}_3'$ and so $H' \subseteq G_5 - \bigcup \mathcal{J}'' = G_5 - G_5^\diamond$ by (2.10.17).) Now let $H_A := H'[A]$, $H_B := H'[B]$, $H := H_A + H_B$. Note that H is the $2r$-regular subgraph of H' obtained by removing all the vertices in V_0. Let

$$s' := rfK_1 + 7r^\diamond \overset{(2.10.4)}{=} Lfr_3 + 7r^\diamond \overset{(2.10.8)}{=} \phi_0^{\rm rob} n.$$

Recall from (β_3') that each of the s' exceptional path systems in $\mathcal{EF}_{CA} + \mathcal{EF}_{PCA}$ contains a unique exceptional system and $\mathcal{J}^{\mathrm{rob}}$ is the set of all these s' exceptional systems. Thus Corollary 2.9.5(ii)(b) implies that $H + G^{\mathrm{rob}}$ has a decomposition into edge-disjoint spanning subgraphs $H_1'', \dots, H_{s'}''$ such that, writing $\mathcal{H}_4 := \{H_1'', \dots, H_{s'}''\}$, we have:

(ζ_1) For each $H_s'' \in \mathcal{H}_4$ there is some exceptional system $J_s'' \in \mathcal{J}^{\mathrm{rob}}$ such that $J_s'' \subseteq H_s''$.

(ζ_2) If J_s'' is a Hamilton exceptional system then H_s'' is a Hamilton cycle on $V(G)$. If J_s'' is a matching exceptional system then H_s'' is the edge-disjoint union of two perfect matchings on $V(G)$.

(ζ_3) Let $\mathcal{H}_4' := H_1'' + \cdots + H_{s'}''$. Then $\mathcal{H}_4$ contains precisely $e_{\mathcal{H}_4'}(A', B')/2$ Hamilton cycles on $V(G)$.

Indeed, (ζ_1) and (ζ_2) follow from Corollary 2.9.5(ii)(b) (recall that if $\mathcal{J}^{\mathrm{rob}}$ contains a matching exceptional system, then $|A'| = |B'| = n/2$ is even by (β_3')). For (ζ_3), note that $e_{\mathcal{H}_4'}(A', B') = e_{G^{\mathrm{rob}}}(A', B') = e_{\bigcup \mathcal{J}^{\mathrm{rob}}}(A', B')$ by (β_2'). Now (ζ_3) follows from (β_3') and (ζ_2).

Note that $\mathcal{H}_1 \cup \mathcal{H}_2 \cup \mathcal{H}_3 \cup \mathcal{H}_4$ corresponds to a decomposition of G into Hamilton cycles and perfect matchings. It remains to show that the proportion of Hamilton cycles in this decomposition is as desired.

First suppose that $e_G(A', B') \geq D$. By (α_3), $\mathcal{H}_1$ consists of Hamilton cycles and one perfect matching if D is odd. By (γ_3), (δ_2) and (ε_2), both $\mathcal{H}_2$ and $\mathcal{H}_3$ consist of Hamilton cycles. By (β_3') and (ζ_2) this also holds for $\mathcal{H}_4$. So $\mathcal{H}_1 \cup \mathcal{H}_2 \cup \mathcal{H}_3 \cup \mathcal{H}_4$ consists of Hamilton cycles and one perfect matching if D is odd.

Next suppose that $e_G(A', B') < D$. Then by (α_3), (δ_3), (ε_3) and (ζ_3) the numbers of Hamilton cycles in $\mathcal{H}_1$, $\mathcal{H}_2$, $\mathcal{H}_3$ and $\mathcal{H}_4$ are precisely $\lfloor e_{\mathcal{H}_1'}(A', B')/2 \rfloor$, $e_{\mathcal{H}_2'}(A', B')/2$, $e_{\mathcal{H}_3'}(A', B')/2$ and $e_{\mathcal{H}_4'}(A', B')/2$. Hence, $\mathcal{H}_1 \cup \mathcal{H}_2 \cup \mathcal{H}_3 \cup \mathcal{H}_4$ contains precisely

$$\left\lfloor \frac{e_{\mathcal{H}_1' \cup \mathcal{H}_2' \cup \mathcal{H}_3' \cup \mathcal{H}_4'}(A', B')}{2} \right\rfloor = \left\lfloor \frac{e_G(A', B')}{2} \right\rfloor \geq \left\lfloor \frac{F}{2} \right\rfloor$$

edge-disjoint Hamilton cycles, where F is the size of the minimum cut in G. Since clearly G cannot have more than $\lfloor F/2 \rfloor$ edge-disjoint Hamilton cycles, it follows that we have equality in the final step, as required. □

CHAPTER 3

Exceptional systems for the two cliques case

In this chapter we prove all the results that were stated in Section 2.7. Recall that the exceptional edges are all those edges incident to A_0 and B_0 as well as all those edges joining A' to B'. The results stated in Section 2.7 generated a decomposition of these exceptional edges into exceptional systems: Each such exceptional system was then extended into a Hamilton cycle. (Recall that actually, the exceptional systems may contain some non-exceptional edges as well.) This is the most difficult part of the construction of the Hamilton cycle decomposition and so forms the heart of the argument for the two clique case.

Let G be a D-regular graph and let A', B' be a partition of $V(G)$. Recall that we say that G is *critical* (with respect to A', B' and D) if both of the following hold:

- $\Delta(G[A', B']) \geq 11D/40$;
- $e(H) \leq 41D/40$ for all subgraphs H of $G[A', B']$ with $\Delta(H) \leq 11D/40$.

Recall that Lemmas 2.7.3–2.7.5 guarantee our desired decomposition of the exceptional edges into exceptional systems. Lemma 2.7.3 covers the non-critical case when $G[A', B']$ contains many edges, Lemma 2.7.4 covers the critical case when $G[A', B']$ contains many edges and Lemma 2.7.5 tackles the case when $G[A', B']$ contains only a few edges.

3.1. Proof of Lemma 2.7.1

The following lemma (which collects some basic properties of critical graphs) immediately implies Lemma 2.7.1.

LEMMA 3.1.1. *Suppose that $0 < 1/n \ll 1$ and that $D, n \in \mathbb{N}$ are such that*

$$D \geq n - 2\lfloor n/4 \rfloor - 1 = \begin{cases} n/2 - 1 & \text{if } n = 0 \pmod 4, \\ (n-1)/2 & \text{if } n = 1 \pmod 4, \\ n/2 & \text{if } n = 2 \pmod 4, \\ (n+1)/2 & \text{if } n = 3 \pmod 4. \end{cases} \tag{3.1.1}$$

Let G be a D-regular graph on n vertices and let A', B' be a partition of $V(G)$ with $|A'|, |B'| \geq D/2$ and $\Delta(G[A', B']) \leq D/2$. Suppose that G is critical. Let W be the set of vertices $w \in V(G)$ such that $d_{G[A',B']}(w) \geq 11D/40$. Then the following properties are satisfied:

(i) $1 \leq |W| \leq 3$.
(ii) *Either $D = (n-1)/2$ and $n = 1 \pmod 4$, or $D = n/2 - 1$ and $n = 0 \pmod 4$. Furthermore, if $n = 1 \pmod 4$, then $|W| = 1$.*
(iii) $e_G(A', B') \leq 17D/10 + 5 < n$.

(iv)

$$e_{G-W}(A',B') \leq \begin{cases} 3D/4+5 & \text{if } |W|=1, \\ 19D/40+5 & \text{if } |W|=2, \\ D/5+5 & \text{if } |W|=3. \end{cases}$$

(v) *There exists a set W' of vertices such that $W \subseteq W'$, $|W'| \leq 3$ and for all $w' \in W'$ and $v \in V(G) \setminus W'$ we have*

$$d_{G[A',B']}(w') \geq \frac{21D}{80},\ d_{G[A',B']}(v) \leq \frac{11D}{40} \text{ and } d_{G[A',B']}(w') - d_{G[A',B']}(v) \geq \frac{D}{240}.$$

Proof. Let $w_1, \dots, w_4$ be vertices of G such that

$$d_{G[A',B']}(w_1) \geq \cdots \geq d_{G[A',B']}(w_4) \geq d_{G[A',B']}(v)$$

for all $v \in V(G)\setminus\{w_1,\dots,w_4\}$. Let $W_4 := \{w_1,\dots,w_4\}$. Suppose that $d_{G[A',B']}(w_4) \geq 21D/80$. Let H be a spanning subgraph of $G[A',B']$ such that $d_H(w_i) = \lceil 21D/80 \rceil$ for all $i \leq 4$ and such that every vertex $v \in V(G) \setminus W_4$ satisfies $N_H(v) \subseteq W_4$. Thus $\Delta(H) = \lceil 21D/80 \rceil$ and so $e(H) \leq 41D/40$ since G is critical. On the other hand, $e(H) \geq 4 \cdot \lceil 21D/80 \rceil - 4$, a contradiction. (Here we subtract four to account for the edges of H' between vertices in W.) Hence, $d_{G[A',B']}(w_4) < 21D/80$ and so $|W| \leq 3$. But $|W| \geq 1$ since G is critical. So (i) holds.

Let j be minimal such that $d_{G[A',B']}(w_j) \leq 21D/80$. So $1 < j \leq 4$. Choose an index i with $1 \leq i < j$ such that $W \subseteq \{w_1,\dots,w_i\}$ and $d_{G[A',B']}(w_i) - d_{G[A',B']}(w_{i+1}) \geq D/240$. Then the set $W' := \{w_1,\dots,w_i\}$ satisfies (v).

Let H' be a spanning subgraph of $G[A',B']$ such that $G[A' \setminus W, B' \setminus W] \subseteq H'$ and $d_{H'}(w) = \lfloor 11D/40 \rfloor$ for all $w \in W$. Similarly as before, $e(H') \leq 41D/40$ since G is critical. Thus

$$\begin{aligned} 41D/40 \geq e(H') &\geq e(H'-W) + \lfloor 11D/40 \rfloor |W| - 2 \\ &= e_{G-W}(A',B') + \lfloor 11D/40 \rfloor |W| - 2. \end{aligned}$$

This in turn implies that

$$e_{G-W}(A',B') \leq (41 - 11|W|)D/40 + 5. \tag{3.1.2}$$

Together with (i) this implies (iv). If $D \geq n/2$, then by Proposition 2.2.3 we have $e_{G-W}(A',B') \geq D - 28$. This contradicts (iv). Thus (3.1.1) implies that $D = (n-1)/2$ and $n = 1 \pmod 4$, or $D = n/2 - 1$ and $n = 0 \pmod 4$. If $n = 1 \pmod 4$ and $D = (n-1)/2$, then Proposition 2.2.3 implies that $e_{G-W}(A',B') \geq D/2 - 28$. Hence, by (iv) we deduce that $|W| = 1$ and so (ii) holds. Since $|W| \leq 3$ and $\Delta(G[A',B']) \leq D/2$, we have

$$e_G(A',B') \leq e_{G-W}(A',B') + \frac{|W|D}{2} \overset{(3.1.2)}{\leq} \frac{(41+9|W|)D}{40} + 5 \leq \frac{17D}{10} + 5 < n.$$

(The last inequality follows from (ii).) This implies (iii). □

3.2. Non-critical Case with $e(A',B') \geq D$

In this section we prove Lemma 2.7.3. Recall that Lemma 2.7.3 gives a decomposition of the exceptional edges into exceptional systems in the non-critical case when $e(A',B') \geq D$. The proof splits into the following four steps:

Step 1 We first decompose $G^\diamond$ into edge-disjoint 'localized' subgraphs $H(i, i')$ and $H'(i, i')$ (where $1 \leq i, i' \leq K$). More precisely, each $H(i, i')$ only contains A_0A_i-edges and $B_0B_{i'}$-edges of $G^\diamond$ while all edges of $H'(i, i')$ lie in $G^\diamond[A_0 \cup A_i, B_0 \cup B_{i'}]$, and all the edges of $G^\diamond$ are distributed evenly amongst the $H(i, i')$ and $H'(i, i')$ (see Lemma 2.5.2). We will then move a small number of $A'B'$-edges between the $H'(i, i')$ in order to obtain graphs $H''(i, i')$ such that $e(H''(i, i'))$ is even (see Lemma 3.2.1).

Step 2 We decompose each $H''(i, i')$ into $(D - \phi n)/(2K^2)$ Hamilton exceptional system candidates (see Lemma 3.2.3).

Step 3 Most of the Hamilton exceptional system candidates constructed in Step 2 will be extended into an (i, i')-HES (see Lemma 3.2.4).

Step 4 The remaining Hamilton exceptional system candidates will be extended into Hamilton exceptional systems, which need not be localized (see Lemma 3.2.5). (Altogether, these will be the λn Hamilton exceptional systems in $\mathcal{J}$ which are not mentioned in Lemma 2.7.3(b).)

3.2.1. Step 1: Constructing the Graphs $H''(i, i')$. Let $H(i, i')$ and $H'(i, i')$ be the graphs obtained by applying Lemma 2.5.2 to $G^\diamond$. We would like to decompose each $H'(i, i')$ into Hamilton exceptional system candidates. In order to do this, $e(H'(i, i'))$ must be even. The next lemma shows that we can ensure this property without destroying the other properties of the $H'(i, i')$ too much by moving a small number of edges between the $H'(i, i')$.

LEMMA 3.2.1. *Suppose that $0 < 1/n \ll \varepsilon_0 \ll \varepsilon \ll \varepsilon' \ll \lambda, 1/K \ll 1$, that $D \geq n/3$, that $0 \leq \phi \ll 1$ and that $D, n, K, m, (D - \phi n)/(2K^2) \in \mathbb{N}$. Define α by*

$$(3.2.1) \qquad 2\alpha n := \frac{D - \phi n}{K^2} \qquad \text{and let} \qquad \gamma := \alpha - \frac{2\lambda}{K^2}.$$

Suppose that the following conditions hold:

(i) *G is a D-regular graph on n vertices.*
(ii) *$\mathcal{P}$ is a (K, m, ε_0)-partition of $V(G)$ such that $D \leq e_G(A', B') \leq \varepsilon_0 n^2$ and $\Delta(G[A', B']) \leq D/2$. Furthermore, G is not critical.*
(iii) *G_0 is a subgraph of G such that $G[A_0] + G[B_0] \subseteq G_0$, $e_{G_0}(A', B') \leq \phi n$ and $d_{G_0}(v) = \phi n$ for all $v \in V_0$.*
(iv) *Let $G^\diamond := G - G[A] - G[B] - G_0$. $e_{G^\diamond}(A', B')$ is even and $(G^\diamond, \mathcal{P})$ is a $(K, m, \varepsilon_0, \varepsilon)$-exceptional scheme.*

Then $G^\diamond$ can be decomposed into edge-disjoint spanning subgraphs $H(i, i')$ and $H''(i, i')$ of $G^\diamond$ (for all $1 \leq i, i' \leq K$) such that the following properties hold, where $G'(i, i') := H(i, i') + H''(i, i')$:

(b_1) *Each $H(i, i')$ contains only A_0A_i-edges and $B_0B_{i'}$-edges.*
(b_2) *$H''(i, i') \subseteq G^\diamond[A', B']$. Moreover, all but at most $\varepsilon' n$ edges of $H''(i, i')$ lie in $G^\diamond[A_0 \cup A_i, B_0 \cup B_{i'}]$.*
(b_3) *$e(H''(i, i'))$ is even and $2\alpha n \leq e(H''(i, i')) \leq 11\varepsilon_0 n^2/(10K^2)$.*
(b_4) *$\Delta(H''(i, i')) \leq 31\alpha n/30$.*
(b_5) *$d_{G'(i,i')}(v) = (2\alpha \pm \varepsilon')n$ for all $v \in V_0$.*
(b_6) *Let $\widetilde{H}$ be any spanning subgraph of $H''(i, i')$ which maximises $e(\widetilde{H})$ under the constraints that $\Delta(\widetilde{H}) \leq 3\gamma n/5$, $H''(i, i')[A_0, B_0] \subseteq \widetilde{H}$ and $e(\widetilde{H})$ is even. Then $e(\widetilde{H}) \geq 2\alpha n$.*

Proof. Since $\phi \ll 1/3 \leq D/n$, we deduce that

(3.2.2)
$$\alpha \geq 1/(7K^2), \qquad (1-14\lambda)\alpha \leq \gamma < \alpha \quad \text{and} \quad \varepsilon \ll \varepsilon' \ll \lambda, 1/K, \alpha, \gamma \ll 1.$$

Note that (ii) and (iii) together imply that

$$e_{G^\diamond}(A', B') \geq D - \phi n \overset{(3.2.1)}{=} 2K^2 \alpha n \overset{(3.2.2)}{\geq} n/4. \tag{3.2.3}$$

By (i) and (iii), each $v \in V_0$ satisfies

$$d_{G^\diamond}(v) = D - \phi n \overset{(3.2.1)}{=} 2K^2 \alpha n. \tag{3.2.4}$$

Apply Lemma 2.5.2 to decompose $G^\diamond$ into subgraphs $H(i,i')$, $H'(i,i')$ (for all $1 \leq i, i' \leq K$) satisfying the following properties, where $G(i,i') := H(i,i') + H'(i,i')$:

(a_1') Each $H(i,i')$ contains only A_0A_i-edges and $B_0B_{i'}$-edges.
(a_2') All edges of $H'(i,i')$ lie in $G^\diamond[A_0 \cup A_i, B_0 \cup B_{i'}]$.
(a_3') $e(H'(i,i')) = (1 \pm 16\varepsilon) e_{G^\diamond}(A', B')/K^2$. In particular,
$$2(1-16\varepsilon)\alpha n \leq e(H'(i,i')) \leq (1+16\varepsilon)\varepsilon_0 n^2/K^2.$$
(a_4') $d_{H'(i,i')}(v) = (d_{G^\diamond[A',B']}(v) \pm 2\varepsilon n)/K^2$ for all $v \in V_0$.
(a_5') $d_{G(i,i')}(v) = (2\alpha \pm 4\varepsilon/K^2)n$ for all $v \in V_0$.

Indeed, (a_3') follows from (3.2.3), Lemma 2.5.2(a_3) and (ii), while (a_5') follows from (3.2.4) and Lemma 2.5.2(a_5). We now move some $A'B'$-edges of $G^\diamond$ between the $H'(i,i')$ such that the graphs $H''(i,i')$ obtained in this way satisfy the following conditions:

- Each $H''(i,i')$ is obtained from $H'(i,i')$ by adding or removing at most $32K^2\varepsilon\alpha n \leq \sqrt{\varepsilon} n$ edges.
- $e(H''(i,i')) \geq 2\alpha n$ and $e(H''(i,i'))$ is even.

Note that this is possible by (a_3') and since $\alpha n \in \mathbb{N}$ and $e_{G^\diamond}(A', B') \geq 2K^2\alpha n$ is even by (iv).

We will show that the graphs $H(i,i')$ and $H''(i,i')$ satisfy conditions (b_1)–(b_6). Clearly both (b_1) and (b_2) hold. (a_3') implies that

(3.2.5)
$$e(H''(i,i')) = (1 \pm 16\varepsilon) e_{G^\diamond}(A', B')/K^2 \pm \sqrt{\varepsilon} n \overset{(3.2.2),(3.2.3)}{=} (1 \pm \varepsilon') e_{G^\diamond}(A', B')/K^2.$$

Together with (ii) and our choice of the $H''(i,i')$ this implies (b_3). (b_5) follows from (a_5') and the fact that $d_{G'(i,i')}(v) = d_{G(i,i')}(v) \pm \sqrt{\varepsilon} n$. Similarly, ($a_4'$) implies that for all $v \in V_0$ we have

$$d_{H''(i,i')}(v) = (d_{G^\diamond[A',B']}(v) \pm \varepsilon' n)/K^2. \tag{3.2.6}$$

Recall that $\Delta(G[A', B']) \leq D/2$ by (ii). Thus

$$\Delta(H''(i,i')) \overset{(3.2.6)}{\leq} \frac{D/2 + \varepsilon' n}{K^2} \overset{(3.2.1)}{=} \left(\alpha + \frac{\phi + 2\varepsilon'}{2K^2}\right) n \overset{(3.2.2)}{\leq} \frac{31\alpha n}{30},$$

so (b_4) holds.

So it remains to verify (b_6). To do this, fix $1 \leq i, i' \leq K$ and set $H'' := H''(i,i')$. Let $\widetilde{H}$ be a subgraph of H'' as defined in (b_6). We need to show that $e(\widetilde{H}) \geq 2\alpha n$. Suppose the contrary that $e(\widetilde{H}) < 2\alpha n$. We will show that this contradicts the assumption that G is not critical. Roughly speaking, the argument will be that if $\widetilde{H}$ is sparse, then so is H''. This in turn implies that $G^\diamond$ is also sparse, and thus any subgraph of $G[A', B']$ of comparatively small maximum degree is also sparse, which leads to a contradiction.

Let X be the set of all those vertices x for which $d_{\widetilde{H}}(x) \geq 3\gamma n/5-2$. So $X \subseteq V_0$ by (iv) and (ESch3). Note that if $X = \emptyset$, then $\widetilde{H} = H''$ and so $e(\widetilde{H}) \geq 2\alpha n$ by (b$_3$). If $|X| \geq 4$, then $e(\widetilde{H}) \geq 4(3\gamma n/5 - 2) - 4 \geq 2\alpha n$ by (3.2.2). Hence $1 \leq |X| \leq 3$. Note that $\widetilde{H} - X$ contains all but at most one edge from $H'' - X$. Together with the fact that $\widetilde{H}[X]$ contains at most two edges (since $|X| \leq 3$ and $\widetilde{H}$ is bipartite) this implies that

$$\begin{aligned} 2\alpha n > e(\widetilde{H}) &\geq e(\widetilde{H} - X) + \left(\sum_{x \in X} d_{\widetilde{H}}(x)\right) - 2 \\ &\geq e(H'' - X) - 1 + |X|(3\gamma n/5 - 2) - 2 \\ &\geq e(H'') - \sum_{x \in X} d_{H''}(x) + |X|(3\gamma n/5 - 2) - 3 \\ &= e(H'') - \sum_{x \in X} (d_{H''}(x) - 3\gamma n/5 + 2) - 3 \end{aligned} \tag{3.2.7}$$

and so

$$e(H'') \overset{(3.2.6)}{<} 2\alpha n + \sum_{x \in X} \left(\frac{d_{G^\diamond[A',B']}(x) + \varepsilon' n}{K^2} - 3\gamma n/5 + 2\right) + 3. \tag{3.2.8}$$

Note that (b$_4$) and (3.2.7) together imply that if $e(H'') \geq 4\alpha n$ then $e(\widetilde{H}) \geq e(H'') - |X|(31\alpha n/30 - 3\gamma n/5 + 2) - 3 \geq 2\alpha n$. Thus $e(H'') < 4\alpha n$ and by (3.2.5) we have $e_{G^\diamond}(A', B') \leq 4K^2\alpha n/(1-\varepsilon') \leq 5K^2\alpha n \leq 3n$. Hence

$$\begin{aligned} e_{G^\diamond}(A', B') &\overset{(3.2.5)}{\leq} K^2 e(H'') + \varepsilon' e_{G^\diamond}(A', B') \leq K^2 e(H'') + 3\varepsilon' n \\ &\overset{(3.2.8)}{\leq} D - \phi n + 7\varepsilon' n + \sum_{x \in X} \left(d_{G^\diamond[A',B']}(x) - K^2(3\gamma n/5)\right). \end{aligned} \tag{3.2.9}$$

Let G' be any subgraph of $G^\diamond[A', B']$ which maximises $e(G')$ under the constraint that $\Delta(G') \leq K^2(3\gamma/5 + 2\varepsilon')n$. Note that if $d_{G^\diamond[A',B']}(v) \geq K^2(3\gamma/5 + 2\varepsilon')n$, then $v \in V_0$ (by (iv) and (ESch3)) and so $d_{H''}(v) > 3\gamma n/5$ by (3.2.6). This in turn implies that $v \in X$. Hence

$$\begin{aligned} e(G') &\leq e_{G^\diamond}(A', B') - \sum_{x \in X} \left(d_{G^\diamond[A',B']}(x) - K^2(3\gamma/5 + 2\varepsilon')n\right) + 2 \\ &\overset{(3.2.9)}{\leq} D - \phi n + 7K^2\varepsilon' n. \end{aligned} \tag{3.2.10}$$

Note that (3.2.6) together with the fact that $X \neq \emptyset$ implies that

$$\Delta(G[A', B']) \geq \Delta(G^\diamond[A', B']) \geq K^2(3\gamma n/5 - 2) - \varepsilon' n \overset{(3.2.1),(3.2.2)}{\geq} 11D/40.$$

Since G is not critical this means that there exists a subgraph G'' of $G[A', B']$ such that $\Delta(G'') \leq 11D/40 \leq K^2(3\gamma/5 + 2\varepsilon')n$ and $e(G'') \geq 41D/40$. Thus

$$D - \phi n + 7K^2\varepsilon' n \overset{(3.2.10)}{\geq} e(G') \geq e(G'') - e_{G_0}(A', B') \geq 41D/40 - \phi n,$$

which is a contradiction. Therefore, we must have $e(\widetilde{H}) \geq 2\alpha n$. Hence (b$_6$) is satisfied. □

3.2.2. Step 2: Decomposing $H''(i,i')$ into Hamilton Exceptional System Candidates. Our next aim is to decompose each $H''(i,i')$ into αn Hamilton exceptional system candidates (this will follow from Lemma 3.2.3). Before we can do this, we need the following result on decompositions of bipartite graphs into 'even matchings'. We say that a matching is *even* if it contains an even number of edges, otherwise it is *odd*.

PROPOSITION 3.2.2. *Suppose that $0 < 1/n \ll \gamma \leq 1$ and that $n, \gamma n \in \mathbb{N}$. Let H be a bipartite graph on n vertices with $\Delta(H) \leq 2\gamma n/3$ and where $e(H) \geq 2\gamma n$ is even. Then H can be decomposed into γn edge-disjoint non-empty even matchings, each of size at most $3e(H)/(\gamma n)$.*

Proof. First note that since $e(H) \geq 2\gamma n$, it suffices to show that H can be decomposed into at most γn edge-disjoint non-empty even matchings, each of size at most $3e(H)/(\gamma n)$. Indeed, by splitting these matchings further if necessary, one can obtain precisely γn non-empty even matchings.

Set $n' := \lfloor 2\gamma n/3 \rfloor$. König's theorem implies that $\chi'(H) \leq n'$. So Proposition 1.4.5 implies that there is a decomposition of H into n' edge-disjoint matchings $M_1, \dots, M_{n'}$ such that $|e(M_s) - e(M_{s'})| \leq 1$ for all $s, s' \leq n'$. Hence we have

$$2 \leq \frac{e(H)}{n'} - 1 \leq e(M_s) \leq \frac{e(H)}{n'} + 1 \leq \frac{3e(H)}{\gamma n}$$

for all $s \leq n'$. Since $e(H)$ is even, there are an even number of odd matchings. Let M_s and $M_{s'}$ be two odd matchings. So $e(M_s), e(M_{s'}) \geq 3$ and thus there exist two disjoint edges $e \in M_s$ and $e' \in M_{s'}$. Hence, $M_s - e$, $M_{s'} - e'$ and $\{e, e'\}$ are three even matchings. Thus, by pairing off the odd matchings and repeating this process, the proposition follows. □

LEMMA 3.2.3. *Suppose that $0 < 1/n \ll \varepsilon_0 \ll \gamma < 1$, that $\gamma + \gamma' < 1$ and that $n, \gamma n, \gamma' n \in \mathbb{N}$. Let H be a bipartite graph on n vertices with vertex classes $A \dot\cup A_0$ and $B \dot\cup B_0$, where $|A_0| + |B_0| \leq \varepsilon_0 n$. Suppose that*

(i) *$e(H)$ is even, $\Delta(H) \leq 16\gamma n/15$ and $\Delta(H[A,B]) < (3\gamma/5 - \varepsilon_0)n$.*

Let H' be a spanning subgraph of H which maximises $e(H')$ under the constraints that $\Delta(H') \leq 3\gamma n/5$, $H[A_0, B_0] \subseteq H'$ and $e(H')$ is even. Suppose that

(ii) *$2(\gamma + \gamma')n \leq e(H') \leq 10\varepsilon_0 \gamma n^2$.*

Then there exists a decomposition of H into edge-disjoint Hamilton exceptional system candidates $F_1, \dots, F_{\gamma n}, F'_1, \dots, F'_{\gamma' n}$ with parameter ε_0 such that $e(F'_s) = 2$ for all $s \leq \gamma' n$.

Since we are in the non-critical case with many edges between A' and B', we will be able to assume that the subgraph H' satisfies (ii).

Roughly speaking, the idea of the proof of Lemma 3.2.3 is to apply the previous proposition to decompose H' into a suitable number of even matchings M_i (using the fact that it has small maximum degree). We then extend these matchings into Hamilton exceptional system candidates to cover all edges of H. The additional edges added to each M_i will be vertex-disjoint from M_i and form vertex-disjoint 2-paths uvw with $v \in V_0$. So the number of connections from A' to B' remains the same (as H is bipartite). Each matching M_i will already be a Hamilton exceptional system candidate, which means that M_i and its extension will have the correct number of connections from A' to B' (which makes this part of the argument simpler than in the critical case).

Proof of Lemma 3.2.3. Set $A' := A_0 \cup A$ and $B' := B_0 \cup B$. We first construct the F'_s. If $\gamma' = 0$, there is nothing to do. So suppose that $\gamma' > 0$. Note that each F'_s has to be a matching of size 2 (this follows from the definition of a Hamilton exceptional system candidate and the fact that $e(F'_s) = 2$). Since H' is bipartite and so

$$\frac{e(H')}{\chi'(H')} = \frac{e(H')}{\Delta(H')} \geq \frac{2(\gamma + \gamma')n}{3\gamma n/5} > \frac{10}{3},$$

we can find a 2-matching F'_1 in H'. Delete the edges in F'_1 from H' and choose another 2-matching F'_2. We repeat this process until we have chosen $\gamma' n$ edge-disjoint 2-matchings $F'_1, \ldots, F'_{\gamma' n}$.

We now construct $F_1, \ldots, F_{\gamma n}$ in two steps: first we construct matchings $M_1, \ldots, M_{\gamma n}$ in H' and then extend each M_i into the desired F_i. Let H_1 and H'_1 be obtained from H and H' by removing all the edges in $F'_1, \ldots, F'_{\gamma' n}$. So now $2\gamma n \leq e(H'_1) \leq 10\varepsilon_0 \gamma n^2$ and both $e(H_1)$ and $e(H'_1)$ are even. Thus Proposition 3.2.2 implies that there is a decomposition of H'_1 into edge-disjoint non-empty even matchings $M_1, \ldots, M_{\gamma n}$, each of size at most $30\varepsilon_0 n$.

Note that each M_i is a Hamilton exceptional system candidate with parameter ε_0. So if $H'_1 = H_1$, then we are done by setting $F_s := M_s$ for each $s \leq \gamma n$. Hence, we may assume that $H'' := H_1 - H'_1 = H - H'$ contains edges. Let X be the set of all those vertices $x \in A_0 \cup B_0$ for which $d_{H''}(x) > 0$. Note that each $x \in X$ satisfies $N_{H''}(x) \subseteq A \cup B$ (since $H[A_0, B_0] \subseteq H'$). This implies that each $x \in X$ satisfies $d_{H'}(x) \geq \lfloor 3\gamma n/5 \rfloor - 1$ or $d_{H''}(x) = 1$. (Indeed, suppose that $d_{H'}(x) \leq \lfloor 3\gamma n/5 \rfloor - 2$ and $d_{H''}(x) \geq 2$. Then we can move two edges incident to x from H'' to H'. The final assumption in (i) and the assumption on $d_{H'}(x)$ together imply that we would still have $\Delta(H') \leq 3\gamma n/5$, a contradiction.) Since $\Delta(H) \leq 16\gamma n/15$ by (i) this in turn implies that $d_{H''}(x) \leq 7\gamma n/15 + 2$ for all $x \in X$.

Let $\mathcal{M}$ be a random subset of $\{M_1, \ldots, M_{\gamma n}\}$ where each M_i is chosen independently with probability $2/3$. By Proposition 1.4.4, with high probability, the following assertions hold:

$$r := |\mathcal{M}| = (2/3 \pm \varepsilon_0)\gamma n$$

$$|\{M_s \in \mathcal{M} : d_{M_s}(v) = 1\}| = 2d_{H'_1}(v)/3 \pm \varepsilon_0 \gamma n \qquad \text{for all } v \in V(H). \tag{3.2.11}$$

By relabeling if necessary, we may assume that $\mathcal{M} = \{M_1, M_2, \ldots, M_r\}$. For each $s \leq r$, we will now extend M_s to a Hamilton exceptional system candidate F_s with parameter ε_0 by adding edges from H''. Suppose that for some $1 \leq s \leq r$ we have already constructed $F_1, \ldots, F_{s-1}$. Set $H''_s := H'' - \sum_{j<s} F_j$. Let W_s be the set of all those vertices $w \in X$ for which $d_{M_s}(w) = 0$ and $d_{H''_s}(w) \geq 32\varepsilon_0 n \geq 2|A_0 \cup B_0| + e(M_s)$. Recall that $X \subseteq A_0 \cup B_0$ and $N_{H''_s}(w) \subseteq N_{H''}(w) \subseteq A \cup B$ for each $w \in X$ and thus also for each $w \in W_s$. Thus there are $|W_s|$ vertex-disjoint 2-paths uwu' with $w \in W_s$ and $u, u' \in N_{H''_s}(w) \setminus V(M_s)$. Assign these 2-paths to M_s and call the resulting graph F_s. Observe that F_s is a Hamilton exceptional system candidate with parameter ε_0. Therefore, we have constructed $F_1, \ldots, F_r$ by extending $M_1, \ldots, M_r$.

We now construct $F_{r+1}, \ldots, F_{\gamma n}$. For this, we first prove that the above construction implies that the current 'leftover' H''_{r+1} has small maximum degree. Indeed, note that if $w \in W_s$, then $d_{H''_{s+1}}(w) = d_{H''_s}(w) - 2$. By (3.2.11), for each

$x \in X$, the number of $M_s \in \mathcal{M}$ with $d_{M_s}(x) = 0$ is

$$\begin{aligned} r - |\{M_s \in \mathcal{M} : d_{M_s}(x) = 1\}| &\geq (2/3 - \varepsilon_0)\gamma n - (2d_{H_1'}(x)/3 + \varepsilon_0 \gamma n) \\ &\geq 2\gamma n/3 - 2d_{H'}(x)/3 - 2\varepsilon_0 \gamma n \\ &\geq 2\gamma n/3 - 2/3 \cdot \lfloor 3\gamma n/5 \rfloor - 2\varepsilon_0 \gamma n \\ &\geq (4/15 - 2\varepsilon_0)\gamma n > d_{H''}(x)/2. \end{aligned}$$

Hence, we have $d_{H''_{r+1}}(x) < 32\varepsilon_0 n$ for all $x \in X$ (as we remove 2 edges at x each time we have $d_{M_s}(x) = 0$ and $d_{H''_s}(x) \geq 32\varepsilon_0 n$). Note that by definition of H', all but at most one edge in H'' must have an endpoint in X. So for $x \notin X$, $d_{H''}(x) \leq |X| + 1 \leq |A_0 \cup B_0| + 1 \leq \varepsilon_0 n + 1$. Therefore, $\Delta(H''_{r+1}) < 32\varepsilon_0 n$.

Let $H''' := H_1 - (F_1 + \cdots + F_r)$. So H''' is the union of H''_{r+1} and all the M_s with $r < s \leq \gamma n$. Since each of H_1 and $F_1, \ldots, F_r$ contains an even number of edges, $e(H''')$ is even. In addition, $M_s \subseteq H'''$ for each $r < s \leq \gamma n$, so $e(H''') \geq 2(\gamma n - r)$. By (3.2.11), since $\Delta(H''_{r+1}) \leq 32\varepsilon_0 n$, we deduce that for every vertex $v \in V(H''')$, we have

$$d_{H'''}(v) \leq \left(\frac{d_{H_1'}(v)}{3} + \varepsilon_0 \gamma n\right) + \Delta(H''_{r+1}) \leq \frac{3\gamma n/5}{3} + \varepsilon_0 \gamma n + 32\varepsilon_0 n \leq \frac{2(\gamma n - r)}{3}$$

In the second inequality, we used that $d_{H_1'}(v) \leq d_{H'}(v)$. Moreover, we have

$$e(H''') = e(H''_{r+1}) + e(M_{r+1} + \cdots + M_{\gamma n}) \leq 32\varepsilon_0 n^2 + 30\varepsilon_0 n(\gamma n - r) \leq 62\varepsilon_0 n^2.$$

Thus, by Proposition 3.2.2 applied with H''' and $\gamma - r/n$ playing the roles of H and γ, there exists a decomposition of H''' into $\gamma n - r$ edge-disjoint non-empty even matchings $F_{r+1}, \ldots, F_{\gamma n}$, each of size at most $3e(H''')/(\gamma n - r) \leq \sqrt{\varepsilon_0}n/2$. Thus each such F_s is a Hamilton exceptional system candidate with parameter ε_0. This completes the proof. □

3.2.3. Step 3: Constructing the Localized Exceptional Systems. The next lemma will be used to extend most of the exceptional system candidates guaranteed by Lemma 3.2.3 into localized exceptional systems. These extensions are required to be 'faithful' in the following sense. Suppose that F is an exceptional system candidate. Then J is a *faithful extension of* F if the following holds:

- J contains F and $F[A', B'] = J[A', B']$.
- If F is a Hamilton exceptional system candidate, then J is a Hamilton exceptional system and the analogue holds if F is a matching exceptional system candidate.

LEMMA 3.2.4. *Suppose that* $0 < 1/n \ll \varepsilon_0 \ll 1$, *that* $0 \leq \gamma \leq 1$ *and that* $n, K, m, \gamma n \in \mathbb{N}$. *Let* $\mathcal{P}$ *be a* (K, m, ε_0)*-partition of a set* V *of* n *vertices. Let* $1 \leq i, i' \leq K$. *Suppose that* H *and* $F_1, \ldots, F_{\gamma n}$ *are pairwise edge-disjoint graphs which satisfy the following conditions:*

(i) $V(H) = V$ *and* H *contains only* A_0A_i*-edges and* $B_0B_{i'}$*-edges.*
(ii) *Each* F_s *is an* (i, i')*-ESC with parameter* ε_0.
(iii) *Each* $v \in V_0$ *satisfies* $d_{H+\sum F_s}(v) \geq (2\gamma + \sqrt{\varepsilon_0})n$.

Then there exist edge-disjoint (i, i')*-ES* $J_1, \ldots, J_{\gamma n}$ *with parameter* ε_0 *in* $H + \sum F_s$ *such that* J_s *is a faithful extension of* F_s *for all* $s \leq \gamma n$.

Proof. For each $s \leq \gamma n$ in turn, we extend F_s into an (i,i')-ES J_s with parameter ε_0 in $H+\sum F_s$ such that J_s and $J_{s'}$ are edge-disjoint for all $s' < s$. Since H does not contain any $A'B'$-edges, the J_s will automatically satisfy $J_s[A',B'] = F_s[A',B']$. Suppose that for some $1 \leq s \leq \gamma n$ we have already constructed $J_1, \dots, J_{s-1}$. Set $H_s := H - \sum_{s'<s} J_{s'}$. Consider any $v \in V_0$. Since v has degree at most 2 in an exceptional system and in an exceptional system candidate, (iii) implies that

$$d_{H_s}(v) \geq d_{H+\sum F_s}(v) - 2\gamma n \geq \sqrt{\varepsilon_0} n.$$

Together with (i) this shows that condition (ii) in Lemma 2.3.2 holds (with H_s playing the role of G). Since $\mathcal{P}$ is a (K, m, ε_0)-partition of V, Lemma 2.3.2(i) holds too. Hence we can apply Lemma 2.3.2 to obtain an exceptional system J_s with parameter ε_0 in $H_s + F_s$ such that J_s is a faithful extension of F_s. (i) and (ii) ensure that J_s is an (i,i')-ES, as required. □

3.2.4. Step 4: Constructing the Remaining Exceptional Systems. Due to condition (iii), Lemma 3.2.4 cannot be used to extend *all* the exceptional system candidates returned by Lemma 3.2.3 into localized exceptional systems. The next lemma will be used to deal with the remaining exceptional system candidates (the resulting exceptional systems will not be localized).

LEMMA 3.2.5. *Suppose that $0 < 1/n \ll \varepsilon_0 \ll \varepsilon' \ll \lambda \ll 1$ and that $n, \lambda n \in \mathbb{N}$. Let A, A_0, B, B_0 be a partition of a set V of n vertices such that $|A_0| + |B_0| \leq \varepsilon_0 n$ and $|A| = |B|$. Suppose that $H, F_1, \dots, F_{\lambda n}$ are pairwise edge-disjoint graphs which satisfy the following conditions:*

(i) *$V(H) = V$ and H contains only A_0A-edges and B_0B-edges.*
(ii) *Each F_s is an exceptional system candidate with parameter ε_0.*
(iii) *For all but at most $\varepsilon' n$ indices $s \leq \lambda n$ the graph F_s is either a matching exceptional system candidate with $e(F_s) = 0$ or a Hamilton exceptional system candidate with $e(F_s) = 2$. In particular, all but at most $\varepsilon' n$ of the F_s satisfy $d_{F_s}(v) \leq 1$ for all $v \in V_0$.*
(iv) *All $v \in V_0$ satisfy $d_{H+\sum F_s}(v) = 2\lambda n$.*
(v) *All $v \in A \cup B$ satisfy $d_{H+\sum F_s}(v) \leq 2\varepsilon_0 n$.*

Then there exists a decomposition of $H+\sum F_s$ into edge-disjoint exceptional systems $J_1, \dots, J_{\lambda n}$ with parameter ε_0 such that J_s is a faithful extension of F_s for all $s \leq \lambda n$.

Proof. Let $V_0 := A_0 \cup B_0$ and let $v_1, \dots, v_{|V_0|}$ denote the vertices of V_0. We will decompose H into graphs J'_s in such a way that the graphs $J_s := J'_s + F_s$ satisfy $d_{J_s}(v_i) = 2$ for all $i \leq |V_0|$ and $d_{J_s}(v) \leq 1$ for all $v \in A \cup B$. Hence each J_s will be an exceptional system with parameter ε_0. Condition (i) guarantees that J_s will be a faithful extension of F_s. Moreover, the J_s will form a decomposition of $H+\sum F_s$. We construct the decomposition of H by considering each vertex v_i of $A_0 \cup B_0$ in turn.

Initially, we set $V(J'_s) = E(J'_s) = \emptyset$ for all $s \leq \lambda n$. Suppose that for some $1 \leq i \leq |V_0|$ we have already assigned (and added) all the edges of H incident with each of $v_1, \dots, v_{i-1}$ to the J'_s. Consider v_i. Without loss of generality assume that $v_i \in A_0$. Note that $N_H(v_i) \subseteq A$ by (i). Define an auxiliary bipartite graph Q_i with vertex classes V_1 and V_2 as follows: $V_1 := N_H(v_i)$ and V_2 consists of $2 - d_{F_s}(v_i)$

copies of F_s for each $s \leq \lambda n$. Moreover, Q_i contains an edge between $v \in V_1$ and $F_s \in V_2$ if and only if $v \notin V(F_s + J'_s)$.

We now show that Q_i contains a perfect matching. For this, note that $|V_1| = 2\lambda n - d_{\sum F_s}(v_i) = |V_2|$ by (iv). (v) implies that for each $v \in V_1 \subseteq A$ we have $d_{\sum(F_s+J'_s)}(v) \leq d_{H+\sum F_s}(v) \leq 2\varepsilon_0 n$. So v lies in at most $2\varepsilon_0 n$ of the graphs $F_s + J'_s$. Therefore, $d_{Q_i}(v) \geq |V_2| - 4\varepsilon_0 n \geq |V_2|/2$ for all $v \in V_1$. (The final inequality follows since (iii) and (iv) together imply that $d_H(v_i) = 2\lambda n - d_{\sum F_s}(v_i) \geq 2\lambda n - (\lambda n - \varepsilon' n) - 2\varepsilon' n \geq \lambda n/2$ and so $|V_2| = |V_1| \geq \lambda n/2$.) On the other hand, since each $F_s + J'_s$ is an exceptional system candidate with parameter ε_0, (ESC3) implies that $|V(F_s + J'_s) \cap A| \leq (\sqrt{\varepsilon_0}/2 + 2\varepsilon_0)n \leq \sqrt{\varepsilon_0} n$ for each $F_s \in V_2$. Therefore $d_{Q_i}(F_s) \geq |V_1| - |V(F_s + J'_s) \cap A| \geq |V_1|/2$ for each $F_s \in V_2$. Thus we can apply Hall's theorem to find a perfect matching M in Q_i. Whenever M contains an edge between v and F_s, we add the edge $v_i v$ to J'_s. This completes the desired assignment of the edges of H at v_i to the J'_s. □

3.2.5. Proof of Lemma 2.7.3. In our proof of Lemma 2.7.3 we will use the following result, which is a consequence of Lemmas 3.2.4 and 3.2.5. Given a suitable set of exceptional system candidates in an exceptional scheme, the lemma extends these into exceptional systems which form a decomposition of the exceptional scheme. We prove the lemma in a slightly more general form than needed for the current case, as we will also use it in the other two cases.

LEMMA 3.2.6. *Suppose that* $0 < 1/n \ll \varepsilon_0 \ll \varepsilon \ll \varepsilon' \ll \lambda, 1/K \ll 1$, *that* $1/(7K^2) \leq \alpha < 1/K^2$ *and that* $n, K, m, \alpha n, \lambda n/K^2 \in \mathbb{N}$. *Let*

$$\gamma := \alpha - \frac{\lambda}{K^2} \qquad \text{and} \qquad \gamma' := \frac{\lambda}{K^2}.$$

Suppose that the following conditions hold:

(i) $(G^*, \mathcal{P})$ *is a* $(K, m, \varepsilon_0, \varepsilon)$*-exceptional scheme with* $|G^*| = n$.
(ii) G^* *is the edge-disjoint union of* $H(i,i')$, $F_1(i,i'), \dots, F_{\gamma n}(i,i')$ *and* $F'_1(i,i'), \dots, F'_{\gamma' n}(i,i')$ *over all* $1 \leq i, i' \leq K$.
(iii) *Each* $H(i,i')$ *contains only* A_0A_i*-edges and* $B_0B_{i'}$*-edges.*
(iv) *Each* $F_s(i,i')$ *is an* (i,i')*-ESC with parameter* ε_0.
(v) *Each* $F'_s(i,i')$ *is an exceptional system candidate with parameter* ε_0. *Moreover, for all but at most* $\varepsilon' n$ *indices* $s \leq \gamma' n$ *the graph* $F'_s(i,i')$ *is either a matching exceptional system candidate with* $e(F'_s(i,i')) = 0$ *or a Hamilton exceptional system candidate with* $e(F'_s(i,i')) = 2$.
(vi) $d_{G^*}(v) = 2K^2\alpha n$ *for all* $v \in V_0$.
(vii) *For all* $1 \leq i, i' \leq K$ *let* $G^*(i,i') := H(i,i') + \sum_{s \leq \gamma n} F_s(i,i') + \sum_{s \leq \gamma' n} F'_s(i,i')$. *Then* $d_{G^*(i,i')}(v) = (2\alpha \pm \varepsilon')n$ *for all* $v \in V_0$.

Then G^* *has a decomposition into* $K^2\alpha n$ *edge-disjoint exceptional systems*

$$J_1(i,i'), \dots, J_{\gamma n}(i,i') \qquad \text{and} \qquad J'_1(i,i'), \dots, J'_{\gamma' n}(i,i')$$

with parameter ε_0, *where* $1 \leq i, i' \leq K$, *such that* $J_s(i,i')$ *is an* (i,i')*-ES which is a faithful extension of* $F_s(i,i')$ *for all* $s \leq \gamma n$ *and* $J'_s(i,i')$ *is a faithful extension of* $F'_s(i,i')$ *for all* $s \leq \gamma' n$.

Proof. Fix any $i, i' \leq K$ and set $H := H(i,i')$ and $F_s := F_s(i,i')$ for all $s \leq \gamma n$. Our first aim is to apply Lemma 3.2.4 in order to extend each of $F_1, \dots, F_{\gamma n}$ into a (i, i')-HES. (iii) and (iv) ensure that conditions (i) and (ii) of Lemma 3.2.4 hold. To verify Lemma 3.2.4(iii), note that by (v) and (vii) each $v \in V_0$ satisfies

$$\begin{aligned} d_{H+\sum F_s}(v) &= d_{G^*(i,i')}(v) - d_{\sum_s F'_s(i,i')}(v) \geq (2\alpha - \varepsilon')n - (\gamma' - \varepsilon')n - 2\varepsilon' n \\ &= (2\alpha - \gamma' - 2\varepsilon')n \geq (2\gamma + \sqrt{\varepsilon_0})n. \end{aligned}$$

(Here the first inequality follows since (v) implies that $d_{F'_s(i,i')}(v) \leq 1$ for all but at most $\varepsilon' n$ indices $s \leq \gamma' n$.) Thus we can indeed apply Lemma 3.2.4 to find edge-disjoint (i, i')-ES $J_1(i, i'), \dots, J_{\gamma n}(i, i')$ with parameter ε_0 in $H + \sum F_s$ such that $J_s(i, i')$ is a faithful extension of F_s for all $s \leq \gamma n$. We repeat this procedure for all $1 \leq i, i' \leq K$ to obtain $K^2\gamma n$ edge-disjoint (localized) exceptional systems.

Our next aim is to apply Lemma 3.2.5 in order to construct the $J'_s(i, i')$. Let H_0 be the union of $H(i, i') - (J_1(i, i') + \dots + J_{\gamma n}(i, i'))$ over all $i, i' \leq K$. Relabel the $F'_s(i, i')$ (for all $s \leq \gamma' n$ and all $i, i' \leq K$) to obtain exceptional system candidates $F'_1, \dots, F'_{\lambda n}$. Note that by (vi) each $v \in V_0$ satisfies

$$d_{H_0+\sum F'_s}(v) = d_{G^*}(v) - 2K^2\gamma n = 2K^2\alpha n - 2K^2\gamma n = 2\lambda n. \tag{3.2.12}$$

Thus condition (iv) of Lemma 3.2.5 holds with H_0, F'_s playing the roles of H, F_s. (iii) and (v) imply that conditions (i)–(iii) of Lemma 3.2.5 hold with $K^2\varepsilon'$ playing the role of ε'. To verify Lemma 3.2.5(v), note that each $v \in A$ satisfies $d_{H_0+\sum F'_s}(v) \leq d_{G^*}(v, A_0) + d_{G^*}(v, B') \leq 2\varepsilon_0 n$ by (iii), (i) and (ESch3). Similarly each $v \in B$ satisfies $d_{H_0+\sum F'_s}(v) \leq 2\varepsilon_0 n$. Thus we can apply Lemma 3.2.5 with $H_0, F'_s, K^2\varepsilon'$ playing the roles of H, F_s, ε' to obtain a decomposition of $H_0 + \sum_s F'_s$ into λn edge-disjoint exceptional systems $J'_1, \dots, J'_{\lambda n}$ with parameter ε_0 such that J'_s is a faithful extension of F'_s for all $s \leq \lambda n$. Recall that each F'_s is a $F'_{s'}(i, i')$ for some $i, i' \leq K$ and some $s' \leq \gamma' n$. Let $J'_{s'}(i, i') := J'_s$. Then all the $J_s(i, i')$ and all the $J'_s(i, i')$ are as required in the lemma. □

We now combine Lemmas 3.2.1, 3.2.3 and 3.2.6 in order to prove Lemma 2.7.3.

Proof of Lemma 2.7.3. Let $G^\diamond$ be as defined in Lemma 2.7.3(iv). Choose a new constant ε' such that $\varepsilon \ll \varepsilon' \ll \lambda, 1/K$. Set

$$2\alpha n := \frac{D - \phi n}{K^2}, \qquad \gamma_1 := \alpha - \frac{2\lambda}{K^2} \qquad \text{and} \qquad \gamma'_1 := \frac{2\lambda}{K^2}. \tag{3.2.13}$$

Similarly as in the proof of Lemma 3.2.1, since $\phi \ll 1/3 \leq D/n$, we have

(3.2.14)

$$\alpha \geq 1/(7K^2), \qquad (1 - 14\lambda)\alpha \leq \gamma_1 < \alpha \qquad \text{and} \qquad \varepsilon \ll \varepsilon' \ll \lambda, 1/K, \alpha, \gamma_1 \ll 1.$$

Apply Lemma 3.2.1 with γ_1 playing the role of γ in order to obtain a decomposition of $G^\diamond$ into edge-disjoint spanning subgraphs $H(i, i')$ and $H''(i, i')$ (for all $1 \leq i, i' \leq K$) which satisfy the following properties, where $G'(i, i') := H(i, i') + H''(i, i')$:

(b_1) Each $H(i, i')$ contains only A_0A_i-edges and $B_0B_{i'}$-edges.
(b_2) $H''(i, i') \subseteq G^\diamond[A', B']$. Moreover, all but at most $\varepsilon' n$ edges of $H''(i, i')$ lie in $G^\diamond[A_0 \cup A_i, B_0 \cup B_{i'}]$.
(b_3) $e(H''(i, i'))$ is even and $2\alpha n \leq e(H''(i, i')) \leq 11\varepsilon_0 n^2/(10K^2)$.
(b_4) $\Delta(H''(i, i')) \leq 31\alpha n/30$.
(b_5) $d_{G'(i,i')}(v) = (2\alpha \pm \varepsilon')n$ for all $v \in V_0$.
(b_6) Let $\widetilde{H}$ any spanning subgraph of $H''(i, i')$ which maximises $e(\widetilde{H})$ under the constraints that $\Delta(\widetilde{H}) \leq 3\gamma_1 n/5$, $H''(i, i')[A_0, B_0] \subseteq \widetilde{H}$ and $e(\widetilde{H})$ is even. Then $e(\widetilde{H}) \geq 2\alpha n$.

Fix any $1 \leq i, i' \leq K$. Set $H := H(i, i')$ and $H'' := H''(i, i')$. Our next aim is to decompose H'' into suitable 'localized' Hamilton exceptional system candidates. For this, we will apply Lemma 3.2.3 with H'', γ_1, γ_1' playing the roles of H, γ, γ'. Note that $\Delta(H'') \leq 31\alpha n/30 \leq 16\gamma_1 n/15$ by (b$_4$) and (3.2.14). Moreover, $\Delta(H''[A, B]) \leq \Delta(G^\diamond[A, B]) \leq \varepsilon_0 n$ by (iv) and (ESch3). Since $e(H'')$ is even by (b$_3$), it follows that condition (i) of Lemma 3.2.3 holds. Condition (ii) of Lemma 3.2.3 follows from (b$_6$) and the fact that any $\widetilde{H}$ as in (b$_6$) satisfies $e(\widetilde{H}) \leq e(H'') \leq 11\varepsilon_0 n^2/(10K^2) \leq 10\varepsilon_0\gamma_1 n^2$ (the last inequality follows from (3.2.14)). Thus we can indeed apply Lemma 3.2.3 in order to decompose H'' into αn edge-disjoint Hamilton exceptional system candidates $F_1, \ldots, F_{\gamma_1 n}, F_1', \ldots, F_{\gamma_1' n}'$ with parameter ε_0 such that $e(F_s') = 2$ for all $s \leq \gamma_1' n$. Next we set

$$\gamma_2 := \alpha - \frac{\lambda}{K^2} \qquad \text{and} \qquad \gamma_2' := \frac{\lambda}{K^2}.$$

Condition (b$_2$) ensures that by relabeling the F_s's and F_s''s we obtain αn edge-disjoint Hamilton exceptional system candidates $F_1(i, i'), \ldots, F_{\gamma_2 n}(i, i'), F_1'(i, i'), \ldots, F_{\gamma_2' n}'(i, i')$ with parameter ε_0 such that properties (a$'$) and (b$'$) hold:

- (a$'$) $F_s(i, i')$ is an (i, i')-HESC for every $s \leq \gamma_2 n$. Moreover, at least $\gamma_2' n$ of the $F_s(i, i')$ satisfy $e(F_s(i, i')) = 2$.
- (b$'$) $e(F_s'(i, i')) = 2$ for all but at most $\varepsilon' n$ of the $F_s'(i, i')$.

Indeed, we can achieve this by relabeling each F_s which is a subgraph of $G^\diamond[A_0 \cup A_i, B_0 \cup B_{i'}]$ as one of the $F_{s'}(i, i')$ and each F_s for which is not the case as one of the $F_{s'}'(i, i')$.

Our next aim is to apply Lemma 3.2.6 with $G^\diamond, \gamma_2, \gamma_2'$ playing the roles of G^*, γ, γ'. Clearly conditions (i) and (ii) of Lemma 3.2.6 hold. (iii) follows from (b$_1$). (iv) and (v) follow from (a$'$) and (b$'$). (vi) follows from Lemma 2.7.3(i),(iii). Finally, (vii) follows from (b$_5$) since $G'(i, i')$ plays the role of $G^*(i, i')$. Thus we can indeed apply Lemma 3.2.6 to obtain a decomposition of $G^\diamond$ into $K^2\alpha n$ edge-disjoint Hamilton exceptional systems $J_1(i, i'), \ldots, J_{\gamma_2 n}(i, i')$ and $J_1'(i, i'), \ldots, J_{\gamma_2' n}'(i, i')$ with parameter ε_0, where $1 \leq i, i' \leq K$, such that $J_s(i, i')$ is an (i, i')-HES which is a faithful extension of $F_s(i, i')$ for all $s \leq \gamma_2 n$ and $J_s'(i, i')$ is a faithful extension of $F_s'(i, i')$ for all $s \leq \gamma_2' n$. Then the set $\mathcal{J}$ of all these Hamilton exceptional systems is as required in Lemma 2.7.3. □

3.3. Critical Case with $e(A', B') \geq D$

The aim of this section is to prove Lemma 2.7.4. Recall that Lemma 2.7.4 gives a decomposition of the exceptional edges into exceptional systems in the critical case when $e(A', B') \geq D$. The overall strategy for the proof is similar to that of Lemma 2.7.3. As before, it consists of four steps. In Step 1, we use Lemma 3.3.1 instead of Lemma 3.2.1. In Step 2, we use Lemma 3.3.3 instead of Lemma 3.2.3. We still use Lemma 3.2.6 which combines Steps 3 and 4.

3.3.1. Step 1: Constructing the Graphs $H''(i, i')$. The next lemma is an analogue of Lemma 3.2.1. We will apply it with the graph $G^\diamond$ from Lemma 2.7.4(iv) playing the role of G. Note that instead of assuming that our graph G given in Lemma 2.7.4 is critical, the lemma assumes that $e_{G^\diamond}(A', B') \leq 2n$. This is a weaker assumption, since if G is critical, then $e_{G^\diamond}(A', B') \leq e_G(A', B') < n$

by Lemma 3.1.1(iii). Using only this weaker assumption has the advantage that we can also apply the lemma in the proof of Lemma 2.7.5, i.e. the case when $e_G(A', B') < D$. (b$_7$) is only used in the latter application.

LEMMA 3.3.1. *Suppose that* $0 < 1/n \ll \varepsilon_0 \ll \varepsilon \ll 1/K \ll 1$ *and that* $n, K, m \in \mathbb{N}$. *Let* $(G, \mathcal{P})$ *be a* $(K, m, \varepsilon_0, \varepsilon)$*-exceptional scheme with* $|G| = n$ *and* $e_G(A_0), e_G(B_0) = 0$. *Let* W_0 *be a subset of* V_0 *of size at most* 2 *such that for each* $w \in W_0$, *we have*

$$K^2 \leq d_{G[A',B']}(w) \leq e_G(A', B')/2. \tag{3.3.1}$$

Suppose that $e_G(A', B') \leq 2n$ *is even. Then* G *can be decomposed into edge-disjoint spanning subgraphs* $H(i, i')$ *and* $H''(i, i')$ *of* G *(for all* $1 \leq i, i' \leq K$*) such that the following properties hold, where* $G'(i, i') := H(i, i') + H''(i, i')$*:*

(b$_1$) *Each* $H(i, i')$ *contains only* A_0A_i*-edges and* $B_0B_{i'}$*-edges.*
(b$_2$) $H''(i, i') \subseteq G[A', B']$. *Moreover, all but at most* $20\varepsilon n/K^2$ *edges of* $H''(i, i')$ *lie in* $G[A_0 \cup A_i, B_0 \cup B_{i'}]$.
(b$_3$) $e(H''(i, i')) = 2\lceil e_G(A', B')/(2K^2)\rceil$ *or* $e(H''(i, i')) = 2\lfloor e_G(A', B')/(2K^2)\rfloor$.
(b$_4$) $d_{H''(i,i')}(v) = (d_{G[A',B']}(v) \pm 25\varepsilon n)/K^2$ *for all* $v \in V_0$.
(b$_5$) $d_{G'(i,i')}(v) = (d_G(v) \pm 25\varepsilon n)/K^2$ *for all* $v \in V_0$.
(b$_6$) *Each* $w \in W_0$ *satisfies* $d_{H''(i,i')}(w) = \lceil d_{G[A',B']}(w)/K^2\rceil$ *or* $d_{H''(i,i')}(w) = \lfloor d_{G[A',B']}(w)/K^2\rfloor$.
(b$_7$) *Each* $w \in W_0$ *satisfies* $2d_{H''(i,i')}(w) \leq e(H''(i, i'))$.

Proof. Since $e_G(A', B')$ is even, there exist unique non-negative integers b and q such that $e_G(A', B') = 2K^2b + 2q$ and $q < K^2$. Hence, for all $1 \leq i, i' \leq K$, there are integers $b_{i,i'} \in \{2b, 2b + 2\}$ such that $\sum_{i,i' \leq K} b_{i,i'} = e_G(A', B')$. In particular, the number of pairs i, i' for which $b_{i,i'} = b + 2$ is precisely q. We will choose the graphs $H''(i, i')$ such that $e(H''(i, i')) = b_{i,i'}$. (In particular, this will ensure that (b$_3$) holds.) The following claim will help to ensure (b$_6$) and (b$_7$).

Claim. *For each* $w \in W_0$ *and all* $i, i' \leq K$ *there is an integer* $a_{i,i'} = a_{i,i'}(w)$ *which satisfies the following properties:*

- $a_{i,i'} = \lceil d_{G[A',B']}(w)/K^2\rceil$ *or* $a_{i,i'} = \lfloor d_{G[A',B']}(w)/K^2\rfloor$.
- $2a_{i,i'} \leq b_{i,i'}$.
- $\sum_{i,i' \leq K} a_{i,i'} = d_{G[A',B']}(w)$.

To prove the claim, note that there are unique non-negative integers a and p such that $d_{G[A',B']}(w) = K^2a + p$ and $p < K^2$. Note that $a \geq 1$ by (3.3.1). Moreover,

$$2(K^2a + p) = 2d_{G[A',B']}(w) \overset{(3.3.1)}{\leq} e_G(A', B') = 2K^2b + 2q. \tag{3.3.2}$$

This implies that $a \leq b$. Recall that $b_{i,i'} \in \{2b, 2b + 2\}$. So if $b > a$, then the claim holds by choosing any $a_{i,i'} \in \{a, a+1\}$ such that $\sum_{i,i' \leq K} a_{i,i'} = d_{G[A',B']}(w)$. Hence we may assume that $a = b$. Then (3.3.2) implies that $p \leq q$. Therefore, the claim holds by setting $a_{i,i'} := a + 1$ for exactly p pairs i, i' for which $b_{i,i'} = 2b + 2$ and setting $a_{i,i'} := a$ otherwise. This completes the proof of the claim.

Apply Lemma 2.5.2 to decompose G into subgraphs $H(i, i')$, $H'(i, i')$ (for all $i, i' \leq K$) satisfying the following properties, where $G(i, i') = H(i, i') + H'(i, i')$:

(a$'_1$) Each $H(i, i')$ contains only A_0A_i-edges and $B_0B_{i'}$-edges.
(a$'_2$) All edges of $H'(i, i')$ lie in $G[A_0 \cup A_i, B_0 \cup B_{i'}]$.

(a_3') $e(H'(i,i')) = (e_G(A',B') \pm 8\varepsilon n)/K^2$.
(a_4') $d_{H'(i,i')}(v) = (d_{G[A',B']}(v) \pm 2\varepsilon n)/K^2$ for all $v \in V_0$.
(a_5') $d_{G(i,i')}(v) = (d_G(v) \pm 4\varepsilon n)/K^2$ for all $v \in V_0$.

Indeed, (a_3') follows from Lemma 2.5.2(a_3) and our assumption that $e_G(A',B') \leq 2n$.

Clearly, (a_1') implies that the graphs $H(i,i')$ satisfy (b_1). We will now move some $A'B'$-edges of G between the $H'(i,i')$ such that the graphs $H''(i,i')$ obtained in this way satisfy the following conditions:

- Each $H''(i,i')$ is obtained from $H'(i,i')$ by adding or removing at most $20\varepsilon n/K^2$ edges of G.
- $e(H''(i,i')) = b_{i,i'}$.
- $d_{H''(i,i')}(w) = a_{i,i'}(w)$ for each $w \in W_0$, where $a_{i,i'}(w)$ are integers satisfying the claim.

Write $W_0 =: \{w_1\}$ if $|W_0| = 1$ and $W_0 =: \{w_1, w_2\}$ if $|W_0| = 2$. If $W_0 \neq \emptyset$, then (a_4') implies that $d_{H'(i,i')}(w_1) = a_{i,i'}(w_1) \pm (2\varepsilon n/K^2 + 1)$. For each $i, i' \leq K$, we add or remove at most $2\varepsilon n/K^2 + 1$ edges incident to w_1 such that the graphs $H''(i,i')$ obtained in this way satisfy $d_{H''(i,i')}(w_1) = a_{i,i'}(w_1)$. Note that since $a_{i,i'}(w_1) \geq \lfloor d_{G[A',B']}(w_1)/K^2 \rfloor \geq 1$ by (3.3.1), we can do this in such a way that we do not move the edge w_1w_2 (if it exists). Similarly, if $|W_0| = 2$, then for each $i, i' \leq K$ we add or remove at most $2\varepsilon n/K^2 + 1$ edges incident to w_2 such that the graphs $H''(i,i')$ obtained in this way satisfy $d_{H''(i,i')}(w_2) = a_{i,i'}(w_2)$. As before, we do this in such a way that we do not move the edge w_1w_2 (if it exists).

Thus $d_{H''(i,i')}(w_1) = a_{i,i'}(w_1)$ and $d_{H''(i,i')}(w_2) = a_{i,i'}(w_2)$ for all $1 \leq i, i' \leq K$ (if w_1, w_2 exist). In particular, together with the claim, this implies that $d_{H''(i,i')}(w_1), d_{H''(i,i')}(w_2) \leq b_{i,i'}/2$. Thus the number of edges of $H''(i,i')$ incident to W_0 is at most

$$\sum_{w \in W_0} d_{H''(i,i')}(w) \leq b_{i,i'}. \tag{3.3.3}$$

(This holds regardless of the size of W_0.) On the other hand, (a_3') implies that for all $i, i' \leq K$ we have

$$e(H''(i,i')) = (e_G(A',B') \pm 8\varepsilon n)/K^2 \pm 2(2\varepsilon n/K^2 + 1) = b_{i,i'} \pm 13\varepsilon n/K^2.$$

Together with (3.3.3) this ensures that we can add or delete at most $13\varepsilon n/K^2$ edges which do not intersect W_0 to or from each $H''(i,i')$ in order to ensure that $e(H''(i,i')) = b_{i,i'}$ for all $i, i' \leq K$. Hence, (b_3), (b_6) and (b_7) hold. Moreover,

$$e(H''(i,i') - H'(i,i')) \leq |W_0|(2\varepsilon n/K^2 + 1) + 13\varepsilon n/K^2 \leq 20\varepsilon n/K^2. \tag{3.3.4}$$

So (b_2) follows from (a_2'). Finally, (b_4) and (b_5) follow from (3.3.4), (a_4') and (a_5'). □

3.3.2. Step 2: Decomposing $H''(i,i')$ into Hamilton Exceptional System Candidates. Before we can prove an analogue of Lemma 3.2.3, we need the following result. It will allow us to distribute the edges incident to the (up to three) vertices w_i of high degree in $G[A',B']$ in a suitable way among the localized Hamilton exceptional system candidates F_j. The degrees of these high degree vertices w_i will play the role of the a_i. The c_j will account for edges (not incident to w_i) which have already been assigned to the F_j. (b) and (c) will be used to ensure (ESC4), i.e. that the total number of 'connections' between A' and B' is even and positive.

LEMMA 3.3.2. *Let $1 \leq q \leq 3$ and $0 \leq \eta < 1$ and $r, \eta r \in \mathbb{N}$. Suppose that $a_1, \dots, a_q \in \mathbb{N}$ and $c_1, \dots, c_r \in \{0,1,2\}$ satisfy the following conditions:*

(i) $c_1 \geq \dots \geq c_r \geq c_1 - 1$.
(ii) $\sum_{i \leq q} a_i + \sum_{j \leq r} c_j = 2(1+\eta)r$.
(iii) $31r/60 \leq a_1, a_2 \leq r$ *and* $31r/60 \leq a_3 \leq 31r/30$.

Then for all $i \leq q$ and all $j \leq r$ there are $a_{i,j} \in \{0,1,2\}$ such that the following properties hold:

(a) $\sum_{j \leq r} a_{i,j} = a_i$ *for all* $i \leq q$.
(b) $c_j + \sum_{i \leq q} a_{i,j} = 4$ *for all* $j \leq \eta r$ *and* $c_j + \sum_{i \leq q} a_{i,j} = 2$ *for all* $\eta r < j \leq r$.
(c) *For all $j \leq r$ there are at least $2 - c_j$ indices $i \leq q$ with $a_{i,j} = 1$.*

Proof. We will choose $a_{i,1}, \dots, a_{i,r}$ for each $i \leq q$ in turn such that the following properties (α_i)–(ρ_i) hold, where we write $c_j^{(i)} := c_j + \sum_{i' \leq i} a_{i',j}$ for each $0 \leq i \leq q$ (so $c_j^{(0)} = c_j$):

(α_i) If $i \geq 1$ then $\sum_{j \leq r} a_{i,j} = a_i$.
(β_i) $4 \geq c_1^{(i)} \geq \dots \geq c_r^{(i)}$.
(γ_i) If $\sum_{j \leq r} c_j^{(i)} < 2r$, then $|c_j^{(i)} - c_{j'}^{(i)}| \leq 1$ for all $j, j' \leq r$.
(δ_i) If $\sum_{j \leq r} c_j^{(i)} \geq 2r$, then $c_j^{(i)} \geq 2$ for all $j \leq \eta r$ and $c_j^{(i)} = 2$ for all $\eta r < j \leq r$.
(ρ_i) If $1 \leq i \leq q$ and $c_j^{(i-1)} < 2$ for some $j \leq r$, then $a_{i,j} \in \{0,1\}$.

We will then show that the $a_{i,j}$ defined in this way are as required in the lemma.

Note that (i) and the fact that $c_1, \dots, c_r \in \{0,1,2\}$ together imply (β_0)–(δ_0). Moreover, (α_0) and (ρ_0) are vacuously true. Suppose that for some $1 \leq i \leq q$ we have already defined $a_{i',j}$ for all $i' < i$ and all $j \leq r$ such that $(\alpha_{i'})$–$(\rho_{i'})$ hold. In order to define $a_{i,j}$ for all $j \leq r$, we distinguish the following cases.

Case 1: $\sum_{j \leq r} c_j^{(i-1)} \geq 2r$.

Recall that in this case $c_j^{(i-1)} \geq 2$ for all $j \leq r$ by (δ_{i-1}). For each $j \leq r$ in turn we choose $a_{i,j} \in \{0,1,2\}$ as large as possible subject to the constraints that

- $a_{i,j} + c_j^{(i-1)} \leq 4$ and
- $\sum_{j' \leq j} a_{i,j'} \leq a_i$.

Since $c_j^{(i)} = a_{i,j} + c_j^{(i-1)}$, (β_i) follows from (β_{i-1}) and our choice of the $a_{i,j}$. (γ_i) is vacuously true. To verify (δ_i), note that $c_j^{(i)} \geq c_j^{(i-1)} \geq 2$ by (δ_{i-1}). Suppose that the second part of (δ_i) does not hold, i.e. that $c_{\eta n+1}^{(i)} > 2$. This means that $a_{i,\eta n+1} > 0$. Together with our choice of the $a_{i,j}$ this implies that $c_j^{(i)} = 4$ for all $j \leq \eta n$. Thus

$$2(1+\eta)r = 4\eta r + 2(r - \eta r) < \sum_{j \leq r} c_j^{(i)} = \sum_{j \leq r} a_{i,j} + \sum_{i' < i} a_{i'} + \sum_{j \leq r} c_j \leq \sum_{i' \leq i} a_{i'} + \sum_{j \leq r} c_j$$

contradicting (ii). Thus the second part of (δ_i) holds too. Moreover, $c_{\eta n+1}^{(i)} = c_{\eta n+1}^{(i-1)} = 2$ also means that $a_{i,\eta n+1} = 0$. So $\sum_{j' \leq \eta n} a_{i,j'} = a_i$, i.e. (α_i) holds. (ρ_i) is vacuously true since $c_j^{(i-1)} \geq 2$ by (δ_{i-1}).

Case 2: $2r - a_i \le \sum_{j\le r} c_j^{(i-1)} < 2r$.

If $i \in \{1, 2\}$ then together with (iii) this implies that

$$\sum_{j\le r} c_j^{(i-1)} \ge r \ge a_i. \tag{3.3.5}$$

If $i = 3$ then

$$\sum_{j\le r} c_j^{(i-1)} \ge \sum_{j\le r} \sum_{i'\le 2} a_{i',j} = a_1 + a_2 \ge \frac{31r}{30} \ge a_3 \tag{3.3.6}$$

by (iii). In particular, in both cases we have $\sum_{j\le r} c_j^{(i-1)} \ge r$. Together with (γ_{i-1}) this implies that $c_j^{(i-1)} \in \{1, 2\}$ for all $j \le r$. Let $0 \le r' \le r$ be the largest integer such that $c_{r'}^{(i-1)} = 2$. So $r' < r$ and $\sum_{j\le r} c_j^{(i-1)} = r + r'$. Together with (3.3.5) and (3.3.6) this in turn implies that $a_i \le r + r'$ (regardless of the value of i).

Set $a_{i,j} := 1$ for all $r' < j \le r$. Note that

$$\sum_{r'<j\le r} a_{i,j} = r - r' = 2r - \sum_{j\le r} c_j^{(i-1)} \le a_i,$$

where the final inequality comes from the assumption of Case 2. Take $a_{i,1}, \dots, a_{i,r'}$ to be a sequence of the form $2, \dots, 2, 0, \dots, 0$ (in the case when $a_i - \sum_{r'<j\le r} a_{i,j}$ is even) or $2, \dots, 2, 1, 0, \dots, 0$ (in the case when $a_i - \sum_{r'<j\le r} a_{i,j}$ is odd) which is chosen in such a way that $\sum_{j\le r'} a_{i,j} = a_i - \sum_{r'<j\le r} a_{i,j} = a_i - r + r'$. This can be done since $a_i \le r + r'$ implies that the right hand side is at most $2r'$.

Clearly, (α_i), (β_i) and (ρ_i) hold. Since $\sum_{j\le r} c_j^{(i)} = a_i + \sum_{j\le r} c_j^{(i-1)} \ge 2r$ as we are in Case 2, (γ_i) is vacuously true. Clearly, our choice of the $a_{i,j}$ guarantees that $c_j^{(i)} \ge 2$ for all $j \le r$. As in Case 1 one can show that $c_j^{(i)} = 2$ for all $\eta r < j \le r$. Thus (δ_i) holds.

Case 3: $\sum_{j\le r} c_j^{(i-1)} < 2r - a_i$.

Note that in this case

$$2r > \sum_{j\le r} c_j^{(i-1)} + a_i = \sum_{i'\le i} a_{i'} + \sum_{j\le r} c_j,$$

and so $i < q$ by (ii). Together with (iii) this implies that $a_i \le r$. Thus for all $j \le r$ we can choose $a_{i,j} \in \{0, 1\}$ such that (α_i)–(γ_i) and (ρ_i) are satisfied. (δ_i) is vacuously true.

This completes the proof of the existence of numbers $a_{i,j}$ (for all $i \le q$ and all $j \le r$) satisfying (α_i)–(ρ_i). It remains to show that these $a_{i,j}$ are as required in the lemma. Clearly, (α_1)–(α_q) imply that (a) holds. Since $c_j^{(q)} = c_j + \sum_{i\le q} a_{i,j}$ the second part of (b) follows from (δ_q). Since $c_j^{(q)} \le 4$ for each $j \le \eta r$ by (β_q), together with (ii) this in turn implies that the first part of (b) must hold too. If $c_j < 2$, then (ρ_1)–(ρ_q) and (b) together imply that for at least $2 - c_j$ indices i we have $a_{i,j} = 1$. Therefore, (c) holds. □

We can now use the previous lemma to decompose the bipartite graph induced by A' and B' into Hamilton exceptional system candidates.

LEMMA 3.3.3. *Suppose that $0 < 1/n \ll \varepsilon_0 \ll \alpha < 1$, that $0 \leq \eta < 199/200$ and that $n, \alpha n/200, \eta\alpha n \in \mathbb{N}$. Let H be a bipartite graph on n vertices with vertex classes $A\dot\cup A_0$ and $B\dot\cup B_0$ where $|A_0| + |B_0| \leq \varepsilon_0 n$. Furthermore, suppose that the following conditions hold:*

- (c_1) *$e(H) = 2(1+\eta)\alpha n$.*
- (c_2) *There is a set $W' \subseteq V(H)$ with $1 \leq |W'| \leq 3$ and such that*

$$e(H - W') \leq 199\alpha n/100 \text{ and } d_H(w) \geq 13\alpha n/25 \text{ for all } w \in W'.$$

- (c_3) *There exists a set $W_0 \subseteq W'$ with $|W_0| = \min\{2, |W'|\}$ and such that $d_H(w) \leq \alpha n$ for all $w \in W_0$ and $d_H(w') \leq 41\alpha n/40$ for all $w' \in W' \setminus W_0$.*
- (c_4) *For all $w \in W'$ and all $v \in V(H) \setminus W'$ we have $d_H(w) - d_H(v) \geq \alpha n/150$.*
- (c_5) *For all $v \in A \cup B$ we have $d_H(v) \leq \varepsilon_0 n$.*

Then there exists a decomposition of H into edge-disjoint Hamilton exceptional system candidates $F_1, \dots, F_{\alpha n}$ such that $e(F_s) = 4$ for all $s \leq \eta\alpha n$ and $e(F_s) = 2$ for all $\eta\alpha n < s \leq \alpha n$. Furthermore, at least $\alpha n/200$ of the F_s satisfy the following two properties:

- *$d_{F_s}(w) = 1$ for all $w \in W_0$,*
- *$e(F_s) = 2$.*

Roughly speaking, the idea of the proof is first to find the F_s which satisfy the final two properties. Let H_1 be the graph obtained from H by removing the edges in all these F_s. We will decompose $H_1 - W'$ into matchings M_j of size at most two. Next, we extend these matchings into Hamilton exceptional system candidates F_j using Lemma 3.3.2. In particular, if $e(M_j) < 2$, then we will use one or more edges incident to W' to ensure that the number of $A'B'$-connections is positive and even, as required by (ESC4). (Note that it does not suffice to ensure that the number of $A'B'$-edges is positive and even for this.)

Proof. Set $H' := H - W'$, $W_0 =: \{w_1, w_{|W_0|}\}$ and $W' =: \{w_1, \dots, w_{|W'|}\}$. Hence, if $|W'| = 3$, then $W' \setminus W_0 = \{w_3\}$. Otherwise $W' = W_0$.

We will first construct $e_H(W')$ Hamilton exceptional system candidates F_s, such that each of them is a matching of size two and together they cover all edges in $H[W']$. So suppose that $e_H(W') > 0$. Thus $|W'| = 2$ or $|W'| = 3$. If $|W'| = 2$, let f denote the unique edge in $H[W']$. Note that

$$e(H') \geq e(H) - (d_H(w_1) + d_H(w_2) - 1) \geq 2(1+\eta)\alpha n - (2\alpha n - 1) \geq 1$$

by (c_1) and (c_3). So there exists an edge f' in H'. Therefore, $M_1' := \{f, f'\}$ is a matching. If $|W'| = 3$, then $e_H(W') \leq 2$ as H is bipartite. Since by (c_2) each $w \in W'$ satisfies $d_H(w) \geq 13\alpha n/25$, it is easy to construct $e_H(W')$ 2-matchings $M_1', M_{e_H(W')}'$ such that $d_{M_s'}(w) = 1$ for all $w \in W'$ and all $s \leq e_H(W')$ and such that $H[W'] \subseteq M_1' \cup M_{e_H(W')}'$. Set $F_{\alpha n - s+1} := M_s'$ for all $s \leq e_H(W')$ (regardless of the size of W').

We now greedily choose $\alpha n/200 - e_H(W')$ additional 2-matchings $F_{199\alpha n/200+1}$, $\dots, F_{\alpha n - e_H(W')}$ in H which are edge-disjoint from each other and from $F_{\alpha n}$, $F_{\alpha n - e_H(W')+1}$ and such that $d_{F_s}(w) = 1$ for all $w \in W_0$ and all $199\alpha n/200 < s \leq \alpha n - e_H(W')$. To see that this can be done, recall that by (c_2) we have

$d_H(w) \geq 13\alpha n/25$ for all $w \in W'$ (and thus for all $w \in W_0$) and that (c_1) and (c_3) together imply that $e(H - W_0) \geq 2(1+\eta)\alpha n - \alpha n > \alpha n$ if $|W_0| = 1$.

Thus $F_{199\alpha n/200+1}, \dots, F_{\alpha n}$ are Hamilton exceptional system candidates satisfying the two properties in the 'furthermore part' of the lemma. Let H_1 and H_1' be the graphs obtained from H and H' by deleting all the $\alpha n/100$ edges in these Hamilton exceptional system candidates. Set

$$r := 199\alpha n/200 \qquad \text{and} \qquad \eta' := \eta\alpha n/r = 200\eta/199. \tag{3.3.7}$$

Thus $0 \leq \eta' < 1$ and we now have

$$H_1[W'] = \emptyset, \quad e(H_1) = e(H) - \alpha n/100 = 2(1+\eta')r \quad \text{and} \quad e(H_1') \leq 2r. \tag{3.3.8}$$

(To verify the last inequality note that $e(H_1') \leq e(H - W') \leq 2r$ by (c_2).) Also, (c_2) and (c_4) together imply that for all $w \in W'$ and all $v \in V(H) \setminus W'$ we have

$$d_{H_1}(w) \geq \alpha n/2 \geq 4\varepsilon_0 n \qquad \text{and} \qquad d_{H_1}(w) - d_{H_1}(v) \geq 2\varepsilon_0 n. \tag{3.3.9}$$

Moreover, by (c_2) and (c_3), each $w \in W_0$ satisfies

$$\begin{aligned} 31r/60 \leq 13\alpha n/25 - \alpha n/200 \leq d_H(w) - d_{H-H_1}(w) = d_{H_1}(w) \\ \leq \alpha n - \alpha n/200 = r. \end{aligned} \tag{3.3.10}$$

Similarly, if $|W'| = 3$ and so w_3 exists, then

$$\begin{aligned} 31r/60 \leq 13\alpha n/25 - \alpha n/200 \leq d_H(w_3) - d_{H-H_1}(w_3) = d_{H_1}(w_3) \\ \leq 41\alpha n/40 \leq 31r/30. \end{aligned} \tag{3.3.11}$$

(3.3.9) and (3.3.10) together imply that $d_{H_1'}(v) \leq d_{H_1}(v) < d_{H_1}(w_1) \leq r$ for all $v \in V(H) \setminus W'$. Thus $\chi'(H_1') \leq \Delta(H_1') \leq r$. Together with Proposition 1.4.5 this implies that H_1' can be decomposed into r edge-disjoint matchings $M_1, \dots, M_r$ such that $|m_j - m_{j'}| \leq 1$ for all $1 \leq j, j' \leq r$, where we set $m_j := e(M_j)$.

Our next aim is to apply Lemma 3.3.2 with $|W'|$, $d_{H_1}(w_i)$, m_j, η' playing the roles of q, a_i, c_j, η (for all $i \leq |W'|$ and all $j \leq r$). Since $\sum_{j \leq r} m_j = e(H_1') \leq 2r$ by (3.3.8) and since $|m_j - m_{j'}| \leq 1$, it follows that $m_j \in \{0,1,2\}$ for all $j \leq r$. Moreover, by relabeling the matchings M_j if necessary, we may assume that $m_1 \geq m_2 \geq \dots \geq m_r$. Thus condition (i) of Lemma 3.3.2 holds. (ii) holds too since $\sum_{i \leq |W'|} d_{H_1}(w_i) + \sum_{j \leq r} m_j = e(H_1) = 2(1+\eta')r$ by (3.3.8). Finally, (iii) follows from (3.3.10) and (3.3.11). Thus we can indeed apply Lemma 3.3.2 in order to obtain numbers $a_{i,j} \in \{0,1,2\}$ (for all $i \leq |W'|$ and $j \leq r$) which satisfy the following properties:

(a$'$) $\sum_{j \leq r} a_{i,j} = d_{H_1}(w_i)$ for all $i \leq |W'|$.

(b$'$) $m_j + \sum_{i \leq |W'|} a_{i,j} = 4$ for all $j \leq \eta' r$ and $m_j + \sum_{i \leq |W'|} a_{i,j} = 2$ for all $\eta' r < j \leq r$.

(c$'$) If $m_j < 2$ then there exist at least $2 - m_j$ indices i such that $a_{i,j} = 1$.

For all $j \leq r$, our Hamilton exceptional system candidate F_j will consist of the edges in M_j as well as of $a_{i,j}$ edges of H_1 incident to w_i (for each $i \leq |W'|$). So let $F_j^0 := M_j$ for all $j \leq r$. For each $i = 1, \dots, |W'|$ in turn, we will now assign the edges of H_1 incident with w_i to $F_1^{i-1}, \dots, F_r^{i-1}$ such that the resulting graphs $F_1^i, \dots, F_r^i$ satisfy the following properties:

(α_i) If $i \geq 1$, then $e(F_j^i) - e(F_j^{i-1}) = a_{i,j}$.

(β_i) F^i_j is a path system. Every vertex $v \in A \cup B$ is incident to at most one edge of F^i_j. For every $v \in V_0 \setminus W'$ we have $d_{F^i_j}(v) \leq 2$. If $e(F^i_j) \leq 2$, we even have $d_{F^i_j}(v) \leq 1$.

(γ_i) Let b^i_j be the number of vertex-disjoint maximal paths in F^i_j with one endpoint in A' and the other in B'. If $a_{i,j} = 1$ and $i \geq 1$, then $b^i_j = b^{i-1}_j + 1$. Otherwise $b^i_j = b^{i-1}_j$.

We assign the edges of H_1 incident with w_i to $F^{i-1}_1, \dots, F^{i-1}_r$ in two steps. In the first step, for each index $j \leq r$ with $a_{i,j} = 2$ in turn, we assign an edge of H_1 between w_i and V_0 to F^{i-1}_j whenever there is such an edge left. More formally, to do this, we set $N_0 := N_{H_1}(w_i)$. For each $j \leq r$ in turn, if $a_{i,j} = 2$ and $N_{j-1} \cap V_0 \neq \emptyset$, then we choose a vertex $v \in N_{j-1} \cap V_0$ and set $F'_j := F^{i-1}_j + w_i v$, $N_j := N_{j-1} \setminus \{v\}$ and $a'_{i,j} := 1$. Otherwise, we set $F'_j := F^{i-1}_j$, $N_j := N_{j-1}$ and $a'_{i,j} := a_{i,j}$.

Therefore, after having dealt with all indices $j \leq r$ in this way, we have that

$$\text{either } a'_{i,j} \leq 1 \text{ for all } j \leq r \text{ or } N_r \cap V_0 = \emptyset \text{ (or both).} \tag{3.3.12}$$

Note that by (b′) we have $e(F'_j) \leq m_j + \sum_{i' \leq i} a_{i',j} \leq 4$ for all $j \leq r$. Moreover, (a′) implies that $|N_r| = \sum_{j \leq r} a'_{i,j}$. Also, $N_r \setminus V_0 = N_{H_1}(w_i) \setminus V_0$, and so $N_{H_1}(w_i) \setminus N_r \subseteq V_0$. Hence

$$|N_r| = |N_{H_1}(w_i)| - |N_{H_1}(w_i) \setminus N_r| \geq d_{H_1}(w_i) - |V_0| \geq d_{H_1}(w_i) - \varepsilon_0 n. \tag{3.3.13}$$

In the second step, we assign the remaining edges of H_1 incident with w_i to $F'_1, \dots, F'_r$. We achieve this by finding a perfect matching M in a suitable auxiliary graph.

Claim. *Define a graph Q with vertex classes N_r and V' as follows: V' consists of $a'_{i,j}$ copies of F'_j for each $j \leq r$. Q contains an edge between $v \in N_r$ and $F'_j \in V'$ if and only v is not an endpoint of an edge in F'_j. Then Q has a perfect matching M.*

To prove the claim, note that

$$|V'| = \sum_{j \leq r} a'_{i,j} = |N_r| \overset{(3.3.13)}{\geq} d_{H_1}(w_i) - \varepsilon_0 n. \tag{3.3.14}$$

Moreover, since $F'_j \subseteq H$ is bipartite and so every edge of F'_j has at most one endpoint in N_r, it follows that

$$d_Q(F'_j) \geq |N_r| - e(F'_j) \geq |N_r| - 4 \tag{3.3.15}$$

for each $F'_j \in V'$. Consider any $v \in N_r$. Clearly, there are at most $d_{H_1}(v)$ indices $j \leq r$ such that v is an endpoint of an edge of F'_j. If $v \in N_r \setminus V_0 \subseteq A \cup B$, then by (c$_5$), v lies in at most $2d_{H_1}(v) \leq 2d_H(v) \leq 2\varepsilon_0 n$ elements of V'. (The factor 2 accounts for the fact that each F'_j occurs in V' precisely $a'_{i,j} \leq 2$ times.) So

$$d_Q(v) \geq |V'| - 2\varepsilon_0 n \overset{(3.3.14)}{\geq} d_{H_1}(w_i) - 3\varepsilon_0 n \overset{(3.3.9)}{\geq} \varepsilon_0 n.$$

If $v \in N_r \cap V_0$, then (3.3.12) implies that $a'_{i,j} \leq 1$ for all $j \leq r$. Thus

$$d_Q(v) \geq |V'| - d_{H_1}(v) \overset{(3.3.14)}{\geq} (d_{H_1}(w_i) - d_{H_1}(v)) - \varepsilon_0 n \overset{(3.3.9)}{\geq} 2\varepsilon_0 n - \varepsilon_0 n = \varepsilon_0 n.$$

To summarize, for all $v \in N_r$ we have $d_Q(v) \geq \varepsilon_0 n$. Together with (3.3.15) and the fact that $|N_r| = |V'|$ by (3.3.14) this implies that Q contains a perfect matching M by Hall's theorem. This proves the claim.

For each $j \le r$, let F_j^i be the graph obtained from F_j' by adding the edge w_iv whenever the perfect matching M (as guaranteed by the claim) contains an edge between v and F_j'.

Let us now verify (α_i)–(γ_i) for all $i \le |W'|$. Clearly, (α_0)–(γ_0) hold and $b_j^0 = m_j$. Now suppose that $i \ge 1$ and that (α_{i-1})–(γ_{i-1}) hold. Clearly, (α_i) holds by our construction of $F_1^i, \dots, F_r^i$. Now consider any $j \le r$. If $a_{i,j} = 0$, then (β_i) and (γ_i) follow from (β_{i-1}) and (γ_{i-1}). If $a_{i,j} = 1$, then the unique edge in $F_j^i - F_j^{i-1}$ is vertex-disjoint from any edge of F_j^{i-1} (by the definition of Q) and so (β_i) holds. Moreover, $b_j^i = b_j^{i-1} + 1$ and so (γ_i) holds. So suppose that $a_{i,j} = 2$. Then the unique two edges in $F_j^i - F_j^{i-1}$ form a path $P = v'w_iv''$ of length two with internal vertex w_i. Moreover, at least one of the edges of P, w_iv'' say, was added to F_j^{i-1} in the second step of our construction of F_j^i. Thus $d_{F_j^i}(v'') = 1$. The other edge w_iv' of P was either added in the first or in the second step. If w_iv' was added in the second step, then $d_{F_j^i}(v') = 1$. Altogether this shows that in this case (γ_i) holds and (β_i) follows from (β_{i-1}). So suppose that w_iv' was added to F_j^{i-1} in the first step of our construction of F_j^i. Thus $v' \in V_0 \setminus W'$. But since $a_{i,j} = 2$, (b′) implies that $e(F_j^{i-1}) = m_j + \sum_{i'<i} a_{i',j} \le 2$. Together with (β_{i-1}) this shows that $d_{F_j^{i-1}}(v) \le 1$ for all $v \in V_0 \setminus W'$. Hence $d_{F_j^{i-1}}(v') \le 1$ and so $d_{F_j^i}(v') \le 2$. Together with (β_{i-1}) this implies (β_i). (Note that if $e(F_j^{i-1}) = 0$, then the above argument actually shows that $d_{F_j^i(v')} \le 1$, as required.) Moreover, the above observations also guarantee that (γ_i) holds. Thus $F_1^i, \dots, F_r^i$ satisfy (α_i)–(γ_i).

After having assigned the edges of H_1 incident with w_i for all $i \le |W'|$, we have obtained graphs $F_1^{|W'|}, \dots, F_r^{|W'|}$. Let $F_j := F_j^{|W'|}$ for all $j \le r$. Note that by $(\gamma_{|W'|})$ for all $j \le r$ the number of vertex-disjoint maximal $A'B'$-paths in F_j is precisely $b_j^{|W'|}$.

We now claim that $b_j^{|W'|}$ is positive and even. To verify this, recall that $b_j^0 = m_j$. Let odd_j be the number of $a_{i,j}$ with $a_{i,j} = 1$ and $i \le |W'|$. So $b_j^{|W'|} = m_j + \mathrm{odd}_j$. Together with (c′) this immediately implies that $b_j^{|W'|} \ge 2$. Moreover, since $a_{i,j} \in \{0, 1, 2\}$ we have

$$b_j^{|W'|} = m_j + \mathrm{odd}_j = m_j + \sum_{i \le |W'|,\ a_{i,j} \text{ is odd}} a_{i,j}.$$

Together with (b′) this now implies that $b_j^{|W'|}$ is even. This proves the claim.

Together with (a′), (b′) and (α_i), (β_i) for all $i \le |W'|$ this in turn shows that $F_1, \dots, F_r$ form a decomposition of H_1 into edge-disjoint Hamilton exceptional system candidates with $e(F_j) = 4$ for all $j \le \eta' r$ and $e(F_j) = 2$ for all $\eta' r < j \le r$. Recall that $\eta' r = \eta \alpha n$ by (3.3.7) and that we have already constructed Hamilton exceptional system candidates $F_{199\alpha n/200+1}, \dots, F_{\alpha n}$ which satisfy the 'furthermore statement' of the lemma, and thus in particular consist of precisely two edges. This completes the proof of the lemma. □

3.3.3. Proof of Lemma 2.7.4. We will now combine Lemmas 3.3.1, 3.3.3 and 3.2.6 in order to prove Lemma 2.7.4. This will complete the construction of the required exceptional sequences in the case when G is both critical and $e(G[A', B']) \geq D$.

Proof of Lemma 2.7.4. Let $G^\diamond$ be as defined in Lemma 2.7.4(iv). Our first aim is to decompose $G^\diamond$ into suitable 'localized' subgraphs via Lemma 3.3.1. Choose a new constant ε' such that $\varepsilon \ll \varepsilon' \ll \lambda, 1/K$ and define α by

$$2\alpha n := \frac{D - \phi n}{K^2}. \tag{3.3.16}$$

Recall from Lemma 2.7.4(ii) that $D = (n-1)/2$ or $D = n/2 - 1$. Together with our assumption that $\phi \ll 1$ this implies that

$$\frac{1 - 2/n - 2\phi}{4K^2} \leq \alpha \leq \frac{1-2\phi}{4K^2} \qquad \text{and} \qquad \varepsilon \ll \varepsilon' \ll \lambda, 1/K, \alpha \ll 1. \tag{3.3.17}$$

Note that by Lemma 2.7.4(ii) and (iii) we have $e_{G^\diamond}(A', B') \geq D - \phi n = 2K^2\alpha n$. Together with Lemma 3.1.1(iii) this implies that

(3.3.18)

$$2K^2\alpha n \leq e_{G^\diamond}(A', B') \leq e_G(A', B') \leq 17D/10 + 5 \overset{(3.3.16)}{\leq} 18K^2\alpha n/5 \overset{(3.3.17)}{<} n.$$

Moreover, recall that by Lemma 2.7.4(i) and (iii) we have

$$d_{G^\diamond}(v) = 2K^2\alpha n \qquad \text{for all } v \in V_0. \tag{3.3.19}$$

Let W be the set of all those vertices $w \in V(G)$ with $d_{G[A',B']}(w) \geq 11D/40$. So W is as defined in Lemma 3.1.1 and $1 \leq |W| \leq 3$ by Lemma 3.1.1(i). Let $W' \subseteq V(G)$ be as guaranteed by Lemma 3.1.1(v). Thus $W \subseteq W'$, $|W'| \leq 3$,

(3.3.20)

$$d_{G[A',B']}(w') \geq \frac{21D}{80}, \ d_{G[A',B']}(v) \leq \frac{11D}{40} \ \text{ and } \ d_{G[A',B']}(w') - d_{G[A',B']}(v) \geq \frac{D}{240}.$$

for all $w' \in W'$ and all $v \in V(G) \setminus W'$. In particular, $W' \subseteq V_0$. (This follows since Lemma 2.7.4(iii),(iv) and (ESch3) together imply that $d_{G[A',B']}(v) = d_{G^\diamond[A',B']}(v) + d_{G_0[A',B']}(v) \leq \varepsilon_0 n + e_{G_0}(A', B') \leq \varepsilon_0 n + \phi n$ for all $v \in A \cup B$.) Let w_1, w_2, w_3 be vertices of G such that

$$d_{G[A',B']}(w_1) \geq d_{G[A',B']}(w_2) \geq d_{G[A',B']}(w_3) \geq d_{G[A',B']}(v)$$

for all $v \in V(G) \setminus \{w_1, w_2, w_3\}$, where w_1 and w_2 are as in Lemma 2.7.4(v). Hence W consists of $w_1, \dots, w_{|W|}$ and W' consists of $w_1, \dots, w_{|W'|}$. Set $W_0 := \{w_1, w_2\} \cap W'$. Since $d_{G_0}(v) = \phi n$ for each $v \in V_0$ (and thus for each $v \in W_0$), each $w \in W_0$ satisfies

$$K^2 \leq 21D/80 - \phi n \overset{(3.3.20)}{\leq} d_{G^\diamond[A',B']}(w) \leq K^2\alpha n \overset{(3.3.18)}{\leq} e_{G^\diamond}(A', B')/2. \tag{3.3.21}$$

(Here the third inequality follows from Lemma 2.7.4(v).) Apply Lemma 3.3.1 to $G^\diamond$ in order to obtain a decomposition of $G^\diamond$ into edge-disjoint spanning subgraphs $H(i,i')$ and $H''(i,i')$ (for all $1 \leq i, i' \leq K$) which satisfy the following properties, where $G'(i,i') := H(i,i') + H''(i,i')$:

- (b$'_1$) Each $H(i,i')$ contains only A_0A_i-edges and $B_0B_{i'}$-edges.
- (b$'_2$) $H''(i,i') \subseteq G^\diamond[A', B']$. Moreover, all but at most $20\varepsilon n/K^2$ edges of $H''(i,i')$ lie in $G^\diamond[A_0 \cup A_i, B_0 \cup B_{i'}]$.
- (b$'_3$) $e(H''(i,i')) = 2\lceil e_{G^\diamond}(A', B')/(2K^2)\rceil$ or $e(H''(i,i')) = 2\lfloor e_{G^\diamond}(A', B')/(2K^2)\rfloor$. In particular, $2\alpha n \leq e(H''(i,i')) \leq 19\alpha n/5$ by (3.3.18).
- (b$'_4$) $d_{H''(i,i')}(v) = (d_{G^\diamond[A',B']}(v) \pm 25\varepsilon n)/K^2$ for all $v \in V_0$.

(b$'_5$) $d_{G'(i,i')}(v) = (d_{G^\diamond}(v) \pm 25\varepsilon n)/K^2 = (2\alpha \pm 25\varepsilon/K^2)n$ for all $v \in V_0$ by (3.3.19).

(b$'_6$) Each $w \in W_0$ satisfies $d_{H''(i,i')}(w) \leq \lceil d_{G^\diamond[A',B']}(w)/K^2 \rceil \leq \alpha n$ by (3.3.21).

Our next aim is to apply Lemma 3.3.3 to each $H''(i,i')$ to obtain suitable Hamilton exceptional system candidates (in particular almost all of them will be 'localized'). So consider any $1 \leq i, i' \leq K$ and let $H'' := H''(i,i')$. We claim that there exists $0 \leq \eta \leq 9/10$ such that H'' satisfies the following conditions (which in turn imply conditions (c$_1$)–(c$_5$) of Lemma 3.3.3):

(c$'_1$) $e(H'') = 2(1+\eta)\alpha n$ and $\eta\alpha n \in \mathbb{N}$.

(c$'_2$) $e(H'' - W') \leq 199\alpha n/100$ and $d_{H''}(w) \geq 13\alpha n/25$ for all $w \in W'$.

(c$'_3$) $d_{H''}(w) \leq \alpha n$ for all $w \in W_0$ and $d_{H''}(w') \leq 41\alpha n/40$ for all $w' \in W' \setminus W_0$.

(c$'_4$) For all $w \in W'$ and all $v \in V(G) \setminus W'$ we have $d_{H''}(w) - d_{H''}(v) \geq \alpha n/150$.

(c$'_5$) For all $v \in A \cup B$ we have $d_{H''}(v) \leq \varepsilon_0 n$.

Clearly, (b$'_3$) implies the first part of (c$'_1$). Since $e(H'')$ is even by (b$'_3$) and $\alpha n \in \mathbb{N}$, it follows that $\eta\alpha n \in \mathbb{N}$. To verify the first part of (c$'_2$), note that (b$'_3$) and (b$'_4$) together imply that

$$\begin{aligned} e(H'' - W') &= e(H'') - \sum_{w \in W'} d_{H''}(w) + e(H''[W']) \\ &\leq 2\lceil e_{G^\diamond}(A',B')/(2K^2) \rceil - \sum_{w \in W'} (d_{G^\diamond[A',B']}(w) - 25\varepsilon n)/K^2 + 3 \\ &\leq (e_{G^\diamond - W'}(A',B') + 80\varepsilon n)/K^2. \end{aligned}$$

Together with Lemma 3.1.1(iv) this implies that

$$e(H''-W') \leq (e_{G-W'}(A',B')+80\varepsilon n)/K^2 \leq ((3D/4+5)+80\varepsilon n)/K^2 \leq 199\alpha n/100.$$

To verify the second part of (c$'_2$), note that by (3.3.20) and Lemma 2.7.4(iii) each $w \in W'$ satisfies $d_{G^\diamond[A',B']}(w) \geq d_{G[A',B']}(w) - \phi n \geq 21D/80 - \phi n$. Together with (b$'_4$) this implies $d_{H''}(w) \geq 26\alpha n/50$. Thus (c$'_2$) holds. By (b$'_6$) we have $d_{H''}(w) \leq \alpha n$ for all $w \in W_0$. If $w' \in W' \setminus W_0$, then Lemma 2.7.4(ii) implies $d_{G[A',B']}(w') \leq D/2 \leq 51K^2\alpha n/50$. Thus, $d_{H''}(w') \leq 41\alpha n/40$ by (b$'_4$). Altogether this shows that (c$'_3$) holds. (c$'_4$) follows from (3.3.20), (b$'_4$) and the fact that $d_{G^\diamond[A',B']}(v) \geq d_{G[A',B']}(v) - \phi n$ for all $v \in V(G)$ by Lemma 2.7.4(iii). (c$'_5$) holds since $d_{H''}(v) \leq d_{G^\diamond[A',B']}(v) \leq \varepsilon_0 n$ for all $v \in A \cup B$ by (ESch3).

Now we apply Lemma 3.3.3 in order to decompose H'' into αn edge-disjoint Hamilton exceptional system candidates $F_1, \dots, F_{\alpha n}$ such that $e(F_s) \in \{2,4\}$ for all $s \leq \alpha n$ and such that at least $\alpha n/200$ of F_s satisfy $e(F_s) = 2$ and $d_{F_s}(w) = 1$ for all $w \in W_0$. Let

$$\gamma := \alpha - \frac{\lambda}{K^2} \qquad \text{and} \qquad \gamma' := \frac{\lambda}{K^2}.$$

Recall that by (b$'_2$) all but at most $20\varepsilon n/K^2 \leq \varepsilon' n$ edges of H'' lie in $G^\diamond[A_0 \cup A_i, B_0 \cup B_{i'}]$. Together with (3.3.17) this ensures that we can relabel the F_s if necessary to obtain αn edge-disjoint Hamilton exceptional system candidates $F_1(i,i'), \dots, F_{\gamma n}(i,i')$ and $F'_1(i,i'), \dots, F'_{\gamma' n}(i,i')$ such that the following properties hold:

(a$'$) $F_s(i,i')$ is an (i,i')-HESC for every $s \leq \gamma n$. Moreover, $\gamma' n$ of the $F_s(i,i')$ satisfy $e(F_s(i,i')) = 2$ and $d_{F_s(i,i')}(w) = 1$ for all $w \in W_0$.

(b$'$) $e(F'_s(i,i')) = 2$ for all but at most $\varepsilon' n$ of the $F'_s(i,i')$.

(c′) $e(F_s(i,i')), e(F'_s(i,i')) \in \{2,4\}$.

For (b′) and the 'moreover' part of (a′), we use that $\alpha n/200 - \varepsilon' n \geq 2\lambda n/K^2 = 2\gamma' n$. Our next aim is to apply Lemma 3.2.6 with $G^\diamond$ playing the role of G^* to extend the above exceptional system candidates into exceptional systems. Clearly conditions (i) and (ii) of Lemma 3.2.6 hold. (iii) follows from (b'_1). (iv) and (v) follow from (a′)–(c′). (vi) follows from Lemma 2.7.4(i),(iii). Finally, (vii) follows from (b'_5) since $G'(i,i')$ plays the role of $G^*(i,i')$. Thus we can indeed apply Lemma 3.2.6 to obtain a decomposition of $G^\diamond$ into $K^2\alpha n$ edge-disjoint Hamilton exceptional systems $J_1(i,i'), \dots, J_{\gamma n}(i,i')$ and $J'_1(i,i'), \dots, J'_{\gamma' n}(i,i')$ with parameter ε_0, where $1 \leq i, i' \leq K$, such that $J_s(i,i')$ is an (i,i')-HES which is a faithful extension of $F_s(i,i')$ for all $s \leq \gamma n$ and $J'_s(i,i')$ is a faithful extension of $F'_s(i,i')$ for all $s \leq \gamma' n$. Then the set $\mathcal{J}$ of all these exceptional systems is as required in Lemma 2.7.4. (Since W_0 contains $\{w_1, w_2\} \cap W$, the 'moreover part' of (a′) implies the 'moreover part' of Lemma 2.7.4(b).) □

3.4. The Case when $e(A', B') < D$

The aim of this section is to prove Lemma 2.7.5. This lemma provides a decomposition of the exceptional edges into exceptional systems in the case when $e(A', B') < D$. In this case, we do not need to prove any auxiliary lemmas first, as we can apply those proved in the other two cases (Lemmas 3.2.6 and 3.3.1).

Proof of Lemma 2.7.5. Let ε' be a new constant such that $\varepsilon \ll \varepsilon' \ll \lambda, 1/K$ and set

$$2\alpha n := \frac{n/2 - 1 - \phi n}{K^2}. \tag{3.4.1}$$

Similarly as in the proof of Lemma 2.7.4 we have

$$\varepsilon \ll \varepsilon' \ll \lambda, 1/K, \alpha \ll 1. \tag{3.4.2}$$

We claim that $G^\diamond$ can be decomposed into edge-disjoint spanning subgraphs $H(i,i')$ and $H''(i,i')$ (for all $1 \leq i, i' \leq K$) which satisfy the following properties, where $G'(i,i') := H(i,i') + H''(i,i')$:

(b'_1) Each $H(i,i')$ contains only A_0A_i-edges and $B_0B_{i'}$-edges.
(b'_2) $H''(i,i') \subseteq G^\diamond[A', B']$. Moreover, all but at most $\varepsilon' n$ edges of $H''(i,i')$ lie in $G^\diamond[A_0 \cup A_i, B_0 \cup B_{i'}]$.
(b'_3) $e(H''(i,i'))$ is even and $e(H''(i,i')) \leq 2\alpha n$.
(b'_4) $\Delta(H''(i,i')) \leq e(H''(i,i'))/2$.
(b'_5) $d_{G'(i,i')}(v) = (2\alpha \pm \varepsilon')n$ for all $v \in V_0$.

To see this, let us first consider the case when $e_{G^\diamond}(A', B') \leq 300\varepsilon n$. Apply Lemma 2.5.2 to $G^\diamond$ in order to obtain a decomposition of $G^\diamond$ into edge-disjoint spanning subgraphs $H(i,i')$ and $H'(i,i')$ (for all $1 \leq i, i' \leq K$) which satisfy Lemma 2.5.2(a_1)–(a_5). Set $H''(1,1) := \bigcup_{i,i' \leq K} H'(i,i') = G^\diamond[A', B']$ and $H''(i,i') := \emptyset$ for all other pairs $1 \leq i, i' \leq K$. Then (b'_1) follows from (a_1). (b'_2) follows from our definition of the $H''(i,i')$ and our assumption that $e_{G^\diamond}(A', B') \leq 300\varepsilon n < \varepsilon' n < \alpha n$. Together with Lemma 2.7.5(iv) this also implies (b'_3). (b'_4) follows from Lemma 2.7.5(v). Note that by Lemma 2.7.5(i) and (iii), every $v \in V_0$ satisfies

$d_{G^\diamond}(v) = n/2 - 1 - \phi n = 2K^2\alpha n$. So, writing $G(i,i') := H(i,i') + H'(i,i')$, ($a_5$) implies that

$$d_{G'(i,i')}(v) = d_{G(i,i')}(v) \pm 300\varepsilon n = (2\alpha \pm 4\varepsilon/K^2)n \pm 300\varepsilon n = (2\alpha \pm \varepsilon')n.$$

Thus (b_5') holds too.

So let us next consider the case when $e_{G^\diamond}(A', B') > 300\varepsilon n$. Let W_0 be the set of all those vertices $v \in V(G)$ for which $d_{G^\diamond[A',B']}(v) \geq 3e_{G^\diamond}(A', B')/8$. Then clearly $|W_0| \leq 2$. Moreover, each $v \in V(G) \setminus W_0$ satisfies

$$(3.4.3) \quad d_{G^\diamond[A',B']}(v) + 26\varepsilon n < 3e_{G^\diamond}(A', B')/8 + e_{G^\diamond}(A', B')/8 = e_{G^\diamond}(A', B')/2.$$

Recall from Lemma 2.7.5(v) that $d_{G^\diamond[A',B']}(w) \leq e_{G^\diamond}(A', B')/2$ for each $w \in W_0$. So we can apply Lemma 3.3.1 to $G^\diamond$ in order to obtain a decomposition of $G^\diamond$ into edge-disjoint spanning subgraphs $H(i,i')$ and $H''(i,i')$ (for all $1 \leq i, i' \leq K$) which satisfy Lemma 3.3.1(b_1)–(b_7). Then (b_1) and (b_2) imply (b_1') and (b_2'). (b_3') follows from (b_3), (3.4.1) and Lemma 2.7.5(v). Note that (b_3), (b_4) and (3.4.3) together imply that

$$(3.4.4) \qquad d_{H''(i,i')}(v) \leq \frac{e_{G^\diamond}(A', B')/2 - \varepsilon n}{K^2} \leq \frac{e(H''(i,i'))}{2}$$

for all $v \in V_0 \setminus W_0$. Note that each $v \in A \cup B$ satisfies $d_{H''(i,i')}(v) \leq d_{G^\diamond[A',B']}(v) \leq \varepsilon_0 n$ by Lemma 2.7.5(iv) and (ESch3). Together with the fact that $e(H''(i,i')) \geq 2\lfloor 300\varepsilon n/(2K^2)\rfloor \geq 2\varepsilon_0 n$ by (b_3), this implies that (3.4.4) also holds for all $v \in A \cup B$. Together with (b_7) this implies (b_4'). (b_5') follows from (b_5) and the fact that by Lemma 2.7.5(i) and (iii) every $v \in V_0$ satisfies $d_{G^\diamond}(v) = n/2 - 1 - \phi n = 2K^2\alpha n$. So ($b_1'$)–($b_5'$) hold in all cases.

We now decompose the localized subgraphs $H''(i,i')$ into exceptional system candidates. For this, fix $i, i' \leq K$ and write H'' for $H''(i,i')$. By (b_4') we have $\Delta(H'') \leq e(H'')/2$ and so $\chi'(H'') \leq e(H'')/2$. Apply Proposition 1.4.5 with $e(H'')/2$ playing the role of m to decompose H'' into $e(H'')/2$ edge-disjoint matchings, each of size 2. Note that $\alpha n - e(H'')/2 \geq 0$ by (b_3'). So we can add some empty matchings to obtain a decomposition of H'' into αn edge-disjoint $M_1, \dots, M_{\alpha n}$ such that each M_s is either empty or has size 2. Let

$$\gamma := \alpha - \frac{\lambda}{K^2} \qquad \text{and} \qquad \gamma' := \frac{\lambda}{K^2}.$$

Recall from (b_2') that all but at most $\varepsilon' n \leq \gamma' n$ edges of H'' lie in $G^\diamond[A_0 \cup A_i, B_0 \cup B_{i'}]$. Hence by relabeling if necessary, we may assume that $M_s \subseteq G^\diamond[A_0 \cup A_i, B_0 \cup B_{i'}]$ for every $s \leq \gamma n$. So by setting $F_s(i,i') := M_s$ for all $s \leq \gamma n$ and $F_s'(i,i') := M_{\gamma n + s}$ for all $s \leq \gamma' n$ we obtain a decomposition of H'' into edge-disjoint exceptional system candidates $F_1(i,i'), \dots, F_{\gamma n}(i,i')$ and $F_1'(i,i'), \dots, F_{\gamma' n}'(i,i')$ such that the following properties hold:

- (a′) $F_s(i,i')$ is an (i,i')-ESC for every $s \leq \gamma n$.
- (b′) Each $F_s(i,i')$ is either a Hamilton exceptional system candidate with $e(F_s(i,i')) = 2$or a matching exceptional system candidate with $e(F_s(i,i')) = 0$. The analogue holds for each $F_{s'}'(i,i')$.

Our next aim is to apply Lemma 3.2.6 with $G^\diamond$ playing the role of G^*, to extend the above exceptional system candidates into exceptional systems. Clearly conditions (i) and (ii) of Lemma 3.2.6 hold. (iii) follows from (b_1'). (iv) and (v) follow from (a′) and (b′). (vi) follows from Lemma 2.7.5(i),(iii). Finally, (vii) follows from (b_5') since $G'(i,i')$ plays the role of $G^*(i,i')$ in Lemma 3.2.6. Thus we can indeed apply

Lemma 3.2.6 to obtain a decomposition of $G^\diamond$ into $K^2\alpha n$ edge-disjoint exceptional systems $J_1(i,i'), \dots, J_{\gamma n}(i,i')$ and $J'_1(i,i'), \dots, J'_{\gamma' n}(i,i')$, where $1 \le i, i' \le K$, such that $J_s(i,i')$ is an (i,i')-ES which is a faithful extension of $F_s(i,i')$ for all $s \le \gamma n$ and $J'_s(i,i')$ is a faithful extension of $F'_s(i,i')$ for all $s \le \gamma' n$. Then the set $\mathcal{J}$ of all these exceptional systems is as required in Lemma 2.7.5. □

CHAPTER 4

The bipartite case

The aim of this chapter is to prove Theorems 1.3.5 and 1.3.8. Recall that Theorem 1.3.8 guarantees many edge-disjoint Hamilton cycles in a graph G when G has large minimum degree and is close to bipartite, whilst Theorem 1.3.5 guarantees a Hamilton decomposition of G when G has sufficiently large minimum degree, is regular and is close to bipartite. In Section 4.1 we give an outline of the proofs. The results from Sections 4.2 and 4.3 are used in both the proofs of Theorems 1.3.5 and 1.3.8. In Sections 4.4 and 4.5 we build up machinery for the proof of Theorem 1.3.5. We then prove Theorem 1.3.8 in Section 4.6 and Theorem 1.3.5 in Section 4.7.

Unlike in the previous chapters, in this chapter we view a matching M as a set of edges. (So $|M|$ for example, denotes the number of edges in M.)

4.1. Overview of the Proofs of Theorems 1.3.5 and 1.3.8

Note that, unlike in Theorem 1.3.5, in Theorem 1.3.8 we do not require a complete decomposition of our graph F into edge-disjoint Hamilton cycles. Therefore, the proof of Theorem 1.3.5 is considerably more involved than the proof of Theorem 1.3.8. Moreover, the ideas in the proof of Theorem 1.3.8 are all used in the proof of Theorem 1.3.5 too.

4.1.1. Proof Overview for Theorem 1.3.8. Let F be a graph on n vertices with $\delta(F) \geq (1/2 - o(1))n$ which is close to the balanced bipartite graph $K_{n/2,n/2}$. Further, suppose that G is a D-regular spanning subgraph of F as in Theorem 1.3.8. Then there is a partition A, B of $V(F)$ such that A and B are of roughly equal size and most edges in F go between A and B. Our ultimate aim is to construct $D/2$ edge-disjoint Hamilton cycles in F.

Suppose first that, in the graph F, both A and B are independent sets of equal size. So F is an almost complete balanced bipartite graph. In this case, the densest spanning even-regular subgraph G of F is also almost complete bipartite. This means that one can extend existing techniques (developed e.g. in [**6**, **7**, **9**, **11**, **31**]) to find an approximate Hamilton decomposition. (In Chapter 5, using such techniques, we prove an approximate decomposition result (Lemma 4.6.1) which is suitable for our purposes. In particular, Lemma 4.6.1 is sufficient to prove Theorem 1.3.8 in this special case.) The real difficulties arise when

(i) F is unbalanced (i.e. $|A| \neq |B|$);
(ii) F has vertices having high degree in both A and B (these are called exceptional vertices).

To illustrate (i), recall the following example due to Babai (which is the extremal construction for Corollary 1.1.5). Consider the graph F on $n = 8k + 2$ vertices consisting of one vertex class A of size $4k+2$ containing a perfect matching and no other edges, one empty vertex class B of size $4k$, and all possible edges between A and B. Thus the minimum degree of F is $4k+1=n/2$. Then one can use Tutte's factor theorem to show that the largest even-regular spanning subgraph G of F has degree $D=2k=(n-2)/4$. Note that to prove Theorem 1.3.8 in this case, each of the $D/2=k$ Hamilton cycles we find must contain exactly two of the $2k+1$ edges in A. In this way, we can 'balance out' the difference in the vertex class sizes.

More generally we will construct our Hamilton cycles in two steps. In the first step, we find a path system J which balances out the vertex class sizes (so in the above example, J would contain two edges in A). Then we extend J into a Hamilton cycle using only AB-edges in F. It turns out that the first step is the difficult one. It is easy to see that a path system J will balance out the sizes of A and B (in the sense that the number of uncovered vertices in A and B is the same) if and only if

$$e_J(A)-e_J(B)=|A|-|B|. \tag{4.1.1}$$

Note that any Hamilton cycle also satisfies this identity. So we need to find a set of $D/2$ path systems J satisfying (4.1.1) (where D is the degree of G). This is achieved (amongst other things) in Sections 4.3.2 and 4.3.3.

As indicated above, our aim is to use Lemma 4.6.1 (our approximate decomposition result for the bipartite case) in order to extend each such J into a Hamilton cycle. To apply Lemma 4.6.1 we also need to extend the balancing path systems J into 'balanced exceptional (path) systems' which contain all the exceptional vertices from (ii). This is achieved in Section 4.3.4. Lemma 4.6.1 also assumes that the path systems are 'localized' with respect to a given subpartition of A,B (i.e. they are induced by a small number of partition classes). Section 4.3.1 prepares the ground for this. The balanced exceptional systems are the analogues of the exceptional systems which we use in the two cliques case (i.e. in Chapter 2).

Finding the balanced exceptional systems is extremely difficult if G contains edges between the set A_0 of exceptional vertices in A and the set B_0 of exceptional vertices in B. So in a preliminary step, we find and remove a small number of edge-disjoint Hamilton cycles covering all A_0B_0-edges in Section 4.2. We put all these steps together in Section 4.6. (Sections 4.4, 4.5 and 4.7 are only relevant for the proof of Theorem 1.3.5.)

4.1.2. Proof Overview for Theorem 1.3.5. The main result of this chapter is Theorem 1.3.5. Suppose that G is a D-regular graph satisfying the conditions of that theorem. Using the approach of the previous subsection, one can obtain an approximate decomposition of G, i.e. a set of edge-disjoint Hamilton cycles covering almost all edges of G. However, one does not have any control over the 'leftover' graph H, which makes a complete decomposition seem infeasible. As in the proof of Theorem 1.3.3, we use the following strategy to overcome this issue and obtain a decomposition of G:

(1) find a (sparse) robustly decomposable graph G^{rob} in G and let G' denote the leftover;
(2) find an approximate Hamilton decomposition of G' and let H denote the (very sparse) leftover;
(3) find a Hamilton decomposition of $G^{\mathrm{rob}}\cup H$.

As before, it is of course far from obvious that such a graph G^{rob} exists. By assumption our graph G can be partitioned into two classes A and B of almost equal size such that almost all the edges in G go between A and B. If both A and B are independent sets of equal size then the 'bipartite' version of the robust decomposition lemma of [**21**] guarantees our desired subgraph G^{rob} of G. Of course, in general our graph G will contain edges in A and B. Our aim is therefore to replace such edges with 'fictive edges' between A and B, so that we can apply this version of the robust decomposition lemma (Lemma 4.5.3). (We note here that Lemma 4.5.3 is designed to deal with bipartite graphs. So its statement is slightly different to the robust decomposition lemma (Lemma 2.9.4) that was applied in the proof of Theorem 1.3.3.)

More precisely, similarly as in the proof of Theorem 1.3.8, we construct a collection of localized balanced exceptional systems. Together these path systems contain all the edges in $G[A]$ and $G[B]$. Again, each balanced exceptional system balances out the sizes of A and B and covers the exceptional vertices in G (i.e. those vertices having high degree into both A and B).

Similarly as in the two cliques case, we now introduce fictive edges. This time, by replacing edges of the balanced exceptional systems with fictive edges, we obtain from G an auxiliary (multi)graph G^* which only contains edges between A and B and which does not contain the exceptional vertices of G. This will allow us to apply the robust decomposition lemma. In particular this ensures that each Hamilton cycle obtained in G^* contains a collection of fictive edges corresponding to a single balanced exceptional system (as before the set-up of the robust decomposition lemma does allow for this). Each such Hamilton cycle in G^* then corresponds to a Hamilton cycle in G.

We now give an example of how we introduce fictive edges. Let m be an integer so that $(m-1)/2$ is even. Set $m' := (m-1)/2$ and $m'' := (m+1)/2$. Define the graph G as follows: Let A and B be disjoint vertex sets of size m. Let A_1, A_2 be a partition of A and B_1, B_2 be a partition of B such that $|A_1| = |B_1| = m''$. Add all edges between A and B. Add a matching $M_1 = \{e_1, \dots, e_{m'/2}\}$ covering precisely the vertices of A_2 and add a matching $M_2 = \{e'_1, \dots, e'_{m'/2}\}$ covering precisely the vertices of B_2. Finally add a vertex v which sends an edge to every vertex in $A_1 \cup B_1$. So G is $(m+1)$-regular (and v would be regarded as an exceptional vertex).

Now pair up each edge e_i with the edge e'_i. Write $e_i = x_{2i-1}x_{2i}$ and $e'_i = y_{2i-1}y_{2i}$ for each $1 \le i \le m'/2$. Let $A_1 = \{a_1, \dots, a_{m''}\}$ and $B_1 = \{b_1, \dots, b_{m''}\}$ and write $f_i := a_ib_i$ for all $1 \le i \le m''$. Obtain G^* from G by deleting v together with the edges in $M_1 \cup M_2$ and by adding the following fictive edges: add f_i for each $1 \le i \le m''$ and add x_jy_j for each $1 \le j \le m'$. Then G^* is a balanced bipartite $(m+1)$-regular multigraph containing only edges between A and B.

First, note that any Hamilton cycle C^* in G^* that contains precisely one fictive edge f_i for some $1 \le i \le m''$ corresponds to a Hamilton cycle C in G, where we replace the fictive edge f_i with a_iv and b_iv. Next, consider any Hamilton cycle C^* in G^* that contains precisely three fictive edges; f_i for some $1 \le i \le m''$ together with $x_{2j-1}y_{2j-1}$ and $x_{2j}y_{2j}$ for some $1 \le j \le m'/2$. Further suppose C^* traverses the vertices $a_i, b_i, x_{2j-1}, y_{2j-1}, x_{2j}, y_{2j}$ in this order. Then C^* corresponds to a Hamilton cycle C in G, where we replace the fictive edges with a_iv, b_iv, e_j and e'_j (see Figure 4.1.1). Here the path system J formed by the edges a_iv, b_iv, e_j and e'_j is an example of a balanced exceptional system. The above ideas are formalized in Section 4.4.

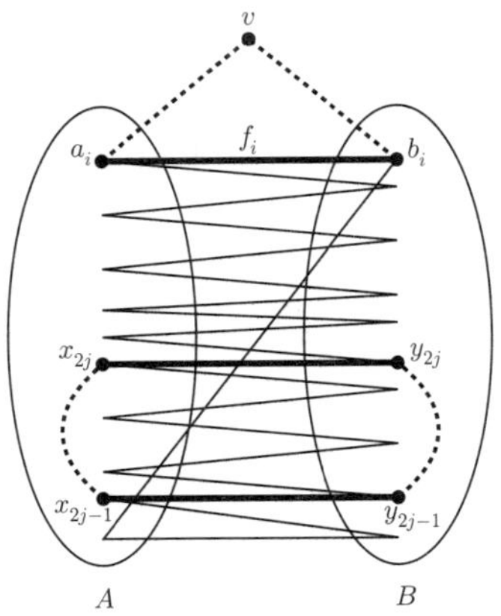

FIGURE 4.1.1. Transforming the problem of finding a Hamilton cycle in G into finding a Hamilton cycle in the balanced bipartite graph G^*

We can now summarize the steps leading to proof of Theorem 1.3.5. In Section 4.2, we find and remove a set of edge-disjoint Hamilton cycles covering all edges in $G[A_0, B_0]$. We can then find the localized balanced exceptional systems in Section 4.3. After this, we need to extend and combine them into certain path systems and factors (which contain fictive edges) in Section 4.4, before we can use them as an 'input' for the robust decomposition lemma in Section 4.5. Finally, all these steps are combined in Section 4.7 to prove Theorem 1.3.5.

4.2. Eliminating Edges between the Exceptional Sets

Suppose that G is a D-regular graph as in Theorem 1.3.5. The purpose of this section is to prove Corollary 4.2.12. Roughly speaking, given $K \in \mathbb{N}$, this corollary states that one can delete a small number of edge-disjoint Hamilton cycles from G to obtain a spanning subgraph G' of G and a partition A, A_0, B, B_0 of $V(G)$ such that (amongst others) the following properties hold:

- almost all edges of G' join $A \cup A_0$ to $B \cup B_0$;
- $|A| = |B|$ is divisible by K;
- every vertex in A has almost all its neighbours in $B \cup B_0$ and every vertex in B has almost all its neighbours in $A \cup A_0$;
- $A_0 \cup B_0$ is small and there are no edges between A_0 and B_0 in G'.

We will call (G', A, A_0, B, B_0) a bi-framework. (The formal definition of a bi-framework is stated before Lemma 4.2.11.) Both A and B will then be split into K clusters of equal size. Our assumption that G is $\varepsilon_{\rm ex}$-bipartite easily implies that there is such a partition A, A_0, B, B_0 which satisfies all these properties apart from the property that there are no edges between A_0 and B_0. So the main part of this section shows that we can cover the collection of all edges between A_0 and B_0 by a small number of edge-disjoint Hamilton cycles.

Since Corollary 4.2.12 will also be used in the proof of Theorem 1.3.8, instead of working with regular graphs we need to consider so-called balanced graphs. We also need to find the above Hamilton cycles in the graph $F \supseteq G$ rather than in G itself (in the proof of Theorem 1.3.5 we will take F to be equal to G).

More precisely, suppose that G is a graph and that A', B' is a partition of $V(G)$, where $A' = A_0 \cup A$, $B' = B_0 \cup B$ and A, A_0, B, B_0 are disjoint. Then we say that G is *D-balanced (with respect to (A, A_0, B, B_0))* if

(B1) $e_G(A') - e_G(B') = (|A'| - |B'|)D/2$;

(B2) all vertices in $A_0 \cup B_0$ have degree exactly D.

Proposition 4.2.1 below implies that whenever A, A_0, B, B_0 is a partition of the vertex set of a D-regular graph H, then H is D-balanced with respect to (A, A_0, B, B_0). Moreover, note that if G is D_G-balanced with respect to (A, A_0, B, B_0) and H is a spanning subgraph of G which is D_H-balanced with respect to (A, A_0, B, B_0), then $G - H$ is $(D_G - D_H)$-balanced with respect to (A, A_0, B, B_0). Furthermore, a graph G is D-balanced with respect to (A, A_0, B, B_0) if and only if G is D-balanced with respect to (B, B_0, A, A_0).

PROPOSITION 4.2.1. *Let H be a graph and let A', B' be a partition of $V(H)$. Suppose that A_0, A is a partition of A' and that B_0, B is a partition of B' such that $|A| = |B|$. Suppose that $d_H(v) = D$ for every $v \in A_0 \cup B_0$ and $d_H(v) = D'$ for every $v \in A \cup B$. Then $e_H(A') - e_H(B') = (|A'| - |B'|)D/2$.*

Proof. Note that

$$\sum_{x \in A'} d_H(x, B') = e_H(A', B') = \sum_{y \in B'} d_H(y, A').$$

Moreover,

$$\begin{aligned} 2e_H(A') &= \sum_{x \in A_0} (D - d_H(x, B')) + \sum_{x \in A} (D' - d_H(x, B')) \\ &= D|A_0| + D'|A| - \sum_{x \in A'} d_H(x, B') \end{aligned}$$

and

$$\begin{aligned} 2e_H(B') &= \sum_{y \in B_0} (D - d_H(y, A')) + \sum_{y \in B} (D' - d_H(y, A')) \\ &= D|B_0| + D'|B| - \sum_{y \in B'} d_H(y, A'). \end{aligned}$$

Therefore

$$2e_H(A') - 2e_H(B') = D(|A_0| - |B_0|) + D'(|A| - |B|) = D(|A_0| - |B_0|) = D(|A'| - |B'|),$$

as desired. □

The following observation states that balancedness is preserved under suitable modifications of the partition.

PROPOSITION 4.2.2. *Let H be D-balanced with respect to (A, A_0, B, B_0). Suppose that A'_0, B'_0 is a partition of $A_0 \cup B_0$. Then H is D-balanced with respect to (A, A'_0, B, B'_0).*

Proof. Observe that the general result follows if we can show that H is D-balanced with respect to (A, A'_0, B, B'_0), where $A'_0 = A_0 \cup \{v\}$, $B'_0 = B_0 \setminus \{v\}$ and $v \in B_0$. (B2) is trivially satisfied in this case, so we only need to check (B1) for the new partition. For this, let $A' := A_0 \cup A$ and $B' := B_0 \cup B$. Now note that (B1) for the original partition implies that

$$\begin{aligned} e_H(A'_0 \cup A) - e_H(B'_0 \cup B) &= e_H(A') + d_H(v, A') - (e_H(B') - d_H(v, B')) \\ &= (|A'| - |B'|)D/2 + D = (|A'_0 \cup A| - |B'_0 \cup B|)D/2. \end{aligned}$$

Thus (B1) holds for the new partition. □

Suppose that G is a graph and A', B' is a partition of $V(G)$. For every vertex $v \in A'$ we call $d_G(v, A')$ the *internal degree of* v in G. Similarly, for every vertex $v \in B'$ we call $d_G(v, B')$ the *internal degree of* v in G.

Given a graph F and a spanning subgraph G of F , we say that (F, G, A, A_0, B, B_0) is an $(\varepsilon, \varepsilon', K, D)$-*weak framework* if the following holds, where $A' := A_0 \cup A$, $B' := B_0 \cup B$ and $n := |G| = |F|$:

(WF1) A, A_0, B, B_0 forms a partition of $V(G) = V(F)$;
(WF2) G is D-balanced with respect to (A, A_0, B, B_0);
(WF3) $e_G(A'), e_G(B') \leq \varepsilon n^2$;
(WF4) $|A| = |B|$ is divisible by K. Moreover, $a + b \leq \varepsilon n$, where $a := |A_0|$ and $b := |B_0|$;
(WF5) all vertices in $A \cup B$ have internal degree at most $\varepsilon' n$ in F;
(WF6) any vertex v has internal degree at most $d_G(v)/2$ in G.

Throughout the chapter, when referring to internal degrees without mentioning the partition, we always mean with respect to the partition A', B', where $A' = A_0 \cup A$ and $B' = B_0 \cup B$. Moreover, a and b will always denote $|A_0|$ and $|B_0|$.

We say that (F, G, A, A_0, B, B_0) is an $(\varepsilon, \varepsilon', K, D)$-*pre-framework* if it satisfies (WF1)–(WF5). The following observation states that pre-frameworks are preserved if we remove suitable balanced subgraphs.

PROPOSITION 4.2.3. *Let $\varepsilon, \varepsilon' > 0$ and $K, D_G, D_H \in \mathbb{N}$. Let (F, G, A, A_0, B, B_0) be an $(\varepsilon, \varepsilon', K, D_G)$-pre framework. Suppose that H is a D_H-regular spanning subgraph of F such that $G \cap H$ is D_H-balanced with respect to (A, A_0, B, B_0). Let $F' := F - H$ and $G' := G - H$. Then (F', G', A, A_0, B, B_0) is an $(\varepsilon, \varepsilon', K, D_G - D_H)$-pre framework.*

Proof. Note that all required properties except possibly (WF2) are not affected by removing edges. But G' satisfies (WF2) since $G \cap H$ is D_H-balanced with respect to (A, A_0, B, B_0). □

LEMMA 4.2.4. *Let $0 < 1/n \ll \varepsilon \ll \varepsilon', 1/K \ll 1$ and let $D \geq n/200$. Suppose that F is a graph on n vertices which is ε-bipartite and that G is a D-regular spanning subgraph of F. Then there is a partition A, A_0, B, B_0 of $V(G) = V(F)$ so that (F, G, A, A_0, B, B_0) is an $(\varepsilon^{1/3}, \varepsilon', K, D)$-weak framework.*

Proof. Let S_1, S_2 be a partition of $V(F)$ which is guaranteed by the assumption that F is ε-bipartite. Let S be the set of all those vertices $x \in S_1$ with $d_F(x, S_1) \geq \sqrt{\varepsilon} n$ together with all those vertices $x \in S_2$ with $d_F(x, S_2) \geq \sqrt{\varepsilon} n$. Since F is ε-bipartite, it follows that $|S| \leq 4\sqrt{\varepsilon} n$.

Given a partition X, Y of $V(F)$, we say that $v \in X$ is *bad for* X, Y if $d_G(v, X) > d_G(v, Y)$ and similarly that $v \in Y$ is *bad for* X, Y if $d_G(v, Y) > d_G(v, X)$. Suppose that there is a vertex $v \in S$ which is bad for S_1, S_2. Then we move v into the class which does not currently contain v to obtain a new partition S'_1, S'_2. We do not change the set S. If there is a vertex $v' \in S$ which is bad for S_1', S_2', then again we move it into the other class.

We repeat this process. After each step, the number of edges in G between the two classes increases, so this process has to terminate with some partition A', B' such that $A' \triangle S_1 \subseteq S$ and $B' \triangle S_2 \subseteq S$. Clearly, no vertex in S is now bad for A', B'. Also, for any $v \in A' \setminus S$ we have

$$\begin{aligned} d_G(v, A') \leq d_F(v, A') \leq d_F(v, S_1) + |S| \leq \sqrt{\varepsilon} n + 4\sqrt{\varepsilon} n < \varepsilon' n \\ < D/2 = d_G(v)/2. \end{aligned} \tag{4.2.1}$$

Similarly, $d_G(v, B') < \varepsilon' n < d_G(v)/2$ for all $v \in B' \setminus S$. Altogether this implies that no vertex is bad for A', B' and thus (WF6) holds. Also note that $e_G(A', B') \geq e_G(S_1, S_2) \geq e(G) - 2\varepsilon n^2$. So

$$e_G(A'), e_G(B') \leq 2\varepsilon n^2. \tag{4.2.2}$$

This implies (WF3).

Without loss of generality we may assume that $|A'| \geq |B'|$. Let A_0' denote the set of all those vertices $v \in A'$ for which $d_F(v, A') \geq \varepsilon' n$. Define $B_0' \subseteq B'$ similarly. We will choose sets $A \subseteq A' \setminus A_0'$ and $A_0 \supseteq A_0'$ and sets $B \subseteq B' \setminus B_0'$ and $B_0 \supseteq B_0'$ such that $|A| = |B|$ is divisible by K and so that A, A_0 and B, B_0 are partitions of A' and B' respectively. We obtain such sets by moving at most $||A' \setminus A_0'| - |B' \setminus B_0'|| + K$ vertices from $A' \setminus A_0'$ to A_0' and at most $||A' \setminus A_0'| - |B' \setminus B_0'|| + K$ vertices from $B' \setminus B_0'$ to B_0'. The choice of A, A_0, B, B_0 is such that (WF1) and (WF5) hold. Further, since $|A| = |B|$, Proposition 4.2.1 implies (WF2).

In order to verify (WF4), it remains to show that $a+b = |A_0 \cup B_0| \leq \varepsilon^{1/3} n$. But (4.2.1) together with its analogue for the vertices in $B' \setminus S$ implies that $A_0' \cup B_0' \subseteq S$. Thus $|A_0'| + |B_0'| \leq |S| \leq 4\sqrt{\varepsilon} n$. Moreover, (WF2), (4.2.2) and our assumption that $D \geq n/200$ together imply that

$$|A'| - |B'| = (e_G(A') - e_G(B'))/(D/2) \leq 2\varepsilon n^2/(D/2) \leq 800\varepsilon n.$$

So altogether, we have

$$\begin{aligned} a + b &\leq |A_0' \cup B_0'| + 2\,||A' \setminus A_0'| - |B' \setminus B_0'|| + 2K \\ &\leq 4\sqrt{\varepsilon} n + 2\,||A'| - |B'| - (|A_0'| - |B_0'|)| + 2K \\ &\leq 4\sqrt{\varepsilon} n + 1600\varepsilon n + 8\sqrt{\varepsilon} n + 2K \leq \varepsilon^{1/3} n. \end{aligned}$$

Thus (WF4) holds. □

Our next goal is to cover the edges of $G[A_0, B_0]$ by edge-disjoint Hamilton cycles. To do this, we will first decompose $G[A_0, B_0]$ into a collection of matchings. We will then extend each such matching into a system of vertex-disjoint paths such that altogether these paths cover every vertex in $G[A_0, B_0]$, each path has its endvertices in $A \cup B$ and the path system is 2-balanced. Since our path system will only contain a small number of nontrivial paths, we can then extend the path system into a Hamilton cycle (see Lemma 4.2.9).

We will call the path systems we are working with A_0B_0-path systems. More precisely, an *A_0B_0-path system (with respect to (A, A_0, B, B_0))* is a path system Q satisfying the following properties:

- Every vertex in $A_0 \cup B_0$ is an internal vertex of a path in Q.

- $A \cup B$ contains the endpoints of each path in Q but no internal vertex of a path in Q.

The following observation (which motivates the use of the word 'balanced') will often be helpful.

PROPOSITION 4.2.5. *Let A_0, A, B_0, B be a partition of a vertex set V. Then an A_0B_0-path system Q with $V(Q) \subseteq V$ is 2-balanced with respect to (A, A_0, B, B_0) if and only if the number of vertices in A which are endpoints of nontrivial paths in Q equals the number of vertices in B which are endpoints of nontrivial paths in Q.*

Proof. Note that by definition any A_0B_0-path system satisfies (B2), so we only need to consider (B1). Let n_A be the number of vertices in A which are endpoints of nontrivial paths in Q and define n_B similarly. Let $a := |A_0|$, $b := |B_0|$, $A' := A \cup A_0$ and $B' := B \cup B_0$. Since $d_Q(v) = 2$ for all $v \in A_0$ and since every vertex in A is either an endpoint of a nontrivial path in Q or has degree zero in Q, we have

$$2e_Q(A') + e_Q(A', B') = \sum_{v \in A'} d_Q(v) = 2a + n_A.$$

So $n_A = 2(e_Q(A') - a) + e_Q(A', B')$, and similarly $n_B = 2(e_Q(B') - b) + e_Q(A', B')$. Therefore, $n_A = n_B$ if and only if $2(e_Q(A') - e_Q(B') - a + b) = 0$ if and only if Q satisfies (B1), as desired. □

The next observation shows that if we have a suitable path system satisfying (B1), we can extend it into a path system which also satisfies (B2).

LEMMA 4.2.6. *Let $0 < 1/n \ll \alpha \ll 1$. Let G be a graph on n vertices such that there is a partition A', B' of $V(G)$ which satisfies the following properties:*

(i) *$A' = A_0 \cup A$, $B' = B_0 \cup B$ and A_0, A, B_0, B are disjoint;*
(ii) *$|A| = |B|$ and $a + b \leq \alpha n$, where $a := |A_0|$ and $b := |B_0|$;*
(iii) *if $v \in A_0$ then $d_G(v, B) \geq 4\alpha n$ and if $v \in B_0$ then $d_G(v, A) \geq 4\alpha n$.*

Let $Q' \subseteq G$ be a path system consisting of at most αn nontrivial paths such that $A \cup B$ contains no internal vertex of a path in Q' and $e_{Q'}(A') - e_{Q'}(B') = a - b$. Then G contains a 2-balanced A_0B_0-path system Q (with respect to (A, A_0, B, B_0)) which extends Q' and consists of at most $2\alpha n$ nontrivial paths. Furthermore, $E(Q) \setminus E(Q')$ consists of A_0B- and AB_0-edges only.

Proof. Since $A \cup B$ contains no internal vertex of a path in Q' and since Q' contains at most αn nontrivial paths, it follows that at most $2\alpha n$ vertices in $A \cup B$ lie on nontrivial paths in Q'. We will now extend Q' into an A_0B_0-path system Q consisting of at most $a + b + \alpha n \leq 2\alpha n$ nontrivial paths as follows:

- for every vertex $v \in A_0$, we join v to $2 - d_{Q'}(v)$ vertices in B;
- for every vertex $v \in B_0$, we join v to $2 - d_{Q'}(v)$ vertices in A.

Condition (iii) and the fact that at most $2\alpha n$ vertices in $A \cup B$ lie on nontrivial paths in Q' together ensure that we can extend Q' in such a way that the endvertices in $A \cup B$ are distinct for different paths in Q. Note that $e_Q(A') - e_Q(B') = e_{Q'}(A') - e_{Q'}(B') = a - b$. Therefore, Q is 2-balanced with respect to (A, A_0, B, B_0). □

The next lemma constructs a small number of 2-balanced A_0B_0-path systems covering the edges of $G[A_0, B_0]$. Each of these path systems will later be extended into a Hamilton cycle.

LEMMA 4.2.7. *Let $0 < 1/n \ll \varepsilon \ll \varepsilon', 1/K \ll \alpha \ll 1$. Let F be a graph on n vertices and let G be a spanning subgraph of F. Suppose that (F, G, A, A_0, B, B_0) is an $(\varepsilon, \varepsilon', K, D)$-weak framework with $\delta(F) \geq (1/4 + \alpha)n$ and $D \geq n/200$. Then for some $r^* \leq \varepsilon n$ the graph G contains r^* edge-disjoint 2-balanced A_0B_0-path systems $Q_1, \dots, Q_{r^*}$ which satisfy the following properties:*

(i) *Together $Q_1, \dots, Q_{r^*}$ cover all edges in $G[A_0, B_0]$;*
(ii) *For each $i \leq r^*$, Q_i contains at most $2\varepsilon n$ nontrivial paths;*
(iii) *For each $i \leq r^*$, Q_i does not contain any edge from $G[A, B]$.*

Proof. (WF4) implies that $|A_0| + |B_0| \leq \varepsilon n$. Thus, by Corollary 1.4.6, there exists a collection $M_1', \dots, M_{r^*}'$ of r^* edge-disjoint matchings in $G[A_0, B_0]$ that together cover all the edges in $G[A_0, B_0]$, where $r^* \leq \varepsilon n$.

We may assume that $a \geq b$ (the case when $b > a$ follows analogously). We will use edges in $G[A']$ to extend each M_i' into a 2-balanced A_0B_0-path system. (WF2) implies that $e_G(A') \geq (a-b)D/2$. Since $d_G(v) = D$ for all $v \in A_0 \cup B_0$ by (WF2), (WF5) and (WF6) imply that $\Delta(G[A']) \leq D/2$. Thus Corollary 1.4.6 implies that $E(G[A'])$ can be decomposed into $\lfloor D/2 \rfloor + 1$ edge-disjoint matchings $M_{A,1}, \dots, M_{A,\lfloor D/2 \rfloor+1}$ such that $||M_{A,i}| - |M_{A,j}|| \leq 1$ for all $i, j \leq \lfloor D/2 \rfloor + 1$.

Notice that at least εn of the matchings $M_{A,i}$ are such that $|M_{A,i}| \geq a - b$. Indeed, otherwise we have that

$$\begin{aligned}(a-b)D/2 \leq e_G(A') &\leq \varepsilon n(a-b) + (a-b-1)(D/2+1-\varepsilon n)\\ &= (a-b)D/2 + a - b - D/2 - 1 + \varepsilon n\\ &< (a-b)D/2 + 2\varepsilon n - D/2 < (a-b)D/2,\end{aligned}$$

a contradiction. (The last inequality follows since $D \geq n/200$.) In particular, this implies that $G[A']$ contains r^* edge-disjoint matchings $M_1'', \dots, M_{r^*}''$ that each consist of precisely $a - b$ edges.

For each $i \leq r^*$, set $M_i := M_i' \cup M_i''$. So for each $i \leq r^*$, M_i is a path system consisting of at most $b + (a-b) = a \leq \varepsilon n$ nontrivial paths such that $A \cup B$ contains no internal vertex of a path in M_i and $e_{M_i}(A') - e_{M_i}(B') = e_{M_i''}(A') = a - b$.

Suppose for some $0 \leq r < r^*$ we have already found a collection $Q_1, \dots, Q_r$ of r edge-disjoint 2-balanced A_0B_0-path systems which satisfy the following properties for each $i \leq r$:

$(\alpha)_i$ Q_i contains at most $2\varepsilon n$ nontrivial paths;
$(\beta)_i$ $M_i \subseteq Q_i$;
$(\gamma)_i$ Q_i and M_j are edge-disjoint for each $j \leq r^*$ such that $i \neq j$;
$(\delta)_i$ Q_i contains no edge from $G[A, B]$.

(Note that $(\alpha)_0$–$(\delta)_0$ are vacuously true.) Let G' denote the spanning subgraph of G obtained from G by deleting the edges lying in $Q_1 \cup \dots \cup Q_r$. (WF2), (WF4) and (WF6) imply that, if $v \in A_0$, $d_{G'}(v, B) \geq D/2 - \varepsilon n - 2r \geq 4\varepsilon n$ and if $v \in B_0$ then $d_{G'}(v, A) \geq 4\varepsilon n$. Thus Lemma 4.2.6 implies that G' contains a 2-balanced A_0B_0-path system Q_{r+1} that satisfies $(\alpha)_{r+1}$–$(\delta)_{r+1}$.

So we can proceed in this way in order to obtain edge-disjoint 2-balanced A_0B_0-path systems $Q_1, \dots, Q_{r^*}$ in G such that $(\alpha)_i$–$(\delta)_i$ hold for each $i \leq r^*$. Note that (i)–(iii) follow immediately from these conditions, as desired. □

The next lemma (Corollary 5.4 in [**20**]) allows us to extend a 2-balanced path system into a Hamilton cycle. Corollary 5.4 concerns so-called '(A, B)-balanced'-path systems rather than 2-balanced A_0B_0-path systems. But the latter satisfies the requirements of the former by Proposition 4.2.5.

LEMMA 4.2.8. *Let $0 < 1/n \ll \varepsilon' \ll \alpha \ll 1$. Let F be a graph and suppose that A_0, A, B_0, B is a partition of $V(F)$ such that $|A| = |B| = n$. Let H be a bipartite subgraph of F with vertex classes A and B such that $\delta(H) \geq (1/2 + \alpha)n$. Suppose that Q is a 2-balanced A_0B_0-path system with respect to (A, A_0, B, B_0) in F which consists of at most $\varepsilon' n$ nontrivial paths. Then F contains a Hamilton cycle C which satisfies the following properties:*

- *$Q \subseteq C$;*
- *$E(C) \setminus E(Q)$ consists of edges from H.*

Now we can apply Lemma 4.2.8 to extend a 2-balanced A_0B_0-path system in a pre-framework into a Hamilton cycle.

LEMMA 4.2.9. *Let $0 < 1/n \ll \varepsilon \ll \varepsilon', 1/K \ll \alpha \ll 1$. Let F be a graph on n vertices and let G be a spanning subgraph of F. Suppose that (F, G, A, A_0, B, B_0) is an $(\varepsilon, \varepsilon', K, D)$-pre-framework, i.e. it satisfies (WF1)–(WF5). Suppose also that $\delta(F) \geq (1/4 + \alpha)n$. Let Q be a 2-balanced A_0B_0-path system with respect to (A, A_0, B, B_0) in G which consists of at most $\varepsilon' n$ nontrivial paths. Then F contains a Hamilton cycle C which satisfies the following properties:*

(i) *$Q \subseteq C$;*
(ii) *$E(C) \setminus E(Q)$ consists of AB-edges;*
(iii) *$C \cap G$ is 2-balanced with respect to (A, A_0, B, B_0).*

Proof. Note that (WF4), (WF5) and our assumption that $\delta(F) \geq (1/4 + \alpha)n$ together imply that every vertex $x \in A$ satisfies

$$d_F(x, B) \geq d_F(x, B') - |B_0| \geq d_F(x) - \varepsilon' n - |B_0| \geq (1/4 + \alpha/2)n \geq (1/2 + \alpha/2)|B|.$$

Similarly, $d_F(x, A) \geq (1/2 + \alpha/2)|A|$ for all $x \in B$. Thus, $\delta(F[A, B]) \geq (1/2 + \alpha/2)|A|$. Applying Lemma 4.2.8 with $F[A, B]$ playing the role of H, we obtain a Hamilton cycle C in F that satisfies (i) and (ii). To verify (iii), note that (ii) and the 2-balancedness of Q together imply that

$$e_{C \cap G}(A') - e_{C \cap G}(B') = e_Q(A') - e_Q(B') = a - b.$$

Since every vertex $v \in A_0 \cup B_0$ satisfies $d_{C \cap G}(v) = d_Q(v) = 2$, (iii) holds. □

We now combine Lemmas 4.2.7 and 4.2.9 to find a collection of edge-disjoint Hamilton cycles covering all the edges in $G[A_0, B_0]$.

LEMMA 4.2.10. *Let $0 < 1/n \ll \varepsilon \ll \varepsilon', 1/K \ll \alpha \ll 1$ and let $D \geq n/100$. Let F be a graph on n vertices and let G be a spanning subgraph of F. Suppose that (F, G, A, A_0, B, B_0) is an $(\varepsilon, \varepsilon', K, D)$-weak framework with $\delta(F) \geq (1/4 + \alpha)n$. Then for some $r^* \leq \varepsilon n$ the graph F contains edge-disjoint Hamilton cycles $C_1, \dots, C_{r^*}$ which satisfy the following properties:*

(i) *Together $C_1, \dots, C_{r^*}$ cover all edges in $G[A_0, B_0]$;*
(ii) *$(C_1 \cup \dots \cup C_{r^*}) \cap G$ is $2r^*$-balanced with respect to (A, A_0, B, B_0).*

Proof. Apply Lemma 4.2.7 to obtain a collection of $r^* \leq \varepsilon n$ edge-disjoint 2-balanced A_0B_0-path systems $Q_1, \dots, Q_{r^*}$ in G which satisfy Lemma 4.2.7(i)–(iii). We will extend each Q_i to a Hamilton cycle C_i.

Suppose that for some $0 \leq r < r^*$ we have found a collection $C_1, \dots, C_r$ of r edge-disjoint Hamilton cycles in F such that the following holds for each $0 \leq i \leq r$:

$(\alpha)_i$ $Q_i \subseteq C_i$;

$(\beta)_i$ $E(C_i) \setminus E(Q_i)$ consists of AB-edges;

$(\gamma)_i$ $G \cap C_i$ is 2-balanced with respect to (A, A_0, B, B_0).

(Note that $(\alpha)_0$–$(\gamma)_0$ are vacuously true.) Let $H_r := C_1 \cup \dots \cup C_r$ (where $H_0 := (V(G), \emptyset)$). So H_r is $2r$-regular. Further, since $G \cap C_i$ is 2-balanced for each $i \leq r$, $G \cap H_r$ is $2r$-balanced. Let $G_r := G - H_r$ and $F_r := F - H_r$. Since (F, G, A, A_0, B, B_0) is an $(\varepsilon, \varepsilon', K, D)$-pre-framework, Proposition 4.2.3 implies that $(F_r, G_r, A, A_0, B, B_0)$ is an $(\varepsilon, \varepsilon', K, D - 2r)$-pre-framework. Moreover, $\delta(F_r) \geq \delta(F) - 2r \geq (1/4 + \alpha/2)n$. Lemma 4.2.7(iii) and $(\beta)_1$–$(\beta)_r$ together imply that Q_{r+1} lies in G_r. Therefore, Lemma 4.2.9 implies that F_r contains a Hamilton cycle C_{r+1} which satisfies $(\alpha)_{r+1}$–$(\gamma)_{r+1}$.

So we can proceed in this way in order to obtain r^* edge-disjoint Hamilton cycles $C_1, \dots, C_{r^*}$ in F such that for each $i \leq r^*$, $(\alpha)_i$–$(\gamma)_i$ hold. Note that this implies that (ii) is satisfied. Further, the choice of $Q_1, \dots, Q_{r^*}$ ensures that (i) holds. □

Given a graph G, we say that (G, A, A_0, B, B_0) is an $(\varepsilon, \varepsilon', K, D)$*-bi-framework* if the following holds, where $A' := A_0 \cup A$, $B' := B_0 \cup B$ and $n := |G|$:

(BFR1) A, A_0, B, B_0 forms a partition of $V(G)$;

(BFR2) G is D-balanced with respect to (A, A_0, B, B_0);

(BFR3) $e_G(A'), e_G(B') \leq \varepsilon n^2$;

(BFR4) $|A| = |B|$ is divisible by K. Moreover, $b \leq a$ and $a + b \leq \varepsilon n$, where $a := |A_0|$ and $b := |B_0|$;

(BFR5) all vertices in $A \cup B$ have internal degree at most $\varepsilon' n$ in G;

(BFR6) $e(G[A_0, B_0]) = 0$;

(BFR7) all vertices $v \in V(G)$ have internal degree at most $d_G(v)/2 + \varepsilon n$ in G.

Note that the main differences to a weak framework are (BFR6) and the fact that a weak framework involves an additional graph F. In particular (BFR1)–(BFR4) imply (WF1)–(WF4). Suppose that $\varepsilon_1 \geq \varepsilon$, $\varepsilon_1' \geq \varepsilon'$ and that K_1 divides K. Then note that every $(\varepsilon, \varepsilon', K, D)$-bi-framework is also an $(\varepsilon_1, \varepsilon_1', K_1, D)$-bi-framework.

LEMMA 4.2.11. *Let $0 < 1/n \ll \varepsilon \ll \varepsilon', 1/K \ll \alpha \ll 1$ and let $D \geq n/100$. Let F be a graph on n vertices and let G be a spanning subgraph of F. Suppose that (F, G, A, A_0, B, B_0) is an $(\varepsilon, \varepsilon', K, D)$-weak framework. Suppose also that $\delta(F) \geq (1/4 + \alpha)n$ and $|A_0| \geq |B_0|$. Then the following properties hold:*

(i) *there is an $(\varepsilon, \varepsilon', K, D_{G'})$-bi-framework (G', A, A_0, B, B_0) such that G' is a spanning subgraph of G with $D_{G'} \geq D - 2\varepsilon n$;*

(ii) *there is a set of $(D - D_{G'})/2 \leq \varepsilon n$ edge-disjoint Hamilton cycles in $F - G'$ containing all edges of $G - G'$. In particular, if D is even then $D_{G'}$ is even.*

Proof. Lemma 4.2.10 implies that there exists some $r^* \leq \varepsilon n$ such that F contains a spanning subgraph H satisfying the following properties:

(a) H is $2r^*$-regular;

(b) H contains all the edges in $G[A_0, B_0]$;
(c) $G \cap H$ is $2r^*$-balanced with respect to (A, A_0, B, B_0);
(d) H has a decomposition into r^* edge-disjoint Hamilton cycles.

Set $G' := G - H$. Then (G', A, A_0, B, B_0) is an $(\varepsilon, \varepsilon', K, D_{G'})$-bi-framework where $D_{G'} := D - 2r^* \geq D - 2\varepsilon n$. Indeed, since (F, G, A, A_0, B, B_0) is an $(\varepsilon, \varepsilon', K, D)$-weak framework, (BFR1) and (BFR3)–(BFR5) follow from (WF1) and (WF3)–(WF5). Further, (BFR2) follows from (WF2) and (c) while (BFR6) follows from (b). (WF6) implies that all vertices $v \in V(G)$ have internal degree at most $d_G(v)/2$ in G. Thus all vertices $v \in V(G')$ have internal degree at most $d_G(v)/2 \leq (d_{G'}(v) + 2r^*)/2 \leq d_{G'}(v)/2 + \varepsilon n$ in G'. So (BFR7) is satisfied. Hence, (i) is satisfied.

Note that by definition of G', H contains all edges of $G - G'$. So since $r^* = (D - D_{G'})/2 \leq \varepsilon n$, (d) implies (ii). □

The following result follows immediately from Lemmas 4.2.4 and 4.2.11.

COROLLARY 4.2.12. *Let* $0 < 1/n \ll \varepsilon \ll \varepsilon^* \ll \varepsilon', 1/K \ll \alpha \ll 1$ *and let* $D \geq n/100$. *Suppose that* F *is an* ε*-bipartite graph on* n *vertices with* $\delta(F) \geq (1/4 + \alpha)n$. *Suppose that* G *is a* D*-regular spanning subgraph of* F. *Then the following properties hold:*

(i) *there is an* $(\varepsilon^*, \varepsilon', K, D_{G'})$*-bi-framework* (G', A, A_0, B, B_0) *such that* G' *is a spanning subgraph of* G, $D_{G'} \geq D - 2\varepsilon^{1/3}n$ *and such that* F *satisfies (WF5) (with respect to the partition* A, A_0, B, B_0*);*
(ii) *there is a set of* $(D - D_{G'})/2 \leq \varepsilon^{1/3}n$ *edge-disjoint Hamilton cycles in* $F - G'$ *containing all edges of* $G - G'$. *In particular, if* D *is even then* $D_{G'}$ *is even.*

4.3. Finding Path Systems which Cover All the Edges within the Classes

The purpose of this section is to prove Corollary 4.3.10 which, given a bi-framework (G, A, A_0, B, B_0), guarantees a set $\mathcal{C}$ of edge-disjoint Hamilton cycles and a set $\mathcal{J}$ of suitable edge-disjoint 2-balanced A_0B_0-path systems such that the graph G^* obtained from G by deleting the edges in all these Hamilton cycles and path systems is bipartite with vertex classes A' and B' and $A_0 \cup B_0$ is isolated in G^*. Each of the path systems in $\mathcal{J}$ will later be extended into a Hamilton cycle by adding suitable edges between A and B. The path systems in $\mathcal{J}$ will need to be 'localized' with respect to a given partition. We prepare the ground for this in the next subsection.

We will call the path systems in $\mathcal{J}$ balanced exceptional systems (see Section 4.3.4 for the definition). These will play a similar role as the exceptional systems in the two cliques case (i.e. in Chapter 2).

Throughout this section, given sets $S, S' \subseteq V(G)$ we often write $E(S)$, $E(S, S')$, $e(S)$ and $e(S, S')$ for $E_G(S)$, $E_G(S, S')$, $e_G(S)$ and $e_G(S, S')$ respectively.

4.3.1. Choosing the Partition and the Localized Slices. Let $K, m \in \mathbb{N}$ and $\varepsilon > 0$. Recall that a (K, m, ε)*-partition* of a set V of vertices is a partition of V into sets $A_0, A_1, \ldots, A_K$ and $B_0, B_1, \ldots, B_K$ such that $|A_i| = |B_i| = m$ for all $1 \leq i \leq K$ and $|A_0 \cup B_0| \leq \varepsilon |V|$. We often write V_0 for $A_0 \cup B_0$ and think of the vertices in V_0 as 'exceptional vertices'. The sets $A_1, \ldots, A_K$ and $B_1, \ldots, B_K$ are called *clusters* of the (K, m, ε_0)-partition and A_0, B_0 are called *exceptional sets.*

Unless stated otherwise, when considering a (K, m, ε)-partition $\mathcal{P}$ we denote the elements of $\mathcal{P}$ by $A_0, A_1, \dots, A_K$ and $B_0, B_1, \dots, B_K$ as above. Further, we will often write A for $A_1 \cup \dots \cup A_K$ and B for $B_1 \cup \dots \cup B_K$.

Suppose that (G, A, A_0, B, B_0) is an $(\varepsilon, \varepsilon', K, D)$-bi-framework with $|G| = n$ and that $\varepsilon_1, \varepsilon_2 > 0$. We say that $\mathcal{P}$ is a $(K, m, \varepsilon, \varepsilon_1, \varepsilon_2)$-*partition for* G if $\mathcal{P}$ satisfies the following properties:

(P1) $\mathcal{P}$ is a (K, m, ε)-partition of $V(G)$ such that the exceptional sets A_0 and B_0 in the partition $\mathcal{P}$ are the same as the sets A_0, B_0 which are part of the bi-framework (G, A, A_0, B, B_0). In particular, $m = |A|/K = |B|/K$;
(P2) $d(v, A_i) = (d(v, A) \pm \varepsilon_1 n)/K$ for all $1 \le i \le K$ and $v \in V(G)$;
(P3) $e(A_i, A_j) = 2(e(A) \pm \varepsilon_2 \max\{n, e(A)\})/K^2$ for all $1 \le i < j \le K$;
(P4) $e(A_i) = (e(A) \pm \varepsilon_2 \max\{n, e(A)\})/K^2$ for all $1 \le i \le K$;
(P5) $e(A_0, A_i) = (e(A_0, A) \pm \varepsilon_2 \max\{n, e(A_0, A)\})/K$ for all $1 \le i \le K$;
(P6) $e(A_i, B_j) = (e(A, B) \pm 3\varepsilon_2 e(A, B))/K^2$ for all $1 \le i, j \le K$;

and the analogous assertions hold if we replace A by B (as well as A_i by B_i etc.) in (P2)–(P5).

Our first aim is to show that for every bi-framework we can find such a partition with suitable parameters.

LEMMA 4.3.1. *Let* $0 < 1/n \ll \varepsilon \ll \varepsilon' \ll \varepsilon_1 \ll \varepsilon_2 \ll 1/K \ll 1$. *Suppose that* (G, A, A_0, B, B_0) *is an* $(\varepsilon, \varepsilon', K, D)$*-bi-framework with* $|G| = n$ *and* $\delta(G) \ge D \ge n/200$. *Suppose that* F *is a graph with* $V(F) = V(G)$. *Then there exists a partition* $\mathcal{P} = \{A_0, A_1, \dots, A_K, B_0, B_1, \dots, B_K\}$ *of* $V(G)$ *so that*

(i) $\mathcal{P}$ *is a* $(K, m, \varepsilon, \varepsilon_1, \varepsilon_2)$*-partition for* G.
(ii) $d_F(v, A_i) = (d_F(v, A) \pm \varepsilon_1 n)/K$ *and* $d_F(v, B_i) = (d_F(v, B) \pm \varepsilon_1 n)/K$ *for all* $1 \le i \le K$ *and* $v \in V(G)$.

Proof. In order to find the required partitions $A_1, \dots, A_K$ of A and $B_1, \dots, B_K$ of B we will apply Lemma 1.4.7 twice, as follows. In the first application we let $U := A$, $R_1 := A_0$, $R_2 := B_0$ and $R_3 := B$. Note that $\Delta(G[U]) \le \varepsilon' n$ by (BFR5) and $d_G(u, R_j) \le |R_j| \le \varepsilon n \le \varepsilon' n$ for all $u \in U$ and $j = 1, 2$ by (BFR4). Moreover, (BFR4) and (BFR7) together imply that $d_G(x, U) \ge D/3 \ge \varepsilon' n$ for each $x \in R_3 = B$. Thus we can apply Lemma 1.4.7 with ε' playing the role of ε to obtain a partition $U_1, \dots, U_K$ of U. We let $A_i := U_i$ for all $i \le K$. Then the A_i satisfy (P2)–(P5) and

$$(4.3.1)\quad e_G(A_i, B) = (e_G(A, B) \pm \varepsilon_2 \max\{n, e_G(A, B)\})/K = (1 \pm \varepsilon_2) e_G(A, B)/K.$$

Further, Lemma 1.4.7(vi) implies that

$$d_F(v, A_i) = (d_F(v, A) \pm \varepsilon_1 n)/K$$

for all $1 \le i \le K$ and $v \in V(G)$.

For the second application of Lemma 1.4.7 we let $U := B$, $R_1 := B_0$, $R_2 := A_0$ and $R_j := A_{j-2}$ for all $3 \le j \le K+2$. As before, $\Delta(G[U]) \le \varepsilon' n$ by (BFR5) and $d_G(u, R_j) \le \varepsilon n \le \varepsilon' n$ for all $u \in U$ and $j = 1, 2$ by (BFR4). Moreover, (BFR4) and (BFR7) together imply that $d_G(x, U) \ge D/3 \ge \varepsilon' n$ for all $3 \le j \le K+2$ and each $x \in R_j = A_{j-2}$. Thus we can apply Lemma 1.4.7 with ε' playing the role of ε to obtain a partition $U_1, \dots, U_K$ of U. Let $B_i := U_i$ for all $i \le K$. Then the B_i satisfy (P2)–(P5) with A replaced by B, A_i replaced by B_i, and so on. Moreover, for all $1 \le i,j \le K$,

$$\begin{aligned} e_G(A_i,B_j) &= (e_G(A_i,B) \pm \varepsilon_2 \max\{n, e_G(A_i,B)\})/K \\ &\overset{(4.3.1)}{=} ((1\pm\varepsilon_2)e_G(A,B) \pm \varepsilon_2(1+\varepsilon_2)e_G(A,B))/K^2 \\ &= (e_G(A,B) \pm 3\varepsilon_2 e_G(A,B))/K^2, \end{aligned}$$

i.e. (P6) holds. Since clearly (P1) holds as well, $A_0, A_1, \dots, A_K$ and $B_0, B_1, \dots, B_K$ together form a $(K, m, \varepsilon, \varepsilon_1, \varepsilon_2)$-partition for G. Further, Lemma 1.4.7(vi) implies that

$$d_F(v, B_i) = (d_F(v,B) \pm \varepsilon_1 n)/K$$

for all $1 \le i \le K$ and $v \in V(G)$. □

The next lemma gives a decomposition of $G[A']$ and $G[B']$ into suitable smaller edge-disjoint subgraphs H^A_{ij} and H^B_{ij}. We say that the graphs H^A_{ij} and H^B_{ij} guaranteed by Lemma 4.3.2 are *localized slices* of G. Note that the order of the indices i and j matters here, i.e. $H^A_{ij} \ne H^A_{ji}$. Also, we allow $i = j$.

LEMMA 4.3.2. *Let $0 < 1/n \ll \varepsilon \ll \varepsilon' \ll \varepsilon_1 \ll \varepsilon_2 \ll 1/K \ll 1$. Suppose that (G, A, A_0, B, B_0) is an $(\varepsilon, \varepsilon', K, D)$-bi-framework with $|G| = n$ and $D \ge n/200$. Let $A_0, A_1, \dots, A_K$ and $B_0, B_1, \dots, B_K$ be a $(K, m, \varepsilon, \varepsilon_1, \varepsilon_2)$-partition for G. Then for all $1 \le i,j \le K$ there are graphs H^A_{ij} and H^B_{ij} with the following properties:*

- (i) *H^A_{ij} is a spanning subgraph of $G[A_0, A_i \cup A_j] \cup G[A_i, A_j] \cup G[A_0]$;*
- (ii) *The sets $E(H^A_{ij})$ over all $1 \le i,j \le K$ form a partition of the edges of $G[A']$;*
- (iii) *$e(H^A_{ij}) = (e(A') \pm 9\varepsilon_2 \max\{n, e(A')\})/K^2$ for all $1 \le i,j \le K$;*
- (iv) *$e_{H^A_{ij}}(A_0, A_i \cup A_j) = (e(A_0, A) \pm 2\varepsilon_2 \max\{n, e(A_0, A)\})/K^2$ for all $1 \le i,j \le K$;*
- (v) *$e_{H^A_{ij}}(A_i, A_j) = (e(A) \pm 2\varepsilon_2 \max\{n, e(A)\})/K^2$ for all $1 \le i,j \le K$;*
- (vi) *For all $1 \le i,j \le K$ and all $v \in A_0$ we have $d_{H^A_{ij}}(v) = d_{H^A_{ij}}(v, A_i \cup A_j) + d_{H^A_{ij}}(v, A_0) = (d(v,A) \pm 4\varepsilon_1 n)/K^2$.*

The analogous assertions hold if we replace A by B, A_i by B_i, and so on.

Proof. In order to construct the graphs H^A_{ij} we perform the following procedure:

- Initially each H^A_{ij} is an empty graph with vertex set $A_0 \cup A_i \cup A_j$.
- For all $1 \le i \le K$ choose a random partition $E(A_0, A_i)$ into K sets U_j of equal size and let $E(H^A_{ij}) := U_j$. (If $E(A_0, A_i)$ is not divisible by K, first distribute up to $K - 1$ edges arbitrarily among the U_j to achieve divisibility.)
- For all $i \le K$, we add all the edges in $E(A_i)$ to H^A_{ii}.
- For all $i, j \le K$ with $i \ne j$, half of the edges in $E(A_i, A_j)$ are added to H^A_{ij} and the other half is added to H^A_{ji} (the choice of the edges is arbitrary).
- The edges in $G[A_0]$ are distributed equally amongst the H^A_{ij}. (So $e_{H^A_{ij}}(A_0) = e(A_0)/K^2 \pm 1$.)

Clearly, the above procedure ensures that properties (i) and (ii) hold. (P5) implies (iv) and (P3) and (P4) imply (v).

Consider any $v \in A_0$. To prove (vi), note that we may assume that $d(v,A) \ge \varepsilon_1 n/K^2$. Let $X := d_{H^A_{ij}}(v, A_i \cup A_j)$. Note that (P2) implies that $\mathbb{E}(X) = (d(v,A) \pm$

$2\varepsilon_1 n)/K^2$ and note that $\mathbb{E}(X) \leq n$. So the Chernoff-Hoeffding bound for the hypergeometric distribution in Proposition 1.4.4 implies that

$$\mathbb{P}(|X - \mathbb{E}(X)| > \varepsilon_1 n/K^2) \leq \mathbb{P}(|X - \mathbb{E}(X)| > \varepsilon_1 \mathbb{E}(X)/K^2) \leq 2e^{-\varepsilon_1^2 \mathbb{E}(X)/3K^4} \leq 1/n^2.$$

Since $d_{H_{ij}^A}(v, A_0) \leq |A_0| \leq \varepsilon_1 n/K^2$, a union bound implies the desired result. Finally, observe that for any $a, b_1, \dots, b_4 > 0$, we have

$$\sum_{i=1}^{4} \max\{a, b_i\} \leq 4 \max\{a, b_1, \dots, b_4\} \leq 4 \max\{a, b_1 + \dots + b_4\}.$$

So (iii) follows from (iv), (v) and the fact that $e_{H_{ij}^A}(A_0) = e(A_0)/K^2 \pm 1$. □

Note that the construction implies that if $i \neq j$, then H_{ij}^A will contain edges between A_0 and A_i but not between A_0 and A_j. However, this additional information is not needed in the subsequent argument.

4.3.2. Decomposing the Localized Slices. Suppose that (G, A, A_0, B, B_0) is an $(\varepsilon, \varepsilon', K, D)$-bi-framework. Recall that $a = |A_0|$, $b = |B_0|$ and $a \geq b$. Since G is D-balanced by (BFR2), we have $e(A') - e(B') = (a - b)D/2$. So there are an integer $q \geq -b$ and a constant $0 \leq c < 1$ such that

$$e(A') = (a + q + c)D/2 \quad \text{and} \quad e(B') = (b + q + c)D/2. \tag{4.3.2}$$

The aim of this subsection is to prove Lemma 4.3.5, which guarantees a decomposition of each localized slice H_{ij}^A into path systems (which will be extended into A_0B_0-path systems in Section 4.3.4) and a sparse (but not too sparse) leftover graph G_{ij}^A.

The following two results will be used in the proof of Lemma 4.3.5.

LEMMA 4.3.3. *Let $0 < 1/n \ll \alpha, \beta, \gamma$ so that $\gamma < 1/2$. Suppose that G is a graph on n vertices such that $\Delta(G) \leq \alpha n$ and $e(G) \geq \beta n$. Then G contains a spanning subgraph H such that $e(H) = \lceil (1 - \gamma) e(G) \rceil$ and $\Delta(G - H) \leq 6\gamma\alpha n/5$.*

Proof. Let H' be a spanning subgraph of G such that

- $\Delta(H') \leq 6\gamma\alpha n/5$;
- $e(H') \geq \gamma e(G)$.

To see that such a graph H' exists, consider a random subgraph of G obtained by including each edge of G with probability $11\gamma/10$. Then $\mathbb{E}(\Delta(H')) \leq 11\gamma\alpha n/10$ and $\mathbb{E}(e(H')) = 11\gamma e(G)/10$. Thus applying Proposition 1.4.4 we have that, with high probability, H' is as desired.

Define H to be a spanning subgraph of G such that $H \supseteq G - H'$ and $e(H) = \lceil (1 - \gamma) e(G) \rceil$. Then $\Delta(G - H) \leq \Delta(H') \leq 6\gamma\alpha n/5$, as required. □

LEMMA 4.3.4. *Suppose that G is a graph such that $\Delta(G) \leq D - 2$ where $D \in \mathbb{N}$ is even. Suppose A_0, A is a partition of $V(G)$ such that $d_G(x) \leq D/2 - 1$ for all $x \in A$ and $\Delta(G[A_0]) \leq D/2 - 1$. Then G has a decomposition into $D/2$ edge-disjoint path systems $P_1, \dots, P_{D/2}$ such that the following conditions hold:*

(i) *For each $i \leq D/2$, any internal vertex on a path in P_i lies in A_0;*
(ii) *$|e(P_i) - e(P_j)| \leq 1$ for all $i, j \leq D/2$.*

Proof. Let G_1 be a maximal spanning subgraph of G under the constraints that $G[A_0] \subseteq G_1$ and $\Delta(G_1) \leq D/2-1$. Note that $G[A_0] \cup G[A] \subseteq G_1$. Set $G_2 := G - G_1$. So G_2 only contains A_0A-edges. Further, since $\Delta(G) \leq D-2$, the maximality of G_1 implies that $\Delta(G_2) \leq D/2-1$.

Define an auxiliary graph G', obtained from G_1 as follows: let $A_0 = \{a_1, \dots, a_m\}$. Add a new vertex set $A'_0 = \{a'_1, \dots, a'_m\}$ to G_1. For each $i \leq m$ and $x \in A$, we add an edge between a'_i and x if and only if $a_i x$ is an edge in G_2.

Thus $G'[A_0 \cup A]$ is isomorphic to G_1 and $G'[A'_0, A]$ is isomorphic to G_2. By construction and since $d_G(x) \leq D/2-1$ for all $x \in A$, we have that $\Delta(G') \leq D/2-1$. Hence, Corollary 1.4.6 implies that $E(G')$ can be decomposed into $D/2$ edge-disjoint matchings $M_1, \dots, M_{D/2}$ such that $||M_i| - |M_j|| \leq 1$ for all $i, j \leq D/2$.

By identifying each vertex $a'_i \in A'_0$ with the corresponding vertex $a_i \in A_0$, $M_1, \dots, M_{D/2}$ correspond to edge-disjoint subgraphs $P_1, \dots, P_{D/2}$ of G such that

- $P_1, \dots, P_{D/2}$ together cover all the edges in G;
- $|e(P_i) - e(P_j)| \leq 1$ for all $i, j \leq D/2$.

Note that $d_{M_i}(x) \leq 1$ for each $x \in V(G')$. Thus $d_{P_i}(x) \leq 1$ for each $x \in A$ and $d_{P_i}(x) \leq 2$ for each $x \in A_0$. This implies that any cycle in P_i must lie in $G[A_0]$. However, M_i is a matching and $G'[A'_0] \cup G'[A_0, A'_0]$ contains no edges. Therefore, P_i contains no cycle, and so P_i is a path system such that any internal vertex on a path in P_i lies in A_0. Hence $P_1, \dots, P_{D/2}$ satisfy (i) and (ii). □

LEMMA 4.3.5. *Let $0 < 1/n \ll \varepsilon \ll \varepsilon' \ll \varepsilon_1 \ll \varepsilon_2 \ll \varepsilon_3 \ll \varepsilon_4 \ll 1/K \ll 1$. Suppose that (G, A, A_0, B, B_0) is an $(\varepsilon, \varepsilon', K, D)$-bi-framework with $|G| = n$ and $D \geq n/200$. Let $A_0, A_1, \dots, A_K$ and $B_0, B_1, \dots, B_K$ be a $(K, m, \varepsilon, \varepsilon_1, \varepsilon_2)$-partition for G. Let H^A_{ij} be a localized slice of G as guaranteed by Lemma 4.3.2. Define c and q as in (4.3.2). Suppose that $t := (1-20\varepsilon_4)D/2K^2 \in \mathbb{N}$. If $e(B') \geq \varepsilon_3 n$, set t^* to be the largest integer which is at most ct and is divisible by K^2. Otherwise, set $t^* := 0$. Define*

$$\ell_a := \begin{cases} 0 & \text{if } e(A') < \varepsilon_3 n; \\ a-b & \text{if } e(A') \geq \varepsilon_3 n \text{ but } e(B') < \varepsilon_3 n; \\ a+q+c & \text{otherwise} \end{cases}$$

and

$$\ell_b := \begin{cases} 0 & \text{if } e(B') < \varepsilon_3 n; \\ b+q+c & \text{otherwise.} \end{cases}$$

Then H^A_{ij} has a decomposition into t edge-disjoint path systems $P_1, \dots, P_t$ and a spanning subgraph G^A_{ij} with the following properties:

(i) *For each $s \leq t$, any internal vertex on a path in P_s lies in A_0;*
(ii) *$e(P_1) = \dots = e(P_{t^*}) = \lceil \ell_a \rceil$ and $e(P_{t^*+1}) = \dots = e(P_t) = \lfloor \ell_a \rfloor$;*
(iii) *$e(P_s) \leq \sqrt{\varepsilon} n$ for every $s \leq t$;*
(iv) *$\Delta(G^A_{ij}) \leq 13\varepsilon_4 D/K^2$.*

The analogous assertion (with ℓ_a replaced by ℓ_b and A_0 replaced by B_0) holds for each localized slice H^B_{ij} of G. Furthermore, $\lceil \ell_a \rceil - \lceil \ell_b \rceil = \lfloor \ell_a \rfloor - \lfloor \ell_b \rfloor = a - b$.

Proof. Note that (4.3.2) and (BFR3) together imply that $\ell_a D/2 \leq (a+q+c)D/2 = e(A') \leq \varepsilon n^2$ and so $\lceil \ell_a \rceil \leq \sqrt{\varepsilon} n$. Thus (iii) will follow from (ii). So it remains to prove (i), (ii) and (iv). We split the proof into three cases.

Case 1. $e(A') < \varepsilon_3 n$

(BFR2) and (BFR4) imply that $e(A') - e(B') = (a-b)D/2 \geq 0$. So $e(B') \leq e(A') < \varepsilon_3 n$. Thus $\ell_a = \ell_b = 0$. Set $G^A_{ij} := H^A_{ij}$ and $G^B_{ij} := H^B_{ij}$. Therefore, (iv) is satisfied as $\Delta(H^A_{ij}) \leq e(A') < \varepsilon_3 n \leq 13\varepsilon_4 D/K^2$. Further, (i) and (ii) are vacuous (i.e. we set each P_s to be the empty graph on $V(G)$).

Note that $a = b$ since otherwise $a > b$ and therefore (BFR2) implies that $e(A') \geq (a-b)D/2 \geq D/2 > \varepsilon_3 n$, a contradiction. Hence, $\lceil \ell_a \rceil - \lceil \ell_b \rceil = \lfloor \ell_a \rfloor - \lfloor \ell_b \rfloor = 0 = a - b$.

Case 2. $e(A') \geq \varepsilon_3 n$ and $e(B') < \varepsilon_3 n$

Since $\ell_b = 0$ in this case, we set $G^B_{ij} := H^B_{ij}$ and each P_s to be the empty graph on $V(G)$. Then as in Case 1, (i), (ii) and (iv) are satisfied with respect to H^B_{ij}. Further, clearly $\lceil \ell_a \rceil - \lceil \ell_b \rceil = \lfloor \ell_a \rfloor - \lfloor \ell_b \rfloor = a - b$.

Note that $a > b$ since otherwise $a = b$ and thus $e(A') = e(B')$ by (BFR2), a contradiction to the case assumptions. Since $e(A') - e(B') = (a-b)D/2$ by (BFR2), Lemma 4.3.2(iii) implies that

$$e(H^A_{ij}) \geq (1 - 9\varepsilon_2)e(A')/K^2 - 9\varepsilon_2 n/K^2 \geq (1 - 9\varepsilon_2)(a-b)D/(2K^2) - 9\varepsilon_2 n/K^2$$
$$\geq (1 - \varepsilon_3)(a-b)D/(2K^2) > (a-b)t. \tag{4.3.3}$$

Similarly, Lemma 4.3.2(iii) implies that

$$e(H^A_{ij}) \leq (1 + \varepsilon_4)(a-b)D/(2K^2). \tag{4.3.4}$$

Therefore, (4.3.3) implies that there exists a constant $\gamma > 0$ such that

$$(1 - \gamma)e(H^A_{ij}) = (a-b)t.$$

Since $(1 - 19\varepsilon_4)(1 - \varepsilon_3) > (1 - 20\varepsilon_4)$, (4.3.3) implies that $\gamma > 19\varepsilon_4 \gg 1/n$. Further, since $(1 + \varepsilon_4)(1 - 21\varepsilon_4) < (1 - 20\varepsilon_4)$, (4.3.4) implies that $\gamma < 21\varepsilon_4$.

Note that (BFR5), (BFR7) and Lemma 4.3.2(vi) imply that

$$\Delta(H^A_{ij}) \leq (D/2 + 5\varepsilon_1 n)/K^2. \tag{4.3.5}$$

Thus Lemma 4.3.3 implies that H^A_{ij} contains a spanning subgraph H such that $e(H) = (1-\gamma)e(H^A_{ij}) = (a-b)t$ and

$$\Delta(H^A_{ij} - H) \leq 6\gamma(D/2 + 5\varepsilon_1 n)/(5K^2) \leq 13\varepsilon_4 D/K^2,$$

where the last inequality follows since $\gamma < 21\varepsilon_4$ and $\varepsilon_1 \ll 1$. Setting $G^A_{ij} := H^A_{ij} - H$ implies that (iv) is satisfied.

Our next task is to decompose H into t edge-disjoint path systems so that (i) and (ii) are satisfied. Note that (4.3.5) implies that

$$\Delta(H) \leq \Delta(H^A_{ij}) \leq (D/2 + 5\varepsilon_1 n)/K^2 < 2t - 2.$$

Further, (BFR4) implies that $\Delta(H[A_0]) \leq |A_0| \leq \varepsilon n < t - 1$ and (BFR5) implies that $d_H(x) \leq \varepsilon' n < t - 1$ for all $x \in A$. Since $e(H) = (a-b)t$, Lemma 4.3.4 implies that H has a decomposition into t edge-disjoint path systems $P_1, \dots, P_t$ satisfying (i) and so that $e(P_s) = a - b = \ell_a$ for all $s \leq t$. In particular, (ii) is satisfied.

Case 3. $e(A'), e(B') \geq \varepsilon_3 n$

By definition of ℓ_a and ℓ_b, we have that $\lceil \ell_a \rceil - \lceil \ell_b \rceil = \lfloor \ell_a \rfloor - \lfloor \ell_b \rfloor = a - b$. Notice that since $e(A') \geq \varepsilon_3 n$ and $\varepsilon_2 \ll \varepsilon_3$, certainly $\varepsilon_3 e(A')/(2K^2) > 9\varepsilon_2 n/K^2$.

Therefore, Lemma 4.3.2(iii) implies that

$$\begin{aligned} e(H^A_{ij}) &\geq (1-9\varepsilon_2)e(A')/K^2 - 9\varepsilon_2 n/K^2 \\ &\geq (1-\varepsilon_3)e(A')/K^2 \\ &\geq \varepsilon_3 n/(2K^2). \end{aligned} \tag{4.3.6}$$

Note that $1/n \ll \varepsilon_3/(2K^2)$. Further, (4.3.2) and (4.3.6) imply that

$$\begin{aligned} e(H^A_{ij}) &\geq (1-\varepsilon_3)e(A')/K^2 \\ &= (1-\varepsilon_3)(a+q+c)D/(2K^2) > (a+q)t + t^*. \end{aligned} \tag{4.3.7}$$

Similarly, Lemma 4.3.2(iii) implies that

$$e(H^A_{ij}) \leq (1+\varepsilon_3)(a+q+c)D/(2K^2). \tag{4.3.8}$$

By (4.3.7) there exists a constant $\gamma > 0$ such that

$$(1-\gamma)e(H^A_{ij}) = (a+q)t + t^*.$$

Note that (4.3.7) implies that $1/n \ll 19\varepsilon_4 < \gamma$ and (4.3.8) implies that $\gamma < 21\varepsilon_4$. Moreover, as in Case 2, (BFR5), (BFR7) and Lemma 4.3.2(vi) together show that

$$\Delta(H^A_{ij}) \leq (D/2 + 5\varepsilon_1 n)/K^2. \tag{4.3.9}$$

Thus (as in Case 2 again), Lemma 4.3.3 implies that H^A_{ij} contains a spanning subgraph H such that $e(H) = (1-\gamma)e(H^A_{ij}) = (a+q)t + t^*$ and

$$\Delta(H^A_{ij} - H) \leq 6\gamma(D/2 + 5\varepsilon_1 n)/(5K^2) \leq 13\varepsilon_4 D/K^2.$$

Setting $G^A_{ij} := H^A_{ij} - H$ implies that (iv) is satisfied. Next we decompose H into t edge-disjoint path systems so that (i) and (ii) are satisfied. Note that (4.3.9) implies that

$$\Delta(H) \leq \Delta(H^A_{ij}) \leq (D/2 + 5\varepsilon_1 n)/K^2 < 2t - 2.$$

Further, (BFR4) implies that $\Delta(H[A_0]) \leq |A_0| \leq \varepsilon n < t-1$ and (BFR5) implies that $d_H(x) \leq \varepsilon' n < t-1$ for all $x \in A$. Since $e(H) = (a+q)t + t^*$, Lemma 4.3.4 implies that H has a decomposition into t edge-disjoint path systems $P_1, \dots, P_t$ satisfying (i) and (ii). An identical argument implies that (i), (ii) and (iv) are satisfied with respect to H^B_{ij} also. □

4.3.3. Decomposing the Global Graph. Let G^A_{glob} be the union of the graphs G^A_{ij} guaranteed by Lemma 4.3.5 over all $1 \leq i, j \leq K$. Define G^B_{glob} similarly. The next lemma gives a decomposition of both G^A_{glob} and G^B_{glob} into suitable path systems. Properties (iii) and (iv) of the lemma guarantee that one can pair up each such path system $Q_A \subseteq G^A_{glob}$ with a different path system $Q_B \subseteq G^B_{glob}$ such that $Q_A \cup Q_B$ is 2-balanced (in particular $e(Q_A) - e(Q_B) = a - b$). This property will then enable us to apply Lemma 4.2.9 to extend $Q_A \cup Q_B$ into a Hamilton cycle using only edges between A' and B'.

LEMMA 4.3.6. *Let $0 < 1/n \ll \varepsilon \ll \varepsilon' \ll \varepsilon_1 \ll \varepsilon_2 \ll \varepsilon_3 \ll \varepsilon_4 \ll 1/K \ll 1$. Suppose that (G, A, A_0, B, B_0) is an $(\varepsilon, \varepsilon', K, D)$-bi-framework with $|G| = n$ and such that $D \geq n/200$ and D is even. Let $A_0, A_1, \dots, A_K$ and $B_0, B_1, \dots, B_K$ be a $(K, m, \varepsilon, \varepsilon_1, \varepsilon_2)$-partition for G. Let G^A_{glob} be the union of the graphs G^A_{ij} guaranteed by Lemma 4.3.5 over all $1 \leq i, j \leq K$. Define G^B_{glob} similarly. Suppose that $k := 10\varepsilon_4 D \in \mathbb{N}$. Then the following properties hold:*

(i) *There is an integer* q' *and a real number* $0 \leq c' < 1$ *so that* $e(G^A_{glob}) = (a + q' + c')k$ *and* $e(G^B_{glob}) = (b + q' + c')k$.

(ii) $\Delta(G^A_{glob}), \Delta(G^B_{glob}) < 3k/2$.

(iii) *Let* $k^* := c'k$. *Then* G^A_{glob} *has a decomposition into* k^* *path systems, each containing* $a+q'+1$ *edges, and* $k-k^*$ *path systems, each containing* $a+q'$ *edges. Moreover, each of these* k *path systems* Q *satisfies* $d_Q(x) \leq 1$ *for all* $x \in A$.

(iv) G^B_{glob} *has a decomposition into* k^* *path systems, each containing* $b+q'+1$ *edges, and* $k - k^*$ *path systems, each containing* $b + q'$ *edges. Moreover, each of these* k *path systems* Q *satisfies* $d_Q(x) \leq 1$ *for all* $x \in B$.

(v) *Each of the path systems guaranteed in (iii) and (iv) contains at most* $\sqrt{\varepsilon}n$ *edges.*

Note that in Lemma 4.3.6 and several later statements the parameter ε_3 is implicitly defined by the application of Lemma 4.3.5 which constructs the graphs G^A_{glob} and G^B_{glob}.

Proof. Let t^* and t be as defined in Lemma 4.3.5. Our first task is to show that (i) is satisfied. If $e(A'), e(B') < \varepsilon_3 n$ then $G^A_{glob} = G[A']$ and $G^B_{glob} = G[B']$. Further, $a = b$ in this case since otherwise (BFR4) implies that $a > b$ and so (BFR2) yields that $e(A') \geq (a - b)D/2 \geq D/2 > \varepsilon_3 n$, a contradiction. Therefore, (BFR2) implies that

$$e(G^A_{glob}) - e(G^B_{glob}) = e(A') - e(B') = (a - b)D/2 = 0 = (a - b)k.$$

If $e(A') \geq \varepsilon_3 n$ and $e(B') < \varepsilon_3 n$ then $G^B_{glob} = G[B']$. Further, G^A_{glob} is obtained from $G[A']$ by removing tK^2 edge-disjoint path systems, each of which contains precisely $a - b$ edges. Thus (BFR2) implies that

$$e(G^A_{glob}) - e(G^B_{glob}) = e(A') - e(B') - tK^2(a - b) = (a - b)(D/2 - tK^2) = (a - b)k.$$

Finally, consider the case when $e(A'), e(B') > \varepsilon_3 n$. Then G^A_{glob} is obtained from $G[A']$ by removing $t^* K^2$ edge-disjoint path systems, each of which contain exactly $a+q+1$ edges, and by removing $(t-t^*)K^2$ edge-disjoint path systems, each of which contain exactly $a + q$ edges. Similarly, G^B_{glob} is obtained from $G[B']$ by removing $t^* K^2$ edge-disjoint path systems, each of which contain exactly $b+q+1$ edges, and by removing $(t - t^*)K^2$ edge-disjoint path systems, each of which contain exactly $b + q$ edges. So (BFR2) implies that

$$e(G^A_{glob}) - e(G^B_{glob}) = e(A') - e(B') - (a - b)tK^2 = (a - b)k.$$

Therefore, in every case,

$$e(G^A_{glob}) - e(G^B_{glob}) = (a - b)k. \tag{4.3.10}$$

Define the integer q' and $0 \leq c' < 1$ by $e(G^A_{glob}) = (a + q' + c')k$. Then (4.3.10) implies that $e(G^B_{glob}) = (b + q' + c')k$. This proves (i). To prove (ii), note that Lemma 4.3.5(iv) implies that $\Delta(G^A_{glob}) \leq 13\varepsilon_4 D < 3k/2$ and similarly $\Delta(G^B_{glob}) < 3k/2$.

Note that (BFR5) implies that $d_{G^A_{glob}}(x) \leq \varepsilon' n < k - 1$ for all $x \in A$ and $\Delta(G^A_{glob}[A_0]) \leq |A_0| \leq \varepsilon n < k - 1$. Thus Lemma 4.3.4 together with (i) implies that (iii) is satisfied. (iv) follows from Lemma 4.3.4 analogously.

(BFR3) implies that $e(G^A_{glob}) \le e_G(A') \le \varepsilon n^2$ and $e(G^B_{glob}) \le e_G(B') \le \varepsilon n^2$. Therefore, each path system from (iii) and (iv) contains at most $\lceil \varepsilon n^2/k \rceil \le \sqrt{\varepsilon} n$ edges. So (v) is satisfied. □

We say that a path system $P \subseteq G[A']$ is (i,j,A)-*localized* if

(i) $E(P) \subseteq E(G[A_0, A_i \cup A_j]) \cup E(G[A_i, A_j]) \cup E(G[A_0])$;
(ii) Any internal vertex on a path in P lies in A_0.

We introduce an analogous notion of (i,j,B)-*localized* for path systems $P \subseteq G[B']$.

The following result is a straightforward consequence of Lemmas 4.3.2, 4.3.5 and 4.3.6. It gives a decomposition of $G[A'] \cup G[B']$ into pairs of paths systems so that most of these are localized and so that each pair can be extended into a Hamilton cycle by adding $A'B'$-edges.

COROLLARY 4.3.7. *Let* $0 < 1/n \ll \varepsilon \ll \varepsilon' \ll \varepsilon_1 \ll \varepsilon_2 \ll \varepsilon_3 \ll \varepsilon_4 \ll 1/K \ll 1$. *Suppose that* (G, A, A_0, B, B_0) *is an* $(\varepsilon, \varepsilon', K, D)$*-bi-framework with* $|G| = n$ *and such that* $D \ge n/200$ *and* D *is even. Let* $A_0, A_1, \dots, A_K$ *and* $B_0, B_1, \dots, B_K$ *be a* $(K, m, \varepsilon, \varepsilon_1, \varepsilon_2)$*-partition for* G. *Let* $t_K := (1-20\varepsilon_4)D/2K^4$ *and* $k := 10\varepsilon_4 D$. *Suppose that* $t_K \in \mathbb{N}$. *Then there are* K^4 *sets* $\mathcal{M}_{i_1 i_2 i_3 i_4}$, *one for each* $1 \le i_1, i_2, i_3, i_4 \le K$, *such that each* $\mathcal{M}_{i_1 i_2 i_3 i_4}$ *consists of* t_K *pairs of path systems and satisfies the following properties:*

(a) *Let* (P, P') *be a pair of path systems which forms an element of* $\mathcal{M}_{i_1 i_2 i_3 i_4}$. *Then*
 (i) P *is an* (i_1, i_2, A)*-localized path system and* P' *is an* (i_3, i_4, B)*-localized path system;*
 (ii) $e(P) - e(P') = a - b$;
 (iii) $e(P), e(P') \le \sqrt{\varepsilon} n$.
(b) *The* $2t_K$ *path systems in the pairs belonging to* $\mathcal{M}_{i_1 i_2 i_3 i_4}$ *are all pairwise edge-disjoint.*
(c) *Let* $G(\mathcal{M}_{i_1 i_2 i_3 i_4})$ *denote the spanning subgraph of* G *whose edge set is the union of all the path systems in the pairs belonging to* $\mathcal{M}_{i_1 i_2 i_3 i_4}$. *Then the* K^4 *graphs* $G(\mathcal{M}_{i_1 i_2 i_3 i_4})$ *are edge-disjoint. Further, each* $x \in A_0$ *satisfies* $d_{G(\mathcal{M}_{i_1 i_2 i_3 i_4})}(x) \ge (d_G(x, A) - 15\varepsilon_4 D)/K^4$ *while each* $y \in B_0$ *satisfies* $d_{G(\mathcal{M}_{i_1 i_2 i_3 i_4})}(y) \ge (d_G(y, B) - 15\varepsilon_4 D)/K^4$.
(d) *Let* G_{glob} *be the subgraph of* $G[A'] \cup G[B']$ *obtained by removing all edges contained in* $G(\mathcal{M}_{i_1 i_2 i_3 i_4})$ *for all* $1 \le i_1, i_2, i_3, i_4 \le K$. *Then* $\Delta(G_{glob}) \le 3k/2$. *Moreover,* G_{glob} *has a decomposition into* k *pairs of path systems* $(Q_{1,A}, Q_{1,B}), \dots, (Q_{k,A}, Q_{k,B})$ *so that*
 (i′) $Q_{i,A} \subseteq G_{glob}[A']$ *and* $Q_{i,B} \subseteq G_{glob}[B']$ *for all* $i \le k$;
 (ii′) $d_{Q_{i,A}}(x) \le 1$ *for all* $x \in A$ *and* $d_{Q_{i,B}}(x) \le 1$ *for all* $x \in B$;
 (iii′) $e(Q_{i,A}) - e(Q_{i,B}) = a - b$ *for all* $i \le k$;
 (iv′) $e(Q_{i,A}), e(Q_{i,B}) \le \sqrt{\varepsilon} n$ *for all* $i \le k$.

Proof. Apply Lemma 4.3.2 to obtain localized slices H^A_{ij} and H^B_{ij} (for all $i, j \le K$). Let $t := K^2 t_K$ and let t^* be as defined in Lemma 4.3.5. Since $t/K^2, t^*/K^2 \in \mathbb{N}$ we have $(t - t^*)/K^2 \in \mathbb{N}$. For all $i_1, i_2 \le K$, let $\mathcal{M}^A_{i_1 i_2}$ be the set of t path systems in $H^A_{i_1 i_2}$ guaranteed by Lemma 4.3.5. We call the t^* path systems in $\mathcal{M}^A_{i_1 i_2}$ of size $\lceil \ell_a \rceil$ *large* and the others *small*. We define $\mathcal{M}^B_{i_3 i_4}$ as well as large and small path systems in $\mathcal{M}^B_{i_3 i_4}$ analogously (for all $i_3, i_4 \le K$).

We now construct the sets $\mathcal{M}_{i_1i_2i_3i_4}$ as follows: For all $i_1, i_2 \le K$, consider a random partition of the set of all large path systems in $\mathcal{M}^A_{i_1i_2}$ into K^2 sets of equal size t^*/K^2 and assign (all the path systems in) each of these sets to one of the $\mathcal{M}_{i_1i_2i_3i_4}$ with $i_3, i_4 \le K$. Similarly, randomly partition the set of small path systems in $\mathcal{M}^A_{i_1i_2}$ into K^2 sets, each containing $(t - t^*)/K^2$ path systems. Assign each of these K^2 sets to one of the $\mathcal{M}_{i_1i_2i_3i_4}$ with $i_3, i_4 \le K$. Proceed similarly for each $\mathcal{M}^B_{i_3i_4}$ in order to assign each of its path systems randomly to some $\mathcal{M}_{i_1i_2i_3i_4}$. Then to each $\mathcal{M}_{i_1i_2i_3i_4}$ we have assigned exactly t^*/K^2 large path systems from both $\mathcal{M}^A_{i_1i_2}$ and $\mathcal{M}^B_{i_3i_4}$. Pair these off arbitrarily. Similarly, pair off the small path systems assigned to $\mathcal{M}_{i_1i_2i_3i_4}$ arbitrarily. Clearly, the sets $\mathcal{M}_{i_1i_2i_3i_4}$ obtained in this way satisfy (a) and (b).

We now verify (c). By construction, the K^4 graphs $G(\mathcal{M}_{i_1i_2i_3i_4})$ are edge-disjoint. So consider any vertex $x \in A_0$ and write $d := d_G(x, A)$. Note that $d_{H^A_{i_1i_2}}(x) \ge (d - 4\varepsilon_1 n)/K^2$ by Lemma 4.3.2(vi). Let $G(\mathcal{M}^A_{i_1i_2})$ be the spanning subgraph of G whose edge set is the union of all the path systems in $\mathcal{M}^A_{i_1i_2}$. Then Lemma 4.3.5(iv) implies that

$$d_{G(\mathcal{M}^A_{i_1i_2})}(x) \ge d_{H^A_{i_1i_2}}(x) - \Delta(G^A_{i_1i_2}) \ge \frac{d - 4\varepsilon_1 n}{K^2} - \frac{13\varepsilon_4 D}{K^2} \ge \frac{d - 14\varepsilon_4 D}{K^2}.$$

So a Chernoff-Hoeffding estimate for the hypergeometric distribution (Proposition 1.4.4) implies that

$$d_{G(\mathcal{M}_{i_1i_2i_3i_4})}(x) \ge \frac{1}{K^2}\left(\frac{d - 14\varepsilon_4 D}{K^2}\right) - \varepsilon n \ge \frac{d - 15\varepsilon_4 D}{K^4}.$$

(Note that we only need to apply the Chernoff-Hoeffding bound if $d \ge \varepsilon n$ say, as (c) is vacuous otherwise.)

It remains to check condition (d). First note that $k \in \mathbb{N}$ since $t_K, D/2 \in \mathbb{N}$. Thus we can apply Lemma 4.3.6 to obtain a decomposition of both G^A_{glob} and G^B_{glob} into path systems. Since $G_{glob} = G^A_{glob} \cup G^B_{glob}$, (d) is an immediate consequence of Lemma 4.3.6(ii)–(v). □

4.3.4. Constructing Localized Balanced Exceptional Systems. The localized path systems obtained from Corollary 4.3.7 do not yet cover all of the exceptional vertices. This is achieved via the following lemma: we extend the path systems to achieve this additional property, while maintaining the property of being balanced. More precisely, let

$$\mathcal{P} := \{A_0, A_1, \dots, A_K, B_0, B_1, \dots, B_K\}$$

be a (K, m, ε)-partition of a set V of n vertices. Given $1 \le i_1, i_2, i_3, i_4 \le K$ and $\varepsilon_0 > 0$, an *(i_1, i_2, i_3, i_4)-balanced exceptional system with respect to $\mathcal{P}$ and parameter* ε_0 is a path system J with $V(J) \subseteq A_0 \cup B_0 \cup A_{i_1} \cup A_{i_2} \cup B_{i_3} \cup B_{i_4}$ such that the following conditions hold:

(BES1) Every vertex in $A_0 \cup B_0$ is an internal vertex of a path in J. Every vertex $v \in A_{i_1} \cup A_{i_2} \cup B_{i_3} \cup B_{i_4}$ satisfies $d_J(v) \le 1$.
(BES2) Every edge of $J[A \cup B]$ is either an $A_{i_1}A_{i_2}$-edge or a $B_{i_3}B_{i_4}$-edge.
(BES3) The edges in J cover precisely the same number of vertices in A as in B.
(BES4) $e(J) \le \varepsilon_0 n$.

To shorten the notation, we will often refer to J as an (i_1, i_2, i_3, i_4)-*BES*. If V is the vertex set of a graph G and $J \subseteq G$, we also say that J is an (i_1, i_2, i_3, i_4)-*BES in* G. Note that (BES2) implies that an (i_1, i_2, i_3, i_4)-BES does not contain edges between A and B. Furthermore, an (i_1, i_2, i_3, i_4)-BES is also, for example, an (i_2, i_1, i_4, i_3)-BES. We will sometimes omit the indices i_1, i_2, i_3, i_4 and just refer to a balanced exceptional system (or a BES for short). We will sometimes also omit the partition $\mathcal{P}$, if it is clear from the context. As mentioned before, balanced exceptional systems will play a similar role as the exceptional systems that we used in the two cliques case (i.e. in Chapter 2).

(BES1) implies that each balanced exceptional system is an A_0B_0-path system as defined before Proposition 4.2.5. (However, the converse is not true since, for example, a 2-balanced A_0B_0-path system need not satisfy (BES4).) So (BES3) and Proposition 4.2.5 imply that each balanced exceptional system is also 2-balanced.

We now extend each set $\mathcal{M}_{i_1i_2i_3i_4}$ obtained from Corollary 4.3.7 into a set $\mathcal{J}_{i_1i_2i_3i_4}$ of (i_1, i_2, i_3, i_4)-BES.

LEMMA 4.3.8. *Let* $0 < 1/n \ll \varepsilon \ll \varepsilon_0 \ll \varepsilon' \ll \varepsilon_1 \ll \varepsilon_2 \ll \varepsilon_3 \ll \varepsilon_4 \ll 1/K \ll 1$. *Suppose that* (G, A, A_0, B, B_0) *is an* $(\varepsilon, \varepsilon', K, D)$*-bi-framework with* $|G| = n$ *and such that* $D \geq n/200$ *and* D *is even. Let* $\mathcal{P} := \{A_0, A_1, \dots, A_K, B_0, B_1, \dots, B_K\}$ *be a* $(K, m, \varepsilon, \varepsilon_1, \varepsilon_2)$*-partition for* G. *Suppose that* $t_K := (1 - 20\varepsilon_4)D/2K^4 \in \mathbb{N}$. *Let* $\mathcal{M}_{i_1i_2i_3i_4}$ *be the sets returned by Corollary* 4.3.7. *Then for all* $1 \leq i_1, i_2, i_3, i_4 \leq K$ *there is a set* $\mathcal{J}_{i_1i_2i_3i_4}$ *which satisfies the following properties:*

- (i) $\mathcal{J}_{i_1i_2i_3i_4}$ *consists of* t_K *edge-disjoint* (i_1, i_2, i_3, i_4)*-BES in* G *with respect to* $\mathcal{P}$ *and with parameter* ε_0.
- (ii) *For each of the* t_K *pairs of path systems* $(P, P') \in \mathcal{M}_{i_1i_2i_3i_4}$, *there is a unique* $J \in \mathcal{J}_{i_1i_2i_3i_4}$ *which contains all the edges in* $P \cup P'$. *Moreover, all edges in* $E(J) \setminus E(P \cup P')$ *lie in* $G[A_0, B_{i_3}] \cup G[B_0, A_{i_1}]$.
- (iii) *Whenever* $(i_1, i_2, i_3, i_4) \neq (i'_1, i'_2, i'_3, i'_4)$, $J \in \mathcal{J}_{i_1i_2i_3i_4}$ *and* $J' \in \mathcal{J}_{i'_1i'_2i'_3i'_4}$, *then* J *and* J' *are edge-disjoint.*

We let $\mathfrak{J}$ denote the union of the sets $\mathcal{J}_{i_1i_2i_3i_4}$ over all $1 \leq i_1, i_2, i_3, i_4 \leq K$.

Proof. We will construct the sets $\mathcal{J}_{i_1i_2i_3i_4}$ greedily by extending each pair of path systems $(P, P') \in \mathcal{M}_{i_1i_2i_3i_4}$ in turn into an (i_1, i_2, i_3, i_4)-BES containing $P \cup P'$. For this, consider some arbitrary ordering of the K^4 4-tuples (i_1, i_2, i_3, i_4). Suppose that we have already constructed the sets $\mathcal{J}_{i'_1i'_2i'_3i'_4}$ for all (i'_1, i'_2, i'_3, i'_4) preceding (i_1, i_2, i_3, i_4) so that (i)–(iii) are satisfied. So our aim now is to construct $\mathcal{J}_{i_1i_2i_3i_4}$. Consider an enumeration $(P_1, P'_1), \dots, (P_{t_K}, P'_{t_K})$ of the pairs of path systems in $\mathcal{M}_{i_1i_2i_3i_4}$. Suppose that for some $i \leq t_K$ we have already constructed edge-disjoint (i_1, i_2, i_3, i_4)-BES $J_1, \dots, J_{i-1}$, so that for each $i' < i$ the following conditions hold:

- $J_{i'}$ contains the edges in $P_{i'} \cup P'_{i'}$;
- all edges in $E(J_{i'}) \setminus E(P_{i'} \cup P'_{i'})$ lie in $G[A_0, B_{i_3}] \cup G[B_0, A_{i_1}]$;
- $J_{i'}$ is edge-disjoint from all the balanced exceptional systems in $\bigcup_{(i'_1,i'_2,i'_3,i'_4)} \mathcal{J}_{i'_1i'_2i'_3i'_4}$, where the union is over all (i'_1, i'_2, i'_3, i'_4) preceding (i_1, i_2, i_3, i_4).

We will now construct $J := J_i$. For this, we need to add suitable edges to $P_i \cup P'_i$ to ensure that all vertices of $A_0 \cup B_0$ have degree two. We start with A_0. Recall that $a = |A_0|$ and write $A_0 = \{x_1, \dots, x_a\}$. Let G' denote the subgraph of $G[A', B']$ obtained by removing all the edges lying in $J_1, \dots, J_{i-1}$ as well as all those edges lying in the balanced exceptional systems belonging to $\bigcup_{(i'_1,i'_2,i'_3,i'_4)} \mathcal{J}_{i'_1i'_2i'_3i'_4}$ (where as before the union is over all (i'_1, i'_2, i'_3, i'_4) preceding (i_1, i_2, i_3, i_4)). We will choose the new edges incident to A_0 in J inside $G'[A_0, B_{i_3}]$.

Suppose we have already found suitable edges for $x_1, \dots, x_{j-1}$ and let $J(j)$ be the set of all these edges. We will first show that the degree of x_j inside $G'[A_0, B_{i_3}]$ is still large. Let $d_j := d_G(x_j, A')$. Consider any (i'_1, i'_2, i'_3, i'_4) preceding (i_1, i_2, i_3, i_4). Let $G(\mathcal{J}_{i'_1i'_2i'_3i'_4})$ denote the union of the t_K balanced exceptional systems belonging to $\mathcal{J}_{i'_1i'_2i'_3i'_4}$. Thus $d_{G(\mathcal{J}_{i'_1i'_2i'_3i'_4})}(x_j) = 2t_K$. However, Corollary 4.3.7(c) implies that $d_{G(\mathcal{M}_{i'_1i'_2i'_3i'_4})}(x_j) \geq (d_j - 15\varepsilon_4 D)/K^4$. So altogether, when constructing (the balanced exceptional systems in) $\mathcal{J}_{i'_1i'_2i'_3i'_4}$, we have added at most $2t_K - (d_j - 15\varepsilon_4 D)/K^4$ new edges at x_j, and all these edges join x_j to vertices in $B_{i'_3}$. Similarly, when constructing $J_1, \dots, J_{i-1}$, we have added at most $2t_K - (d_j - 15\varepsilon_4 D)/K^4$ new edges at x_j. Since the number of 4-tuples (i'_1, i'_2, i'_3, i'_4) with $i'_3 = i_3$ is K^3, it follows that

$$\begin{aligned} d_G(x_j, B_{i_3}) - d_{G'}(x_j, B_{i_3}) &\leq K^3 \left(2t_K - \frac{d_j - 15\varepsilon_4 D}{K^4} \right) \\ &= \frac{1}{K} \left((1 - 20\varepsilon_4)D - d_j + 15\varepsilon_4 D \right) \\ &= \frac{1}{K} \left(D - d_j - 5\varepsilon_4 D \right). \end{aligned}$$

Also, (P2) with A replaced by B implies that

$$d_G(x_j, B_{i_3}) \geq \frac{d_G(x_j, B) - \varepsilon_1 n}{K} \geq \frac{d_G(x_j) - d_G(x_j, A') - \varepsilon_1 n}{K} = \frac{D - d_j - \varepsilon_1 n}{K},$$

where here we use (BFR2) and (BFR6). So altogether, we have

$$d_{G'}(x_j, B_{i_3}) \geq (5\varepsilon_4 D - \varepsilon_1 n)/K \geq \varepsilon_4 n / 50K.$$

Let B'_{i_3} be the set of vertices in B_{i_3} not covered by the edges of $J(j) \cup P'_i$. Note that $|B'_{i_3}| \geq |B_{i_3}| - 2|A_0| - 2e(P'_i) \geq |B_{i_3}| - 3\sqrt{\varepsilon} n$ since $a = |A_0| \leq \varepsilon n$ by (BFR4) and $e(P'_i) \leq \sqrt{\varepsilon} n$ by Corollary 4.3.7(a)(iii). So $d_{G'}(x_j, B'_{i_3}) \geq \varepsilon_4 n / 51K$. We can add up to two of these edges to J in order to ensure that x_j has degree two in J. This completes the construction of the edges of J incident to A_0. The edges incident to B_0 are found similarly.

Let J be the graph on $A_0 \cup B_0 \cup A_{i_1} \cup A_{i_2} \cup B_{i_3} \cup B_{i_4}$ whose edge set is constructed in this way. By construction, J satisfies (BES1) and (BES2) since P_j and P'_j are (i_1, i_2, A)-localized and (i_3, i_4, B)-localized respectively. We now verify (BES3). As mentioned before the statement of the lemma, (BES1) implies that J is an A_0B_0-path system (as defined before Proposition 4.2.5). Moreover, Corollary 4.3.7(a)(ii) implies that $P_i \cup P'_i$ is a path system which satisfies (B1) in the definition of 2-balanced. Since J was obtained by adding only $A'B'$-edges, (B1) is preserved in J. Since by construction J satisfies (B2), it follows that J is 2-balanced. So Proposition 4.2.5 implies (BES3).

Finally, we verify (BES4). For this, note that Corollary 4.3.7(a)(iii) implies that $e(P_i), e(P'_i) \leq \sqrt{\varepsilon} n$. Moreover, the number of edges added to $P_i \cup P'_i$ when constructing J is at most $2(|A_0| + |B_0|)$, which is at most $2\varepsilon n$ by (BFR4). Thus $e(J) \leq 2\sqrt{\varepsilon} n + 2\varepsilon n \leq \varepsilon_0 n$. □

4.3.5. Covering G_{glob} by Edge-disjoint Hamilton Cycles. We now find a set of edge-disjoint Hamilton cycles covering the edges of the 'leftover' graph obtained from $G - G[A, B]$ by deleting all those edges lying in balanced exceptional systems belonging to $\mathfrak{J}$.

LEMMA 4.3.9. *Let $0 < 1/n \ll \varepsilon \ll \varepsilon_0 \ll \varepsilon' \ll \varepsilon_1 \ll \varepsilon_2 \ll \varepsilon_3 \ll \varepsilon_4 \ll 1/K \ll 1$. Suppose that (G, A, A_0, B, B_0) is an $(\varepsilon, \varepsilon', K, D)$-bi-framework with $|G| = n$ and such that $D \geq n/200$ and D is even. Let $\mathcal{P} := \{A_0, A_1, \dots, A_K, B_0, B_1, \dots, B_K\}$ be a $(K, m, \varepsilon, \varepsilon_1, \varepsilon_2)$-partition for G. Suppose that $t_K := (1 - 20\varepsilon_4)D/2K^4 \in \mathbb{N}$. Let $\mathfrak{J}$ be as defined after Lemma 4.3.8 and let $G(\mathfrak{J}) \subseteq G$ be the union of all the balanced exceptional systems lying in $\mathfrak{J}$. Let $G^* := G - G(\mathfrak{J})$, let $k := 10\varepsilon_4 D$ and let $(Q_{1,A}, Q_{1,B}), \dots, (Q_{k,A}, Q_{k,B})$ be as in Corollary 4.3.7(d).*

(a) *The graph $G^* - G^*[A, B]$ can be decomposed into k A_0B_0-path systems $Q_1, \dots, Q_k$ which are 2-balanced and satisfy the following properties:*
  (i) *Q_i contains all edges of $Q_{i,A} \cup Q_{i,B}$;*
  (ii) *$Q_1, \dots, Q_k$ are pairwise edge-disjoint;*
  (iii) *$e(Q_i) \leq 3\sqrt{\varepsilon} n$.*

(b) *Let $Q_1, \dots, Q_k$ be as in (a). Suppose that F is a graph on $V(G)$ such that $G \subseteq F$, $\delta(F) \geq 2n/5$ and such that F satisfies (WF5) with respect to ε'. Then there are edge-disjoint Hamilton cycles $C_1, \dots, C_k$ in $F - G(\mathfrak{J})$ such that $Q_i \subseteq C_i$ and $C_i \cap G$ is 2-balanced for each $i \leq k$.*

Proof. We first prove (a). The argument is similar to that of Lemma 4.3.6. Roughly speaking, we will extend each $Q_{i,A}$ into a path system $Q'_{i,A}$ by adding suitable A_0B-edges which ensure that every vertex in A_0 has degree exactly two in $Q'_{i,A}$. Similarly, we will extend each $Q_{i,B}$ into $Q'_{i,B}$ by adding suitable AB_0-edges. We will ensure that no vertex is an endvertex of both an edge in $Q'_{i,A}$ and an edge in $Q'_{i,B}$ and take Q_i to be the union of these two path systems. We first construct all the $Q'_{i,A}$.

Claim 1. *$G^*[A'] \cup G^*[A_0, B]$ has a decomposition into edge-disjoint path systems $Q'_{1,A}, \dots, Q'_{k,A}$ such that*

- *$Q_{i,A} \subseteq Q'_{i,A}$ and $E(Q'_{i,A}) \setminus E(Q_{i,A})$ consists of A_0B-edges in G^* (for each $i \leq k$);*
- *$d_{Q'_{i,A}}(x) = 2$ for every $x \in A_0$ and $d_{Q'_{i,A}}(x) \leq 1$ for every $x \notin A_0$;*
- *no vertex is an endvertex of both an edge in $Q'_{i,A}$ and an edge in $Q_{i,B}$ (for each $i \leq k$).*

To prove Claim 1, let G_{glob} be as defined in Corollary 4.3.7(d). Thus $G_{glob}[A'] = Q_{1,A} \cup \dots \cup Q_{k,A}$. On the other hand, Lemma 4.3.8(ii) implies that $G^*[A'] = G_{glob}[A']$. Hence,

$$G^*[A'] = G_{glob}[A'] = Q_{1,A} \cup \dots \cup Q_{k,A}. \tag{4.3.11}$$

Similarly, $G^*[B'] = G_{glob}[B'] = Q_{1,B} \cup \dots \cup Q_{k,B}$. Moreover, $G_{glob} = G^*[A'] \cup G^*[B']$. Consider any vertex $x \in A_0$. Let $d_{glob}(x)$ denote the degree of x in $Q_{1,A} \cup \dots \cup Q_{k,A}$. So $d_{glob}(x) = d_{G^*}(x, A')$ by (4.3.11). Let

$$d_{loc}(x) := d_G(x, A') - d_{glob}(x) \tag{4.3.12}$$

$$= d_G(x, A') - d_{G^*}(x, A') = d_{G(\mathfrak{J})}(x, A'). \tag{4.3.13}$$

Then

$$d_{loc}(x) + d_G(x, B') + d_{glob}(x) \stackrel{(4.3.12)}{=} d_G(x) = D, \tag{4.3.14}$$

where the final equality follows from (BFR2). Recall that $\mathfrak{J}$ consists of $K^4 t_K$ edge-disjoint balanced exceptional systems. Since x has two neighbours in each of these balanced exceptional systems, the degree of x in $G(\mathfrak{J})$ is $2K^4 t_K = D - 2k$. Altogether this implies that

$$\begin{aligned} d_{G^*}(x, B') &= d_G(x, B') - d_{G(\mathfrak{J})}(x, B') \\ &= d_G(x, B') - (d_{G(\mathfrak{J})}(x) - d_{G(\mathfrak{J})}(x, A')) \\ &\stackrel{(4.3.13)}{=} d_G(x, B') - (D - 2k - d_{loc}(x)) \stackrel{(4.3.14)}{=} 2k - d_{glob}(x). \end{aligned} \tag{4.3.15}$$

Note that this is precisely the total number of edges at x which we need to add to $Q_{1,A}, \dots, Q_{k,A}$ in order to obtain $Q'_{1,A}, \dots, Q'_{k,A}$ as in Claim 1.

We can now construct the path systems $Q'_{i,A}$. For each $x \in A_0$, let $n_i(x) = 2 - d_{Q_{i,A}}(x)$. So $0 \le n_i(x) \le 2$ for all $i \le k$. Recall that $a := |A_0|$ and consider an ordering $x_1, \dots, x_a$ of the vertices in A_0. Let $G^*_j := G^*[\{x_1, \dots, x_j\}, B]$. Assume that for some $0 \le j < a$, we have already found a decomposition of G^*_j into edge-disjoint path systems $Q_{1,j}, \dots, Q_{k,j}$ satisfying the following properties (for all $i \le k$):

- (i′) no vertex is an endvertex of both an edge in $Q_{i,j}$ and an edge in $Q_{i,B}$;
- (ii′) $x_{j'}$ has degree $n_i(x_{j'})$ in $Q_{i,j}$ for all $j' \le j$ and all other vertices have degree at most one in $Q_{i,j}$.

We call this assertion $\mathcal{A}_j$. We will show that $\mathcal{A}_{j+1}$ holds (i.e. the above assertion also holds with j replaced by $j+1$). This in turn implies Claim 1 if we let $Q'_{i,A} := Q_{i,a} \cup Q_{i,A}$ for all $i \le k$.

To prove $\mathcal{A}_{j+1}$, consider the following bipartite auxiliary graph H_{j+1}. The vertex classes of H_{j+1} are $N_{j+1} := N_{G^*}(x_{j+1}) \cap B$ and Z_{j+1}, where Z_{j+1} is a multiset whose elements are chosen from $Q_{1,B}, \dots, Q_{k,B}$. Each $Q_{i,B}$ is included exactly $n_i(x_{j+1})$ times in Z_{j+1}. Note that $N_{j+1} = N_{G^*}(x_{j+1}) \cap B'$ since $e(G[A_0, B_0]) = 0$ by (BFR6). Altogether this implies that

$$\begin{aligned} |Z_{j+1}| &= \sum_{i=1}^{k} n_i(x_{j+1}) = 2k - \sum_{i=1}^{k} d_{Q_{i,A}}(x_{j+1}) = 2k - d_{glob}(x_{j+1}) \\ &\stackrel{(4.3.15)}{=} d_{G^*}(x_{j+1}, B') = |N_{j+1}| \ge k/2. \end{aligned} \tag{4.3.16}$$

The final inequality follows from (4.3.15) since

$$d_{glob}(x_{j+1}) \stackrel{(4.3.11)}{\le} \Delta(G_{glob}[A']) \le 3k/2$$

by Corollary 4.3.7(d). We include an edge in H_{j+1} between $v \in N_{j+1}$ and $Q_{i,B} \in Z_{j+1}$ if v is not an endvertex of an edge in $Q_{i,B} \cup Q_{i,j}$.

Claim 2. *H_{j+1} has a perfect matching M'_{j+1}.*

Given the perfect matching guaranteed by the claim, we construct $Q_{i,j+1}$ from $Q_{i,j}$ as follows: the edges of $Q_{i,j+1}$ incident to x_{j+1} are precisely the edges $x_{j+1}v$ where $vQ_{i,B}$ is an edge of M'_{j+1} (note that there are up to two of these). Thus Claim 2 implies that $\mathcal{A}_{j+1}$ holds. (Indeed, (i′)–(ii′) are immediate from the definition of H_{j+1}.)

To prove Claim 2, consider any vertex $v \in N_{j+1}$. Since $v \in B$, the number of path systems $Q_{i,B}$ containing an edge at v is at most $d_G(v, B')$. The number of indices i for which $Q_{i,j}$ contains an edge at v is at most $d_G(v, A_0) \leq |A_0|$. Since each path system $Q_{i,B}$ occurs at most twice in the multiset Z_{j+1}, it follows that the degree of v in H_{j+1} is at least $|Z_{j+1}| - 2d_G(v, B') - 2|A_0|$. Moreover, $d_G(v, B') \leq \varepsilon' n \leq k/16$ (say) by (BFR5). Also, $|A_0| \leq \varepsilon n \leq k/16$ by (BFR4). So v has degree at least $|Z_{j+1}| - k/4 \geq |Z_{j+1}|/2$ in H_{j+1}.

Now consider any path system $Q_{i,B} \in Z_{j+1}$. Recall that $e(Q_{i,B}) \leq \sqrt{\varepsilon} n \leq k/16$ (say), where the first inequality follows from Corollary 4.3.7(d)(iv′). Moreover, $e(Q_{i,j}) \leq 2|A_0| \leq 2\varepsilon n \leq k/8$, where the second inequality follows from (BFR4). Thus the degree of $Q_{i,B}$ in H_{j+1} is at least

$$|N_{j+1}| - 2e(Q_{i,B}) - e(Q_{i,j}) \geq |N_{j+1}| - k/4 \geq |N_{j+1}|/2.$$

Altogether this implies that H_{j+1} has a perfect matching M'_{j+1}, as required.

This completes the construction of $Q'_{1,A}, \dots, Q'_{k,A}$. Next we construct $Q'_{1,B}, \dots, Q'_{k,B}$ using the same approach.

Claim 3. *$G^*[B'] \cup G^*[B_0, A]$ has a decomposition into edge-disjoint path systems $Q'_{1,B}, \dots, Q'_{k,B}$ such that*

- *$Q_{i,B} \subseteq Q'_{i,B}$ and $E(Q'_{i,B}) \setminus E(Q_{i,B})$ consists of B_0A-edges in G^* (for each $i \leq k$);*
- *$d_{Q'_{i,B}}(x) = 2$ for every $x \in B_0$ and $d_{Q'_{i,B}}(x) \leq 1$ for every $x \notin B_0$;*
- *no vertex is an endvertex of both an edge in $Q'_{i,A}$ and an edge in $Q'_{i,B}$ (for each $i \leq k$).*

The proof of Claim 3 is similar to that of Claim 1. The only difference is that when constructing $Q'_{i,B}$, we need to avoid the endvertices of all the edges in $Q'_{i,A}$ (not just the edges in $Q_{i,A}$). However, $e(Q'_{i,A} - Q_{i,A}) \leq 2|A_0|$, so this does not affect the calculations significantly.

We now take $Q_i := Q'_{i,A} \cup Q'_{i,B}$ for all $i \leq k$. Then the Q_i are pairwise edge-disjoint and

$$e(Q_i) \leq e(Q_{i,A}) + e(Q_{i,B}) + 2|A_0 \cup B_0| \leq 2\sqrt{\varepsilon} n + 2\varepsilon n \leq 3\sqrt{\varepsilon} n$$

by Corollary 4.3.7(d)(iv′) and (BFR4). Moreover, Corollary 4.3.7(d)(iii′) implies that

$$e_{Q_i}(A') - e_{Q_i}(B') = e(Q_{i,A}) - e(Q_{i,B}) = a - b.$$

Thus each Q_i is a 2-balanced A_0B_0-path system. Further, $Q_1, \dots, Q_k$ form a decomposition of

$$G^*[A'] \cup G^*[A_0, B] \cup G^*[B'] \cup G^*[B_0, A] = G^* - G^*[A, B].$$

(The last equality follows since $e(G[A_0, B_0]) = 0$ by (BFR6).) This completes the proof of (a).

To prove (b), note that (F, G, A, A_0, B, B_0) is an $(\varepsilon, \varepsilon', D)$-pre-framework, i.e. it satisfies (WF1)–(WF5). Indeed, recall that (BFR1)–(BFR4) imply (WF1)–(WF4) and that (WF5) holds by assumption. So we can apply Lemma 4.2.9 (with Q_1 playing the role of Q) to extend Q_1 into a Hamilton cycle C_1. Moreover, Lemma 4.2.9(iii) implies that $C_1 \cap G$ is 2-balanced, as required. (Lemma 4.2.9(ii) guarantees that C_1 is edge-disjoint from $Q_2, \dots, Q_k$ and $G(\mathfrak{J})$.)

Let $G_1 := G - C_1$ and $F_1 := F - C_1$. Proposition 4.2.3 (with C_1 playing the role of H) implies that $(F_1, G_1, A, A_0, B, B_0)$ is an $(\varepsilon, \varepsilon', D-2)$-pre-framework. So we can now apply Lemma 4.2.9 to $(F_1, G_1, A, A_0, B, B_0)$ to extend Q_2 into a Hamilton cycle C_2, where $C_2 \cap G$ is also 2-balanced.

We can continue this way to find $C_3, \dots, C_k$. Indeed, suppose that we have found $C_1, \dots, C_i$ for $i < k$. Then we can still apply Lemma 4.2.9 since $\delta(F) - 2i \geq \delta(F) - 2k \geq n/3$. Moreover, $C_j \cap G$ is 2-balanced for all $j \leq i$, so $(C_1 \cup \dots \cup C_i) \cap G$ is $2i$-balanced. This in turn means that Proposition 4.2.3 (applied with $C_1 \cup \dots \cup C_i$ playing the role of H) implies that after removing $C_1, \dots, C_i$, we still have an $(\varepsilon, \varepsilon', D-2i)$-pre-framework and can find C_{i+1}. □

We can now put everything together to find a set of localized balanced exceptional systems and a set of Hamilton cycles which altogether cover all edges of G outside $G[A, B]$. The localized balanced exceptional systems will be extended to Hamilton cycles later on.

COROLLARY 4.3.10. *Let $0 < 1/n \ll \varepsilon \ll \varepsilon_0 \ll \varepsilon' \ll \varepsilon_1 \ll \varepsilon_2 \ll \varepsilon_3 \ll \varepsilon_4 \ll 1/K \ll 1$. Suppose that (G, A, A_0, B, B_0) is an $(\varepsilon, \varepsilon', K, D)$-bi-framework with $|G| = n$ and so that $D \geq n/200$ and D is even. Let $\mathcal{P} := \{A_0, A_1, \dots, A_K, B_0, B_1, \dots, B_K\}$ be a $(K, m, \varepsilon, \varepsilon_1, \varepsilon_2)$-partition for G. Suppose that $t_K := (1 - 20\varepsilon_4)D/2K^4 \in \mathbb{N}$ and let $k := 10\varepsilon_4 D$. Suppose that F is a graph on $V(G)$ such that $G \subseteq F$, $\delta(F) \geq 2n/5$ and such that F satisfies (WF5) with respect to ε'. Then there are k edge-disjoint Hamilton cycles $C_1, \dots, C_k$ in F and for all $1 \leq i_1, i_2, i_3, i_4 \leq K$ there is a set $\mathcal{J}_{i_1 i_2 i_3 i_4}$ such that the following properties are satisfied:*

- (i) *$\mathcal{J}_{i_1 i_2 i_3 i_4}$ consists of t_K (i_1, i_2, i_3, i_4)-BES in G with respect to $\mathcal{P}$ and with parameter ε_0 which are edge-disjoint from each other and from $C_1 \cup \dots \cup C_k$.*
- (ii) *Whenever $(i_1, i_2, i_3, i_4) \neq (i'_1, i'_2, i'_3, i'_4)$, $J \in \mathcal{J}_{i_1 i_2 i_3 i_4}$ and $J' \in \mathcal{J}_{i'_1 i'_2 i'_3 i'_4}$, then J and J' are edge-disjoint.*
- (iii) *Given any $i \leq k$ and $v \in A_0 \cup B_0$, the two edges incident to v in C_i lie in G.*
- (iv) *Let $G^\diamond$ be the subgraph of G obtained by deleting the edges of all the C_i and all the balanced exceptional systems in $\mathcal{J}_{i_1 i_2 i_3 i_4}$ (for all $1 \leq i_1, i_2, i_3, i_4 \leq K$). Then $G^\diamond$ is bipartite with vertex classes A', B' and $V_0 = A_0 \cup B_0$ is an isolated set in $G^\diamond$.*

Proof. This follows immediately from Lemmas 4.3.8 and 4.3.9(b). Indeed, clearly (i)–(iii) are satisfied. To check (iv), note that $G^\diamond$ is obtained from the graph G^* defined in Lemma 4.3.9 by deleting all the edges of the Hamilton cycles C_i. But Lemma 4.3.9 implies that the C_i together cover all the edges in $G^* - G^*[A, B]$. Thus this implies that $G^\diamond$ is bipartite with vertex classes A', B' and V_0 is an isolated set in $G^\diamond$. □

4.4. Special Factors and Balanced Exceptional Factors

As discussed in the proof sketch, the proof of Theorem 1.3.5 proceeds as follows. First we find an approximate decomposition of the given graph G and finally we find a decomposition of the (sparse) leftover from the approximate decomposition (with the aid of a 'robustly decomposable' graph we removed earlier). Both the approximate decomposition as well as the actual decomposition steps assume that we work with a bipartite graph on $A \cup B$ (with $|A| = |B|$). So in both steps, we would need $A_0 \cup B_0$ to be empty, which we clearly cannot assume. On the other hand, in both steps, one can specify 'balanced exceptional path systems' (BEPS) in G with the following crucial property: one can replace each BEPS with a path system BEPS* so that

(α_1) BEPS* is bipartite with vertex classes A and B;
(α_2) a Hamilton cycle C^* in $G^* := G[A, B] + \text{BEPS}^*$ which contains BEPS* corresponds to a Hamilton cycle C in G which contains BEPS (see Section 4.4.1).

Each BEPS will contain one of the balanced exceptional sequences BES constructed in Section 4.3. BEPS* will then be obtained by replacing the edges in BES by suitable 'fictive' edges (i.e. which are not necessarily contained in G).

So, roughly speaking, this allows us to work with G^* rather than G in the two steps. Similarly as in the two clique case, a convenient way of specifying and handling these balanced exceptional path systems is to combine them into 'balanced exceptional factors' BF (see Section 4.4.3 for the definition). (The balanced exceptional path systems and balanced exceptional factors are analogues of the exceptional path systems and exceptional factors considered in Chapter 2.)

As before, one complication is that the 'robust decomposition lemma' (Lemma 4.5.3) we use from [**21**] deals with digraphs rather than undirected graphs. So to be able to apply it, we again need a suitable orientation of the edges of G and so we will actually consider directed path systems $\text{BEPS}^*_{\text{dir}}$ instead of BEPS* above (whereas the path systems BEPS are undirected).

Rather than guaranteeing (α_2) directly, the (bipartite) robust decomposition lemma assumes the existence of certain directed 'special paths systems' SPS which are combined into 'special factors' SF. (Recall that these notions were used in the proof of Theorem 1.3.3; see Section 2.8.1. In this chapter, we use slight variants of these definitions which are introduced in Section 4.4.2.) Each of the Hamilton cycles produced by the lemma then contains exactly one of these special path systems. So to apply the lemma, it suffices to check separately that each $\text{BEPS}^*_{\text{dir}}$ satisfies the conditions required of a special path system and that it also satisfies (α_2).

4.4.1. Constructing the Graphs J^* from the Balanced Exceptional Systems J. Suppose that J is a balanced exceptional system in a graph G with respect to a (K, m, ε_0)-partition $\mathcal{P} = \{A_0, A_1, \dots, A_K, B_0, B_1, \dots, B_K\}$ of $V(G)$. We will now use J to define an auxiliary matching J^*. Every edge of J^* will have one endvertex in A and its other endvertex in B. We will regard J^* as being edge-disjoint from the original graph G. So even if both J^* and G have an edge between the same pair of endvertices, we will regard these as different edges. The edges of such a J^* will be called *fictive edges*. Proposition 4.4.1(ii) below shows that a Hamilton cycle in $G[A \cup B] + J^*$ containing all edges of J^* in a suitable order will correspond to a Hamilton cycle in G which contains J. So when finding our Hamilton cycles, this property will enable us to ignore all the vertices in $V_0 = A_0 \cup B_0$ and to consider a bipartite (multi-)graph between A and B instead.

We construct J^* in two steps. First we will construct a matching J^*_{AB} on $A \cup B$ and then J^*. Since each maximal path in J has endpoints in $A \cup B$ and internal vertices in V_0 by (BES1), a balanced exceptional system J naturally induces a matching J^*_{AB} on $A \cup B$. More precisely, if $P_1, \dots, P_{\ell'}$ are the non-trivial paths

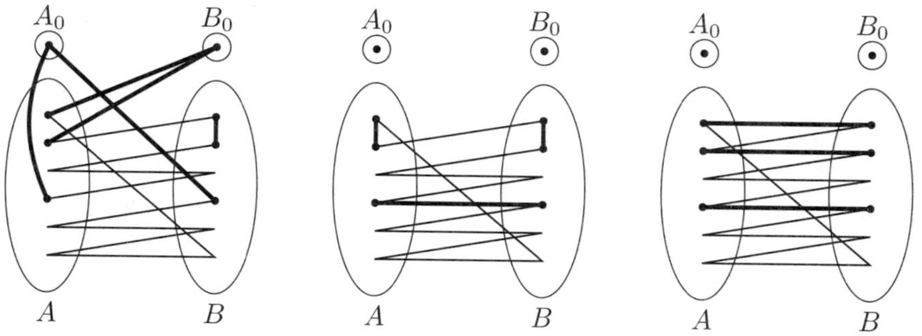

FIGURE 4.4.1. The thick lines illustrate the edges of J, J^*_{AB} and J^* respectively.

in J and x_i, y_i are the endpoints of P_i, then we define $J^*_{AB} := \{x_i y_i : i \le \ell'\}$. Thus J^*_{AB} is a matching by (BES1) and $e(J^*_{AB}) \le e(J)$. Moreover, J^*_{AB} and $E(J)$ cover exactly the same vertices in A. Similarly, they cover exactly the same vertices in B. So (BES3) implies that $e(J^*_{AB}[A]) = e(J^*_{AB}[B])$. We can write $E(J^*_{AB}[A]) = \{x_1x_2, \dots, x_{2s-1}x_{2s}\}$, $E(J^*_{AB}[B]) = \{y_1y_2, \dots, y_{2s-1}y_{2s}\}$ and $E(J^*_{AB}[A, B]) = \{x_{2s+1}y_{2s+1}, \dots, x_{s'}y_{s'}\}$, where $x_i \in A$ and $y_i \in B$. Define $J^* := \{x_iy_i : 1 \le i \le s'\}$. Note that $e(J^*) = e(J^*_{AB}) \le e(J)$. All edges of J^* are called *fictive edges*.

As mentioned before, we regard J^* as being edge-disjoint from the original graph G. Suppose that P is an orientation of a subpath of (the multigraph) $G[A \cup B] + J^*$. We say that P is *consistent with* J^* if P contains all the edges of J^* and P traverses the vertices $x_1, y_1, x_2, \dots, y_{s'-1}, x_{s'}, y_{s'}$ in this order. (This ordering will be crucial for the vertices $x_1, y_1, \dots, x_{2s}, y_{2s}$, but it is also convenient to have an ordering involving all vertices of J^*.) Similarly, we say that a cycle D in $G[A \cup B] + J^*$ is *consistent with* J^* if D contains all the edges of J^* and there exists some orientation of D which traverses the vertices $x_1, y_1, x_2, \dots, y_{s'-1}, x_{s'}, y_{s'}$ in this order.

The next result shows that if J is a balanced exceptional system and C is a Hamilton cycle on $A \cup B$ which is consistent with J^*, then the graph obtained from C by replacing J^* with J is a Hamilton cycle on $V(G)$ which contains J, see Figure 4.4.1. When choosing our Hamilton cycles, this property will enable us ignore all the vertices in V_0 and edges in A and B and to consider the (almost complete) bipartite graph with vertex classes A and B instead.

PROPOSITION 4.4.1. *Let $\mathcal{P} = \{A_0, A_1, \dots, A_K, B_0, B_1, \dots, B_K\}$ be a (K, m, ε)-partition of a vertex set V. Let G be a graph on V and let J be a balanced exceptional system with respect to $\mathcal{P}$.*

(i) *Assume that P is an orientation of a subpath of $G[A \cup B] + J^*$ such that P is consistent with J^*. Then the graph obtained from $P - J^* + J$ by ignoring the orientations of the edges is a path on $V(P) \cup V_0$ whose endvertices are the same as those of P.*
(ii) *If $J \subseteq G$ and D is a Hamilton cycle of $G[A \cup B] + J^*$ which is consistent with J^*, then $D - J^* + J$ is a Hamilton cycle of G.*

Proof. We first prove (i). Let $s := e(J^*_{AB}[A]) = e(J^*_{AB}[B])$ and $J^\diamond := \{x_1y_1, \dots, x_{2s}y_{2s}\}$ (where the x_i and y_i are as in the definition of J^*). So $J^* := J^\diamond \cup \{x_{2s+1}y_{2s+1}, \dots, x_{s'}y_{s'}\}$, where $s' := e(J^*)$. Let P^c denote the path obtained from $P = z_1 \dots z_2$ by reversing its direction. (So $P^c = z_2 \dots z_1$ traverses the vertices $y_{s'}, x_{s'}, y_{2s'-1}, \dots, x_2, y_1, x_1$ in this order.) First note

$$P' := z_1 P x_1 x_2 P^c y_1 y_2 P x_3 x_4 P^c y_3 y_4 \dots x_{2s-1} x_{2s} P^c y_{2s-1} y_{2s} P z_2$$

is a path on $V(P)$. Moreover, the underlying undirected graph of P' is precisely

$$P - J^\diamond + (J^*_{AB}[A] \cup J^*_{AB}[B]) = P - J^* + J^*_{AB}.$$

In particular, P' contains J^*_{AB}. Now recall that if w_1w_2 is an edge in J^*_{AB}, then the vertices w_1 and w_2 are the endpoints of some path P^* in J (where the internal vertices on P^* lie in V_0). Clearly, $P' - w_1w_2 + P^*$ is also a path. Repeating this step for every edge w_1w_2 of J^*_{AB} gives a path P'' on $V(P) \cup V_0$. Moreover, $P'' = P - J^* + J$. This completes the proof of (i).

(ii) now follows immediately from (i). □

4.4.2. Special Path Systems and Special Factors. As mentioned earlier, in order to apply Lemma 4.5.3, we first need to prove the existence of certain 'special path systems'. These are defined below. Note that the definitions given in this section are slight variants of the corresponding definitions used in Chapter 2.

Suppose that

$$\mathcal{P} = \{A_0, A_1, \dots, A_K, B_0, B_1, \dots, B_K\}$$

is a (K, m, ε_0)-partition of a vertex set V and $L, m/L \in \mathbb{N}$. Recall that we say that $(\mathcal{P}, \mathcal{P}')$ is a *(K, L, m, ε_0)-partition of V* if $\mathcal{P}'$ is obtained from $\mathcal{P}$ by partitioning A_i into L sets $A_{i,1}, \dots, A_{i,L}$ of size m/L for all $1 \le i \le K$ and partitioning B_i into L sets $B_{i,1}, \dots, B_{i,L}$ of size m/L for all $1 \le i \le K$. (So $\mathcal{P}'$ consists of the exceptional sets A_0, B_0, the KL clusters $A_{i,j}$ and the KL clusters $B_{i,j}$.) Unless stated otherwise, whenever considering a (K, L, m, ε_0)-partition $(\mathcal{P}, \mathcal{P}')$ of a vertex set V we use the above notation to denote the elements of $\mathcal{P}$ and $\mathcal{P}'$.

Let $(\mathcal{P}, \mathcal{P}')$ be a (K, L, m, ε_0)-partition of V. Consider a spanning cycle $C = A_1B_1 \dots A_KB_K$ on the clusters of $\mathcal{P}$. Given an integer f dividing K, the *canonical interval partition $\mathcal{I}$ of C into f intervals* consists of the intervals

$$A_{(i-1)K/f+1} B_{(i-1)K/f+1} A_{(i-1)K/f+2} \dots B_{iK/f} A_{iK/f+1}$$

for all $i \le f$. (Here $A_{K+1} := A_1$.)

Suppose that G is a digraph on $V \setminus V_0$ and $h \le L$. Let $I = A_jB_jA_{j+1} \dots A_{j'}$ be an interval in $\mathcal{I}$. A *special path system SPS of style h in G spanning the interval I* consists of precisely m/L (non-trivial) vertex-disjoint directed paths $P_1, \dots, P_{m/L}$ such that the following conditions hold:

- (SPS1) Every P_s has its initial vertex in $A_{j,h}$ and its final vertex in $A_{j',h}$.
- (SPS2) SPS contains a matching $\mathrm{Fict}(SPS)$ such that all the edges in $\mathrm{Fict}(SPS)$ avoid the endclusters A_j and $A_{j'}$ of I and such that $E(P_s) \setminus \mathrm{Fict}(SPS) \subseteq E(G)$.
- (SPS3) The vertex set of SPS is $A_{j,h} \cup B_{j,h} \cup A_{j+1,h} \cup \dots \cup B_{j'-1,h} \cup A_{j',h}$.

The edges in $\mathrm{Fict}(SPS)$ are called *fictive edges of SPS*.

Let $\mathcal{I} = \{I_1, \dots, I_f\}$ be the canonical interval partition of C into f intervals. A *special factor SF with parameters (L, f) in G (with respect to C, $\mathcal{P}'$)* is a 1-regular digraph on $V \setminus V_0$ which is the union of Lf digraphs $SPS_{j,h}$ (one for all $j \le f$

and $h \leq L$) such that each $SPS_{j,h}$ is a special path system of style h in G which spans I_j. We write $\mathrm{Fict}(SF)$ for the union of the sets $\mathrm{Fict}(SPS_{j,h})$ over all $j \leq f$ and $h \leq L$ and call the edges in $\mathrm{Fict}(SF)$ *fictive edges of SF*.

We will always view fictive edges as being distinct from each other and from the edges in other digraphs. So if we say that special factors $SF_1, \dots, SF_r$ are pairwise edge-disjoint from each other and from some digraph Q on $V \setminus V_0$, then this means that Q and all the $SF_i - \mathrm{Fict}(SF_i)$ are pairwise edge-disjoint, but for example there could be an edge from x to y in Q as well as in $\mathrm{Fict}(SF_i)$ for several indices $i \leq r$. But these are the only instances of multiedges that we allow, i.e. if there is more than one edge from x to y, then all but at most one of these edges are fictive edges.

4.4.3. Balanced Exceptional Path Systems and Balanced Exceptional Factors. We now define balanced exceptional path systems BEPS. It will turn out that they (or rather their bipartite directed versions $\mathrm{BEPS}^*_{\mathrm{dir}}$ involving fictive edges) will satisfy the conditions of the special path systems defined above. Moreover, Hamilton cycles that respect the partition A, B and which contain $\mathrm{BEPS}^*_{\mathrm{dir}}$ correspond to Hamilton cycles in the 'original' graph G (see Proposition 4.4.2).

Let $(\mathcal{P}, \mathcal{P}')$ be a (K, L, m, ε_0)-partition of a vertex set V. Suppose that $K/f \in \mathbb{N}$ and $h \leq L$. Consider a spanning cycle $C = A_1B_1 \dots A_KB_K$ on the clusters of $\mathcal{P}$. Let $\mathcal{I}$ be the canonical interval partition of C into f intervals of equal size. Suppose that G is an oriented bipartite graph with vertex classes A and B. Suppose that $I = A_jB_j \dots A_{j'}$ is an interval in $\mathcal{I}$. A *balanced exceptional path system $BEPS$ of style h for G spanning I* consists of precisely m/L (non-trivial) vertex-disjoint undirected paths $P_1, \dots, P_{m/L}$ such that the following conditions hold:

(BEPS1) Every P_s has one endvertex in $A_{j,h}$ and its other endvertex in $A_{j',h}$.

(BEPS2) $J := BEPS - BEPS[A, B]$ is a balanced exceptional system with respect to $\mathcal{P}$ such that P_1 contains all edges of J and so that the edge set of J is disjoint from $A_{j,h}$ and $A_{j',h}$. Let $P_{1,\mathrm{dir}}$ be the path obtained by orienting P_1 towards its endvertex in $A_{j',h}$ and let J_{dir} be the orientation of J obtained in this way. Moreover, let J^*_{dir} be obtained from J^* by orienting every edge in J^* towards its endvertex in B. Then $P^*_{1,\mathrm{dir}} := P_{1,\mathrm{dir}} - J_{\mathrm{dir}} + J^*_{\mathrm{dir}}$ is a directed path from $A_{j,h}$ to $A_{j',h}$ which is consistent with J^*.

(BEPS3) The vertex set of $BEPS$ is $V_0 \cup A_{j,h} \cup B_{j,h} \cup A_{j+1,h} \cup \dots \cup B_{j'-1,h} \cup A_{j',h}$.

(BEPS4) For each $2 \leq s \leq m/L$, define $P_{s,\mathrm{dir}}$ similarly as $P_{1,\mathrm{dir}}$. Then $E(P_{s,\mathrm{dir}}) \setminus E(J_{\mathrm{dir}}) \subseteq E(G)$ for every $1 \leq s \leq m/L$.

Let $BEPS^*_{\mathrm{dir}}$ be the path system consisting of $P^*_{1,\mathrm{dir}}, P_{2,\mathrm{dir}}, \dots, P_{m/L,\mathrm{dir}}$. Then $BEPS^*_{\mathrm{dir}}$ is a special path system of style h in G which spans the interval I and such that $\mathrm{Fict}(BEPS^*_{\mathrm{dir}}) = J^*_{\mathrm{dir}}$.

Let $\mathcal{I} = \{I_1, \dots, I_f\}$ be the canonical interval partition of C into f intervals. A *balanced exceptional factor BF with parameters (L, f) for G (with respect to C, $\mathcal{P}'$)* is the union of Lf undirected graphs $BEPS_{j,h}$ (one for all $j \leq f$ and $h \leq L$) such that each $BEPS_{j,h}$ is a balanced exceptional path system of style h for G which spans I_j. We write BF^*_{dir} for the union of $BEPS^*_{j,h,\mathrm{dir}}$ over all $j \leq f$ and $h \leq L$. Note that BF^*_{dir} is a special factor with parameters (L, f) in G (with respect to C, $\mathcal{P}'$) such that $\mathrm{Fict}(BF^*_{\mathrm{dir}})$ is the union of $J^*_{j,h,\mathrm{dir}}$ over all $j \leq f$ and $h \leq L$, where $J_{j,h} = BEPS_{j,h} - BEPS_{j,h}[A, B]$ is the balanced exceptional system contained in $BEPS_{j,h}$ (see condition (BEPS2)). In particular, BF^*_{dir} is a 1-regular digraph on

$V \setminus V_0$ while BF is an undirected graph on V with

(4.4.1) $\quad d_{BF}(v) = 2$ for all $v \in V \setminus V_0$ and $d_{BF}(v) = 2Lf$ for all $v \in V_0$.

Given a balanced exceptional path system $BEPS$, let J be as in (BEPS2) and let $BEPS^* := BEPS - J + J^*$. So $BEPS^*$ consists of $P_1^* := P_1 - J + J^*$ as well as $P_2, \dots, P_{m/L}$. The following is an immediate consequence of (BEPS2) and Proposition 4.4.1.

PROPOSITION 4.4.2. *Let $(\mathcal{P}, \mathcal{P}')$ be a (K, L, m, ε_0)-partition of a vertex set V. Suppose that G is a graph on $V \setminus V_0$, that $G_{\rm dir}$ is an orientation of $G[A, B]$ and that $BEPS$ is a balanced exceptional path system for $G_{\rm dir}$. Let J be as in (BEPS2). Let C be a Hamilton cycle of $G + J^*$ which contains $BEPS^*$. Then $C - BEPS^* + BEPS$ is a Hamilton cycle of $G \cup J$.*

Proof. Note that $C - BEPS^* + BEPS = C - J^* + J$. Moreover, (BEPS2) implies that C contains all edges of J^* and is consistent with J^*. So the proposition follows from Proposition 4.4.1(ii) applied with $G \cup J$ playing the role of G. □

4.4.4. Finding Balanced Exceptional Factors in a Bi-scheme. The following definition of a 'bi-scheme' captures the 'non-exceptional' part of the graphs we are working with. For example, this will be the structure within which we find the edges needed to extend a balanced exceptional system into a balanced exceptional path system.

Given an oriented graph G and partitions $\mathcal{P}$ and $\mathcal{P}'$ of a vertex set V, we call $(G, \mathcal{P}, \mathcal{P}')$ a $[K, L, m, \varepsilon_0, \varepsilon]$-*bi-scheme* if the following properties hold:

(BSch1′) $(\mathcal{P}, \mathcal{P}')$ is a (K, L, m, ε_0)-partition of V. Moreover, $V(G) = A \cup B$.

(BSch2′) Every edge of G has one endvertex in A and its other endvertex in B.

(BSch3′) $G[A_{i,j}, B_{i',j'}]$ and $G[B_{i',j'}, A_{i,j}]$ are $[\varepsilon, 1/2]$-superregular for all $i, i' \leq K$ and all $j, j' \leq L$. Further, $G[A_i, B_j]$ and $G[B_j, A_i]$ are $[\varepsilon, 1/2]$-superregular for all $i, j \leq K$.

(BSch4′) $|N_G^+(x) \cap N_G^-(y) \cap B_{i,j}| \geq (1 - \varepsilon)m/5L$ for all distinct $x, y \in A$, all $i \leq K$ and all $j \leq L$. Similarly, $|N_G^+(x) \cap N_G^-(y) \cap A_{i,j}| \geq (1 - \varepsilon)m/5L$ for all distinct $x, y \in B$, all $i \leq K$ and all $j \leq L$.

If $L = 1$ (and so $\mathcal{P} = \mathcal{P}'$), then (BSch1′) just says that $\mathcal{P}$ is a (K, m, ε_0)-partition of $V(G)$.

The next lemma allows us to extend a suitable balanced exceptional system into a balanced exceptional path system. Given $h \leq L$, we say that an (i_1, i_2, i_3, i_4)-BES J has *style* h *(with respect to the* (K, L, m, ε_0)*-partition* $(\mathcal{P}, \mathcal{P}')$*)* if all the edges of J have their endvertices in $V_0 \cup A_{i_1,h} \cup A_{i_2,h} \cup B_{i_3,h} \cup B_{i_4,h}$.

LEMMA 4.4.3. *Suppose that $K, L, n, m/L \in \mathbb{N}$, that $0 < 1/n \ll \varepsilon, \varepsilon_0 \ll 1$ and $\varepsilon_0 \ll 1/K, 1/L$. Let $(G, \mathcal{P}, \mathcal{P}')$ be a $[K, L, m, \varepsilon_0, \varepsilon]$-bi-scheme with $|V(G) \cup V_0| = n$. Consider a spanning cycle $C = A_1 B_1 \dots A_K B_K$ on the clusters of $\mathcal{P}$ and let $I = A_j B_j A_{j+1} \dots A_{j'}$ be an interval on C of length at least 10. Let J be an (i_1, i_2, i_3, i_4)-BES of style $h \leq L$ with parameter ε_0 (with respect to $(\mathcal{P}, \mathcal{P}')$), for some $i_1, i_2, i_3, i_4 \in \{j+1, \dots, j'-1\}$. Then there exists a balanced exceptional path system of style h for G which spans the interval I and contains all edges in J.*

Proof. For each $k \leq 4$, let m_k denote the number of vertices in $A_{i_k,h} \cup B_{i_k,h}$ which are incident to edges of J. We only consider the case when i_1, i_2, i_3 and i_4

are distinct and $m_k > 0$ for each $k \leq 4$, as the other cases can be proved by similar arguments. Clearly $m_1 + \cdots + m_4 \leq 2\varepsilon_0 n$ by (BES4). For every vertex $x \in A$, we define $B(x)$ to be the cluster $B_{i,h} \in \mathcal{P}'$ such that A_i contains x. Similarly, for every $y \in B$, we define $A(y)$ to be the cluster $A_{i,h} \in \mathcal{P}'$ such that B_i contains y.

Let $x_1y_1, \dots, x_{s'}y_{s'}$ be the edges of J^*, with $x_i \in A$ and $y_i \in B$ for all $i \leq s'$. (Recall that the ordering of these edges is fixed in the definition of J^*.) Thus $s' = (m_1 + \cdots + m_4)/2 \leq \varepsilon_0 n$. Moreover, our assumption that $\varepsilon_0 \ll 1/K, 1/L$ implies that $\varepsilon_0 n \leq m/100L$ (say). Together with (BSch4′) this in turn ensures that for every $r \leq s'$, we can pick vertices $w_r \in B(x_r)$ and $z_r \in A(y_r)$ such that w_rx_r, y_rz_r and z_rw_{r+1} are (directed) edges in G and such that all the $4s'$ vertices x_r, y_r, w_r, z_r (for $r \leq s'$) are distinct from each other. Let P_1' be the path $w_1x_1y_1z_1w_2x_2y_2z_2w_3 \dots y_{s'}z_{s'}$. Thus P_1' is a directed path from B to A in $G + J^*_{\mathrm{dir}}$ which is consistent with J^*. (Here J^*_{dir} is obtained from J^* by orienting every edge towards B.) Note that $|V(P_1') \cap A_{i_k,h}| = m_k = |V(P_1') \cap B_{i_k,h}|$ for all $k \leq 4$. (This follows from our assumption that i_1, i_2, i_3 and i_4 are distinct.) Moreover, $V(P_1') \cap (A_i \cup B_i) = \emptyset$ for all $i \notin \{i_1, i_2, i_3, i_4\}$.

Pick a vertex z' in $A_{j,h}$ so that $z'w_1$ is an edge of G. Find a path P_1'' from $z_{s'}$ to $A_{j',h}$ in G such that the vertex set of P_1'' consists of $z_{s'}$ and precisely one vertex in each $A_{i,h}$ for all $i \in \{j+1, \dots, j'\} \setminus \{i_1, i_2, i_3, i_4\}$ and one vertex in each $B_{i,h}$ for all $i \in \{j, \dots, j'-1\} \setminus \{i_1, i_2, i_3, i_4\}$ and no other vertices. (BSch4′) ensures that this can be done greedily. Define $P^*_{1,\mathrm{dir}}$ to be the concatenation of $z'w_1$, P_1' and P_1''. Note that $P^*_{1,\mathrm{dir}}$ is a directed path from $A_{j,h}$ to $A_{j',h}$ in $G + J^*_{\mathrm{dir}}$ which is consistent with J^*. Moreover, $V(P^*_{1,\mathrm{dir}}) \subseteq \bigcup_{i \leq K} A_{i,h} \cup B_{i,h}$,

$$|V(P^*_{1,\mathrm{dir}}) \cap A_{i,h}| = \begin{cases} 1 & \text{for } i \in \{j, \dots, j'\} \setminus \{i_1, i_2, i_3, i_4\}, \\ m_k & \text{for } i = i_k \text{ and } k \leq 4, \\ 0 & \text{otherwise,} \end{cases}$$

while

$$|V(P^*_{1,\mathrm{dir}}) \cap B_{i,h}| = \begin{cases} 1 & \text{for } i \in \{j, \dots, j'-1\} \setminus \{i_1, i_2, i_3, i_4\}, \\ m_k & \text{for } i = i_k \text{ and } k \leq 4, \\ 0 & \text{otherwise.} \end{cases}$$

(BSch4′) ensures that for each $k \leq 4$, there exist $m_k - 1$ (directed) paths $P_1^k, \dots, P_{m_k-1}^k$ in G such that

- P_r^k is a path from $A_{j,h}$ to $A_{j',h}$ for each $r \leq m_k - 1$ and $k \leq 4$;
- each P_r^k contains precisely one vertex in $A_{i,h}$ for each $i \in \{j, \dots, j'\} \setminus \{i_k\}$, one vertex in $B_{i,h}$ for each $i \in \{j, \dots, j'-1\} \setminus \{i_k\}$ and no other vertices;
- $P^*_{1,\mathrm{dir}}, P_1^1, \dots, P_{m_1-1}^1, P_1^2, \dots, P_{m_4-1}^4$ are vertex-disjoint.

Let Q be the union of $P^*_{1,\mathrm{dir}}$ and all the P_r^k over all $k \leq 4$ and $r \leq m_k - 1$. Thus Q is a path system consisting of $m_1 + \cdots + m_4 - 3$ vertex-disjoint directed paths from $A_{j,h}$ to $A_{j',h}$. Moreover, $V(Q)$ consists of precisely $m_1 + \cdots + m_4 - 3 \leq 2\varepsilon_0 n$ vertices in $A_{i,h}$ for every $j \leq i \leq j'$ and precisely $m_1 + \cdots + m_4 - 3$ vertices in $B_{i,h}$ for every $j \leq i < j'$. Set $A'_{i,h} := A_{i,h} \setminus V(Q)$ and $B'_{i,h} := B_{i,h} \setminus V(Q)$ for all $i \leq K$. Note that, for all $j \leq i \leq j'$,

$$(4.4.2) \quad |A'_{i,h}| = \frac{m}{L} - (m_1 + \cdots + m_4 - 3) \geq \frac{m}{L} - 2\varepsilon_0 n \geq \frac{m}{L} - 5\varepsilon_0 mK \geq (1 - \sqrt{\varepsilon_0})\frac{m}{L}$$

since $\varepsilon_0 \ll 1/K, 1/L$. Similarly, $|B'_{i,h}| \geq (1-\sqrt{\varepsilon_0})m/L$ for all $j \leq i < j'$. Pick a new constant ε' such that $\varepsilon, \varepsilon_0 \ll \varepsilon' \ll 1$. Then (BSch3$'$) and (4.4.2) together with Proposition 1.4.1 imply that $G[A'_{i,h}, B'_{i,h}]$ is still $[\varepsilon', 1/2]$-superregular and so we can find a perfect matching in $G[A'_{i,h}, B'_{i,h}]$ for all $j \leq i < j'$. Similarly, we can find a perfect matching in $G[B'_{i,h}, A'_{i+1,h}]$ for all $j \leq i < j'$. The union Q' of all these matchings forms $m/L - (m_1 + \cdots + m_4) + 3$ vertex-disjoint directed paths.

Let P_1 be the undirected graph obtained from $P^*_{1,\mathrm{dir}} - J^*_{\mathrm{dir}} + J$ by ignoring the directions of all the edges. Proposition 4.4.1(i) implies that P_1 is a path on $V(P^*_{1,\mathrm{dir}}) \cup V_0$ with the same endvertices as $P^*_{1,\mathrm{dir}}$. Consider the path system obtained from $(Q \cup Q') \setminus \{P^*_{1,\mathrm{dir}}\}$ by ignoring the directions of the edges on all the paths. Let $BEPS$ be the union of this path system and P_1. Then $BEPS$ is a balanced exceptional path system for G, as required. □

The next lemma shows that we can obtain many edge-disjoint balanced exceptional factors by extending balanced exceptional systems with suitable properties.

LEMMA 4.4.4. *Suppose that $L, f, q, n, m/L, K/f \in \mathbb{N}$, that $K/f \geq 10$, that $0 < 1/n \ll \varepsilon, \varepsilon_0 \ll 1$, that $\varepsilon_0 \ll 1/K, 1/L$ and $Lq/m \ll 1$. Let $(G, \mathcal{P}, \mathcal{P}')$ be a $[K, L, m, \varepsilon_0, \varepsilon]$-bi-scheme with $|V(G) \cup V_0| = n$. Consider a spanning cycle $C = A_1B_1 \dots A_KB_K$ on the clusters of $\mathcal{P}$. Suppose that there exists a set $\mathcal{J}$ of Lfq edge-disjoint balanced exceptional systems with parameter ε_0 such that*

- *for all $i \leq f$ and all $h \leq L$, $\mathcal{J}$ contains precisely q (i_1, i_2, i_3, i_4)-BES of style h (with respect to $(\mathcal{P}, \mathcal{P}')$) for which $i_1, i_2, i_3, i_4 \in \{(i-1)K/f + 2, \dots, iK/f\}$.*

Then there exist q edge-disjoint balanced exceptional factors with parameters (L, f) for G (with respect to C, $\mathcal{P}'$) covering all edges in $\bigcup \mathcal{J}$.

Recall that the canonical interval partition $\mathcal{I}$ of C into f intervals consists of the intervals

$$A_{(i-1)K/f+1}B_{(i-1)K/f+1}A_{(i-1)K/f+2} \dots A_{iK/f+1}$$

for all $i \leq f$. So the condition on $\mathcal{J}$ ensures that for each interval $I \in \mathcal{I}$ and each $h \leq L$, the set $\mathcal{J}$ contains precisely q balanced exceptional systems of style h whose edges are only incident to vertices in V_0 and vertices belonging to clusters in the interior of I. We will use Lemma 4.4.3 to extend each such balanced exceptional system into a balanced exceptional path system of style h spanning I.

Proof of Lemma 4.4.4. Choose a new constant ε' with $\varepsilon, Lq/m \ll \varepsilon' \ll 1$. Let $\mathcal{J}_1, \dots, \mathcal{J}_q$ be a partition of $\mathcal{J}$ such that for all $j \leq q$, $h \leq L$ and $i \leq f$, the set $\mathcal{J}_j$ contains precisely one (i_1, i_2, i_3, i_4)-BES of style h with $i_1, i_2, i_3, i_4 \in \{(i-1)K/f + 2, \dots, iK/f\}$. Thus each $\mathcal{J}_j$ consists of Lf balanced exceptional systems. For each $j \leq q$ in turn, we will choose a balanced exceptional factor EF_j with parameters (L, f) for G such that BF_j and $BF_{j'}$ are edge-disjoint for all $j' < j$ and BF_j contains all edges of the balanced exceptional systems in $\mathcal{J}_j$. Assume that we have already constructed $BF_1, \dots, BF_{j-1}$. In order to construct BF_j, we will choose the Lf balanced exceptional path systems forming BF_j one by one, such that each of these balanced exceptional path systems is edge-disjoint from $BF_1, \dots, BF_{j-1}$ and contains precisely one of the balanced exceptional systems in $\mathcal{J}_j$. Suppose that we have already chosen some of these balanced exceptional path systems and that next we wish to choose a balanced exceptional path system of style h which spans the interval $I \in \mathcal{I}$ of C and contains $J \in \mathcal{J}_j$. Let G' be the oriented graph obtained from G by deleting all the edges in the balanced path systems already chosen for BF_j as well as deleting all the edges in $BF_1, \dots, BF_{j-1}$. Recall from (BSch1′) that $V(G) = A \cup B$. Thus $\Delta(G - G') \leq 2j < 3q$ by (4.4.1). Together with Proposition 1.4.1 this implies that $(G', \mathcal{P}, \mathcal{P}')$ is still a $[K, L, m, \varepsilon_0, \varepsilon']$-bi-scheme. (Here we use that $\Delta(G - G') < 3q = 3Lq/m \cdot m/L$ and $\varepsilon, Lq/m \ll \varepsilon' \ll 1$.) So we can apply Lemma 4.4.3 with ε' playing the role of ε to obtain a balanced exceptional path system of style h for G' (and thus for G) which spans I and contains all edges of J. This completes the proof of the lemma. □

4.5. The Robust Decomposition Lemma

The purpose of this section is to derive the version of the robust decomposition lemma (Corollary 4.5.4) that we will use in this chapter to prove Theorem 1.3.5. (Recall from Section 4.1 that we will not use it in the proof of Theorem 1.3.8.) Similarly as in the two cliques case, Corollary 4.5.4 allows us to transform an approximate Hamilton decomposition into an exact one. In the next subsection, we introduce the necessary concepts. In particular, Corollary 4.5.4 relies on the existence of a so-called bi-universal walk (which is a 'bipartite version' of the universal walk introduced in Section 2.9.1). The (proof of the) robust decomposition lemma then uses edges guaranteed by this bi-universal walk to 'balance out' edges of the graph H when constructing the Hamilton decomposition of $G^{\mathrm{rob}} + H$.

4.5.1. Chord Sequences and Bi-universal Walks. Let R be a digraph whose vertices are $V_1, \dots, V_k$ and suppose that $C = V_1 \dots V_k$ is a Hamilton cycle of R. (Later on the vertices of R will be clusters. So we denote them by capital letters.)

Recall from Section 2.9.1 that a *chord sequence* $CS(V_i, V_j)$ from V_i to V_j in R is an ordered sequence of edges of the form

$$CS(V_i, V_j) = (V_{i_1 - 1}V_{i_2}, V_{i_2 - 1}V_{i_3}, \dots, V_{i_t - 1}V_{i_{t+1}}),$$

where $V_{i_1} = V_i$, $V_{i_{t+1}} = V_j$ and the edge $V_{i_s - 1}V_{i_{s+1}}$ belongs to R for each $s \leq t$.

As before, if $i = j$ then we consider the empty set to be a chord sequence from V_i to V_j and we may assume that $CS(V_i, V_j)$ does not contain any edges of C.

A closed walk U in R is a *bi-universal walk for C with parameter ℓ'* if the following conditions hold:

- (BU1) The edge set of U has a partition into U_{odd} and U_{even}. For every $1 \leq i \leq k$ there is a chord sequence $ECS^{\mathrm{bi}}(V_i, V_{i+2})$ from V_i to V_{i+2} such that U_{even} contains all edges of all these chord sequences for even i (counted with multiplicities) and U_{odd} contains all edges of these chord sequences for odd i. All remaining edges of U lie on C.
- (BU2) Each $ECS^{\mathrm{bi}}(V_i, V_{i+2})$ consists of at most $\sqrt{\ell'}/2$ edges.
- (BU3) U_{even} enters every cluster V_i exactly $\ell'/2$ times and it leaves every cluster V_i exactly $\ell'/2$ times. The same assertion holds for U_{odd}.

Note that condition (BU1) means that if an edge $V_iV_j \in E(R) \setminus E(C)$ occurs in total 5 times (say) in $ECS^{\mathrm{bi}}(V_1, V_3), \dots, ECS^{\mathrm{bi}}(V_k, V_2)$ then it occurs precisely 5 times in U. We will identify each occurrence of V_iV_j in $ECS^{\mathrm{bi}}(V_1, V_3), \dots, ECS^{\mathrm{bi}}(V_k, V_2)$ with a (different) occurrence of V_iV_j in U. Note that the edges of $ECS^{\mathrm{bi}}(V_i, V_{i+2})$ are allowed to appear in a different order within U.

LEMMA 4.5.1. *Let R be a digraph with vertices $V_1, \dots, V_k$ where $k \geq 4$ is even. Suppose that $C = V_1 \dots V_k$ is a Hamilton cycle of R and that $V_{i-1}V_{i+2} \in E(R)$ for every $1 \leq i \leq k$. Let $\ell' \geq 4$ be an even integer. Let $U_{\mathrm{bi},\ell'}$ denote the multiset obtained from $\ell' - 1$ copies of $E(C)$ by adding $V_{i-1}V_{i+2} \in E(R)$ for every $1 \leq i \leq k$. Then the edges in $U_{\mathrm{bi},\ell'}$ can be ordered so that the resulting sequence forms a bi-universal walk for C with parameter ℓ'.*

In the remainder of the chapter, we will also write $U_{\mathrm{bi},\ell'}$ for the bi-universal walk guaranteed by Lemma 4.5.1.

Proof. Let us first show that the edges in $U_{\mathrm{bi},\ell'}$ can be ordered so that the resulting sequence forms a closed walk in R. To see this, consider the multidigraph U obtained from $U_{\mathrm{bi},\ell'}$ by deleting one copy of $E(C)$. Then U is $(\ell' - 1)$-regular and thus has a decomposition into 1-factors. We order the edges of $U_{\mathrm{bi},\ell'}$ as follows: We first traverse all cycles of the 1-factor decomposition of U which contain the cluster V_1. Next, we traverse the edge V_1V_2 of C. Next we traverse all those cycles of the 1-factor decomposition which contain V_2 and which have not been traversed so far. Next we traverse the edge V_2V_3 of C and so on until we reach V_1 again.

Recall that, for each $1 \leq i \leq k$, the edge $V_{i-1}V_{i+2}$ is a chord sequence from V_i to V_{i+2}. Thus we can take $ECS^{\mathrm{bi}}(V_i, V_{i+2}) := V_{i-1}V_{i+2}$. Then $U_{\mathrm{bi},\ell'}$ satisfies (BU1)–(BU3). Indeed, (BU2) is clearly satisfied. Partition one of the copies of $E(C)$ in $U_{\mathrm{bi},\ell'}$ into E_{even} and E_{odd} where $E_{\mathrm{even}} = \{V_iV_{i+1} |\ i \text{ even}\}$ and $E_{\mathrm{odd}} = \{V_iV_{i+1} |\ i \text{ odd}\}$. Note that the union of E_{even} together with all $ECS^{\mathrm{bi}}(V_i, V_{i+2})$ for even i is a 1-factor in R. Add $\ell'/2 - 1$ of the remaining copies of $E(C)$ to this 1-factor to obtain U_{even}. Define U_{odd} to be $E(U_{\mathrm{bi},\ell'}) \setminus U_{\mathrm{even}}$. By construction of U_{even} and U_{odd}, (BU1) and (BU3) are satisfied. □

4.5.2. Bi-setups and the Robust Decomposition Lemma. The aim of this subsection is to state the 'bipartite version' of the robust decomposition lemma (Lemma 4.5.3, proved in [**21**]) and derive Corollary 4.5.4, which we shall use later on in our proof of Theorem 1.3.5. Lemma 4.5.3 guarantees the existence of a 'robustly decomposable' digraph $G^{\mathrm{rob}}_{\mathrm{dir}}$ within a 'bi-setup'. Roughly speaking, a bi-setup is a digraph G together with its 'reduced digraph' R, which contains a Hamilton cycle C and a bi-universal walk U. (So a bi-setup is a 'bipartite analogue' of a setup that was introduced in Section 2.9.2.) In our application, $G[A, B]$ will play the role of G and R will be the complete bipartite digraph.

To define a bi-setup formally, we first need to recall the following definitions. Given a digraph G and a partition $\mathcal{P}$ of $V(G)$ into k clusters $V_1, \dots, V_k$ of equal size, recall that a partition $\mathcal{P}'$ of $V(G)$ is an *ℓ'-refinement of $\mathcal{P}$* if $\mathcal{P}'$ is obtained by splitting each V_i into ℓ' subclusters of equal size. (So $\mathcal{P}'$ consists of $\ell'k$ clusters.) Recall also that $\mathcal{P}'$ is an *ε-uniform ℓ'-refinement* of $\mathcal{P}$ if it is an ℓ'-refinement of $\mathcal{P}$ which satisfies the following condition: Whenever x is a vertex of G, V is a cluster in $\mathcal{P}$ and $|N^+_G(x) \cap V| \geq \varepsilon|V|$ then $|N^+_G(x) \cap V'| = (1 \pm \varepsilon)|N^+_G(x) \cap V|/\ell'$ for each cluster $V' \in \mathcal{P}'$ with $V' \subseteq V$. The inneighbourhoods of the vertices of G satisfy an analogous condition.

We will need the following definition from [**21**], which describes the structure within which the robust decomposition lemma finds the robustly decomposable graph. $(G, \mathcal{P}, \mathcal{P}', R, C, U, U')$ is called an *$(\ell', k, m, \varepsilon, d)$-bi-setup* if the following properties are satisfied:

(BST1) G and R are digraphs. $\mathcal{P}$ is a partition of $V(G)$ into k clusters of size m where k is even. The vertex set of R consists of these clusters.

(BST2) For every edge VW of R, the corresponding pair $G[V, W]$ is $(\varepsilon, \geq d)$-regular.

(BST3) $C = V_1 \dots V_k$ is a Hamilton cycle of R and for every edge V_iV_{i+1} of C the corresponding pair $G[V_i, V_{i+1}]$ is $[\varepsilon, \geq d]$-superregular.

(BST4) U is a bi-universal walk for C in R with parameter ℓ' and $\mathcal{P}'$ is an ε-uniform ℓ'-refinement of $\mathcal{P}$.

(BST5) Let $V_j^1, \dots, V_j^{\ell'}$ denote the clusters in $\mathcal{P}'$ which are contained in V_j (for each $1 \leq j \leq k$). Then U' is a closed walk on the clusters in $\mathcal{P}'$ which is obtained from U as follows: When U visits V_j for the ath time, we let U' visit the subcluster V_j^a (for all $1 \leq a \leq \ell'$).

(BST6) For every edge $V_i^j V_{i'}^{j'}$ of U' the corresponding pair $G[V_i^j, V_{i'}^{j'}]$ is $[\varepsilon, \geq d]$-superregular.

In [**21**], in a bi-setup, the digraph G could also contain an exceptional set, but since we are only using the definition in the case when there is no such exceptional set, we have only stated it in this special case.

Suppose that $(G, \mathcal{P}, \mathcal{P}')$ is a $[K, L, m, \varepsilon_0, \varepsilon]$-bi-scheme and that $C = A_1B_1 \dots A_KB_K$ is a spanning cycle on the clusters of $\mathcal{P}$. Let $\mathcal{P}_{\rm bi} := \{A_1, \dots, A_K, B_1, \dots, B_K\}$. Suppose that $\ell', m/\ell' \in \mathbb{N}$ with $\ell' \geq 4$. Let $\mathcal{P}''_{\rm bi}$ be an ε-uniform ℓ'-refinement of $\mathcal{P}_{\rm bi}$ (which exists by Lemma 2.9.2). Let $C_{\rm bi}$ be the directed cycle obtained from C in which the edge A_1B_1 is oriented towards B_1 and so on. Let $R_{\rm bi}$ be the complete bipartite digraph whose vertex classes are $\{A_1, \dots, A_K\}$ and $\{B_1, \dots, B_K\}$. Let $U_{{\rm bi},\ell'}$ be a bi-universal walk for C with parameter ℓ' as defined in Lemma 4.5.1. Let $U'_{{\rm bi},\ell'}$ be the closed walk obtained from $U_{{\rm bi},\ell'}$ as described in (BST5). We will call

$$(G, \mathcal{P}_{\rm bi}, \mathcal{P}''_{\rm bi}, R_{\rm bi}, C_{\rm bi}, U_{{\rm bi},\ell'}, U'_{{\rm bi},\ell'})$$

the *bi-setup associated to* $(G, \mathcal{P}, \mathcal{P}')$. The following lemma shows that it is indeed a bi-setup.

LEMMA 4.5.2. *Suppose that $K, L, m/L, \ell', m/\ell' \in \mathbb{N}$ with $\ell' \geq 4$, $K \geq 2$ and $0 < 1/m \ll 1/K, \varepsilon \ll \varepsilon', 1/\ell'$. Suppose that $(G, \mathcal{P}, \mathcal{P}')$ is a $[K, L, m, \varepsilon_0, \varepsilon]$-bi-scheme and that $C = A_1B_1 \dots A_KB_K$ is a spanning cycle on the clusters of $\mathcal{P}$. Then*

$$(G, \mathcal{P}_{\rm bi}, \mathcal{P}''_{\rm bi}, R_{\rm bi}, C_{\rm bi}, U_{{\rm bi},\ell'}, U'_{{\rm bi},\ell'})$$

is an $(\ell', 2K, m, \varepsilon', 1/2)$-bi-setup.

Proof. Clearly, $(G, \mathcal{P}_{\rm bi}, \mathcal{P}''_{\rm bi}, R_{\rm bi}, C_{\rm bi}, U_{{\rm bi},\ell'}, U'_{{\rm bi},\ell'})$ satisfies (BST1). (BSch3′) implies that (BST2) and (BST3) hold. Lemma 4.5.1 implies (BST4). (BST5) follows from the definition of $U'_{{\rm bi},\ell'}$. Finally, (BST6) follows from (BSch3′) and Lemma 2.9.2 since $\mathcal{P}''_{\rm bi}$ is an ε-uniform ℓ'-refinement of $\mathcal{P}_{\rm bi}$. □

We now state the 'bipartite version' of the robust decomposition lemma which was proved in [**21**]. It is an analogue of the robust decomposition lemma (Lemma 2.9.4) used in Chapter 2 and works for bi-setups rather than setups. As before, the lemma guarantees the existence of a 'robustly decomposable' digraph $G^{\rm rob}_{\rm dir}$, whose crucial property is that $H + G^{\rm rob}_{\rm dir}$ has a Hamilton decomposition for any sparse bipartite regular digraph H which is edge-disjoint from $G^{\rm rob}_{\rm dir}$.

Again, $G^{\rm rob}_{\rm dir}$ consists of digraphs $CA_{\rm dir}(r)$ (the 'chord absorber') and $PCA_{\rm dir}(r)$ (the 'parity extended cycle switcher') together with some special factors. $G^{\rm rob}_{\rm dir}$ is constructed in two steps: given a suitable set $\mathcal{SF}$ of special factors, the lemma first 'constructs' $CA_{\rm dir}(r)$ and then, given another suitable set $\mathcal{SF}'$ of special factors, the lemma 'constructs' $PCA_{\rm dir}(r)$.

LEMMA 4.5.3. *Suppose that* $0 < 1/m \ll 1/k \ll \varepsilon \ll 1/q \ll 1/f \ll r_1/m \ll d \ll 1/\ell', 1/g \ll 1$ *where* ℓ' *is even and that* $rk^2 \leq m$. *Let*

$$r_2 := 96\ell' g^2 kr, \quad r_3 := rfk/q, \quad r^\diamond := r_1 + r_2 + r - (q-1)r_3, \quad s' := rfk + 7r^\diamond$$

and suppose that $k/14, k/f, k/g, q/f, m/4\ell', fm/q, 2fk/3g(g-1) \in \mathbb{N}$. *Suppose that* $(G, \mathcal{P}, \mathcal{P}', R, C, U, U')$ *is an* $(\ell', k, m, \varepsilon, d)$*-bi-setup and* $C = V_1 \dots V_k$. *Suppose that* $\mathcal{P}^*$ *is a* (q/f)*-refinement of* $\mathcal{P}$ *and that* $SF_1, \dots, SF_{r_3}$ *are edge-disjoint special factors with parameters* $(q/f, f)$ *with respect to* C, $\mathcal{P}^*$ *in* G. *Let* $\mathcal{SF} := SF_1 + \dots + SF_{r_3}$. *Then there exists a digraph* $CA_{\rm dir}(r)$ *for which the following holds:*

(i) $CA_{\rm dir}(r)$ *is an* $(r_1 + r_2)$*-regular spanning subdigraph of* G *which is edge-disjoint from* $\mathcal{SF}$.

(ii) *Suppose that* $SF'_1, \dots, SF'_{r^\diamond}$ *are special factors with parameters* $(1, 7)$ *with respect to* C, $\mathcal{P}$ *in* G *which are edge-disjoint from each other and from* $CA_{\rm dir}(r) + \mathcal{SF}$. *Let* $\mathcal{SF}' := SF'_1 + \dots + SF'_{r^\diamond}$. *Then there exists a digraph* $PCA_{\rm dir}(r)$ *for which the following holds:*
   (a) $PCA_{\rm dir}(r)$ *is a* $5r^\diamond$*-regular spanning subdigraph of* G *which is edge-disjoint from* $CA_{\rm dir}(r) + \mathcal{SF} + \mathcal{SF}'$.
   (b) *Let* $\mathcal{SPS}$ *be the set consisting of all the* s' *special path systems contained in* $\mathcal{SF} + \mathcal{SF}'$. *Let* $V_{\rm even}$ *denote the union of all* V_i *over all even* $1 \leq i \leq k$ *and define* $V_{\rm odd}$ *similarly. Suppose that* H *is an* r*-regular bipartite digraph on* $V(G)$ *with vertex classes* $V_{\rm even}$ *and* $V_{\rm odd}$ *which is edge-disjoint from* $G^{\rm rob}_{\rm dir} := CA_{\rm dir}(r) + PCA_{\rm dir}(r) + \mathcal{SF} + \mathcal{SF}'$. *Then* $H + G^{\rm rob}_{\rm dir}$ *has a decomposition into* s' *edge-disjoint Hamilton cycles* $C_1, \dots, C_{s'}$. *Moreover,* C_i *contains one of the special path systems from* $\mathcal{SPS}$, *for each* $i \leq s'$.

Recall from Section 4.4.2 that we always view fictive edges in special factors as being distinct from each other and from the edges in other graphs. So for example, saying that $CA_{\rm dir}(r)$ and $\mathcal{SF}$ are edge-disjoint in Lemma 4.5.3 still allows for a fictive edge xy in $\mathcal{SF}$ to occur in $CA_{\rm dir}(r)$ as well (but $CA_{\rm dir}(r)$ will avoid all non-fictive edges in $\mathcal{SF}$).

We will use the following 'undirected' consequence of Lemma 4.5.3.

COROLLARY 4.5.4. *Suppose that* $0 < 1/m \ll \varepsilon_0, 1/K \ll \varepsilon \ll 1/L \ll 1/f \ll r_1/m \ll 1/\ell', 1/g \ll 1$ *where* ℓ' *is even and that* $4rK^2 \leq m$. *Let*

$$r_2 := 192\ell' g^2 Kr, \quad r_3 := 2rK/L, \quad r^\diamond := r_1 + r_2 + r - (Lf-1)r_3, \quad s' := 2rfK + 7r^\diamond$$

and suppose that $L, K/7, K/f, K/g, m/4\ell', m/L, 4fK/3g(g-1) \in \mathbb{N}$. *Suppose that* $(G_{\rm dir}, \mathcal{P}, \mathcal{P}')$ *is a* $[K, L, m, \varepsilon_0, \varepsilon]$*-bi-scheme and let* G' *denote the underlying undirected graph of* $G_{\rm dir}$. *Let* $C = A_1 B_1 \dots A_K B_K$ *be a spanning cycle on the clusters in* $\mathcal{P}$. *Suppose that* $BF_1, \dots, BF_{r_3}$ *are edge-disjoint balanced exceptional factors with parameters* (L, f) *for* $G_{\rm dir}$ *(with respect to* C, $\mathcal{P}'$*). Let* $\mathcal{BF} := BF_1 + \dots + BF_{r_3}$. *Then there exists a graph* $CA(r)$ *for which the following holds:*

(i) $CA(r)$ *is a* $2(r_1+r_2)$*-regular spanning subgraph of* G' *which is edge-disjoint from* $\mathcal{BF}$.

(ii) *Suppose that* $BF'_1, \dots, BF'_{r^\diamond}$ *are balanced exceptional factors with parameters* $(1,7)$ *for* $G_{\rm dir}$ *(with respect to* C, $\mathcal{P}$*) which are edge-disjoint from each other and from* $CA(r) + \mathcal{BF}$. *Let* $\mathcal{BF}' := BF'_1 + \dots + BF'_{r^\diamond}$. *Then there exists a graph* $PCA(r)$ *for which the following holds:*
   (a) $PCA(r)$ *is a* $10r^\diamond$*-regular spanning subgraph of* G' *which is edge-disjoint from* $CA(r) + \mathcal{BF} + \mathcal{BF}'$.
   (b) *Let* $\mathcal{BEPS}$ *be the set consisting of all the* s' *balanced exceptional path systems contained in* $\mathcal{BF}+\mathcal{BF}'$. *Suppose that* H *is a* $2r$*-regular bipartite graph on* $V(G_{\rm dir})$ *with vertex classes* $\bigcup_{i=1}^K A_i$ *and* $\bigcup_{i=1}^K B_i$ *which is edge-disjoint from* $G^{\rm rob} := CA(r) + PCA(r) + \mathcal{BF} + \mathcal{BF}'$. *Then* $H + G^{\rm rob}$ *has a decomposition into* s' *edge-disjoint Hamilton cycles* $C_1, \dots, C_{s'}$. *Moreover,* C_i *contains one of the balanced exceptional path systems from* $\mathcal{BEPS}$, *for each* $i \le s'$.

We remark that we write $A_1, \dots, A_K, B_1, \dots, B_K$ for the clusters in $\mathcal{P}$. Note that the vertex set of each of $\mathcal{EF}$, $\mathcal{EF}'$, $G^{\rm rob}$ includes V_0 while that of $G_{\rm dir}$, $CA(r)$, $PCA(r)$, H does not. Here $V_0 = A_0 \cup B_0$, where A_0 and B_0 are the exceptional sets of $\mathcal{P}$.

Proof. Choose new constants ε' and d such that $\varepsilon \ll \varepsilon' \ll 1/L$ and $r_1/m \ll d \ll 1/\ell', 1/g$. Consider the bi-setup $(G_{\rm dir}, \mathcal{P}_{\rm bi}, \mathcal{P}''_{\rm bi}, R_{\rm bi}, C_{\rm bi}, U_{{\rm bi},\ell'}, U'_{{\rm bi},\ell'})$ associated to $(G_{\rm dir}, \mathcal{P}, \mathcal{P}')$. By Lemma 4.5.2, $(G_{\rm dir}, \mathcal{P}_{\rm bi}, \mathcal{P}''_{\rm bi}, R_{\rm bi}, C_{\rm bi}, U_{{\rm bi},\ell'}, U'_{{\rm bi},\ell'})$ is an $(\ell', 2K, m, \varepsilon', 1/2)$-bi-setup and thus also an $(\ell', 2K, m, \varepsilon', d)$-bi-setup. Let $BF^*_{i,{\rm dir}}$ be as defined in Section 4.4.3. Recall from there that, for each $i \le r_3$, $BF^*_{i,{\rm dir}}$ is a special factor with parameters (L, f) with respect to C, $\mathcal{P}'$ in $G_{\rm dir}$ such that ${\rm Fict}(BF^*_{i,{\rm dir}})$ consists of all the edges in the J^* for all the Lf balanced exceptional systems J contained in BF_i. Thus we can apply Lemma 4.5.3 to $(G_{\rm dir}, \mathcal{P}_{\rm bi}, \mathcal{P}''_{\rm bi}, R_{\rm bi}, C_{\rm bi}, U_{{\rm bi},\ell'}, U'_{{\rm bi},\ell'})$ with $2K$, Lf, ε' playing the roles of k, q, ε in order to obtain a spanning subdigraph $CA_{\rm dir}(r)$ of $G_{\rm dir}$ which satisfies Lemma 4.5.3(i). Hence the underlying undirected graph $CA(r)$ of $CA_{\rm dir}(r)$ satisfies Corollary 4.5.4(i). Indeed, to check that $CA(r)$ and $\mathcal{BF}$ are edge-disjoint, by Lemma 4.5.3(i) it suffices to check that $CA(r)$ avoids all edges in all the balanced exceptional systems J contained in BF_i (for all $i \le r_3$). But this follows since $E(G_{\rm dir}) \supseteq E(CA(r))$ consists only of AB-edges by (BSch2$'$) and since no balanced exceptional system contains an AB-edge by (BES2).

Now let $BF'_1, \dots, BF'_{r^\diamond}$ be balanced exceptional factors as described in Corollary 4.5.4(ii). Similarly as before, for each $i \le r^\diamond$, $(BF'_i)^*_{\rm dir}$ is a special factor with parameters $(1,7)$ with respect to C, $\mathcal{P}$ in $G_{\rm dir}$ such that ${\rm Fict}((BF'_i)^*_{\rm dir})$ consists of all the edges in the J^* over all the 7 balanced exceptional systems J contained in BF'_i. Thus we can apply Lemma 4.5.3 to obtain a spanning subdigraph $PCA_{\rm dir}(r)$ of $G_{\rm dir}$ which satisfies Lemma 4.5.3(ii)(a) and (ii)(b). Hence the underlying undirected graph $PCA(r)$ of $PCA_{\rm dir}(r)$ satisfies Corollary 4.5.4(ii)(a).

It remains to check that Corollary 4.5.4(ii)(b) holds too. Thus let H be as described in Corollary 4.5.4(ii)(b). Let $H_{\rm dir}$ be an r-regular orientation of H. (To see that such an orientation exists, apply Petersen's theorem to obtain a decomposition of H into 2-factors and then orient each 2-factor to obtain a (directed) 1-factor.) Let $\mathcal{BF}^*_{\rm dir}$ be the union of the $BF^*_{i,{\rm dir}}$ over all $i \le r_3$ and let $(\mathcal{BF}')^*_{\rm dir}$

be the union of the $(BF_i')^*_{\rm dir}$ over all $i \le r^\diamond$. Then Lemma 4.5.3(ii)(b) implies that $H_{\rm dir} + CA_{\rm dir}(r) + PCA_{\rm dir}(r) + \mathcal{BF}^*_{\rm dir} + (\mathcal{BF}')^*_{\rm dir}$ has a decomposition into s' edge-disjoint (directed) Hamilton cycles $C_1', \dots, C_{s'}'$ such that each C_i' contains $BEPS^*_{i,\rm dir}$ for some balanced exceptional path system $BEPS_i$ from $\mathcal{BEPS}$. Let C_i be the undirected graph obtained from $C_i' - BEPS^*_{i,\rm dir} + BEPS_i$ by ignoring the directions of all the edges. Then Proposition 4.4.2 (applied with G' playing the role of G) implies that $C_1, \dots, C_{s'}$ is a decomposition of $H + G^{\rm rob} = H + CA(r) + PCA(r) + \mathcal{BF} + \mathcal{BF}'$ into edge-disjoint Hamilton cycles. □

4.6. Proof of Theorem 1.3.8

The proof of Theorem 1.3.8 is similar to that of Theorem 1.3.5 except that we do not need to apply the robust decomposition lemma in the proof of Theorem 1.3.8. For both results, we will need an approximate decomposition result (Lemma 4.6.1), which is stated below and proved in Chapter 5. Lemma 4.6.1 is a bipartite analogue of Lemma 2.5.4. It extends a suitable set of balanced exceptional systems into a set of edge-disjoint Hamilton cycles covering most edges of an almost complete and almost balanced bipartite graph.

LEMMA 4.6.1. *Suppose that $0 < 1/n \ll \varepsilon_0 \ll 1/K \ll \rho \ll 1$ and $0 \le \mu \ll 1$, where $n, K \in \mathbb{N}$ and K is even. Suppose that G is a graph on n vertices and $\mathcal{P}$ is a (K, m, ε_0)-partition of $V(G)$. Furthermore, suppose that the following conditions hold:*

- (a) *$d(w, B_i) = (1 - 4\mu \pm 4/K)m$ and $d(v, A_i) = (1 - 4\mu \pm 4/K)m$ for all $w \in A$, $v \in B$ and $1 \le i \le K$.*
- (b) *There is a set $\mathcal{J}$ which consists of at most $(1/4 - \mu - \rho)n$ edge-disjoint balanced exceptional systems with parameter ε_0 in G.*
- (c) *$\mathcal{J}$ has a partition into K^4 sets $\mathcal{J}_{i_1,i_2,i_3,i_4}$ (one for all $1 \le i_1, i_2, i_3, i_4 \le K$) such that each $\mathcal{J}_{i_1,i_2,i_3,i_4}$ consists of precisely $|\mathcal{J}|/K^4$ (i_1, i_2, i_3, i_4)-BES with respect to $\mathcal{P}$.*
- (d) *Each $v \in A \cup B$ is incident with an edge in J for at most $2\varepsilon_0 n$ $J \in \mathcal{J}$.*

Then G contains $|\mathcal{J}|$ edge-disjoint Hamilton cycles such that each of these Hamilton cycles contains some $J \in \mathcal{J}$.

To prove Theorem 1.3.8, we find a bi-framework via Corollary 4.2.12. Then we choose suitable balanced exceptional systems using Corollary 4.3.10. Finally, we extend these into Hamilton cycles using Lemma 4.6.1.

Proof of Theorem 1.3.8. **Step 1: Choosing the constants and a bi-framework.** By making α smaller if necessary, we may assume that $\alpha \ll 1$. Define new constants such that

$$0 < 1/n_0 \ll \varepsilon_{\rm ex} \ll \varepsilon_0 \ll \varepsilon_0' \ll \varepsilon' \ll \varepsilon_1 \ll \varepsilon_2 \ll \varepsilon_3 \ll \varepsilon_4 \ll 1/K \ll \alpha \ll \varepsilon \ll 1,$$

where $K \in \mathbb{N}$ and K is even.

Let G, F and D be as in Theorem 1.3.8. Apply Corollary 4.2.12 with $\varepsilon_{\rm ex}$, ε_0 playing the role of ε, ε^* to find a set $\mathcal{C}_1$ of at most $\varepsilon_{\rm ex}^{1/3} n$ edge-disjoint Hamilton cycles in F so that the graph G_1 obtained from G by deleting all the edges in these Hamilton cycles forms part of an $(\varepsilon_0, \varepsilon', K, D_1)$-bi-framework (G_1, A, A_0, B, B_0) with $D_1 \ge D - 2\varepsilon_{\rm ex}^{1/3} n$. Moreover, F satisfies (WF5) with respect to ε' and

$$|\mathcal{C}_1| = (D - D_1)/2. \qquad (4.6.1)$$

In particular, this implies that $\delta(G_1) \geq D_1$ and that D_1 is even (since D is even). Let F_1 be the graph obtained from F by deleting all those edges lying on Hamilton cycles in $\mathcal{C}_1$. Then

$$\delta(F_1) \geq \delta(F) - 2|\mathcal{C}_1| \geq (1/2 - 3\varepsilon_{\rm ex}^{1/3})n. \tag{4.6.2}$$

Let

$$m := \frac{|A|}{K} = \frac{|B|}{K} \quad \text{and} \quad t_K := \frac{(1-20\varepsilon_4)D_1}{2K^4}.$$

By changing ε_4 slightly, we may assume that $t_K \in \mathbb{N}$.

Step 2: Choosing a (K, m, ε_0)-partition $\mathcal{P}$. Apply Lemma 4.3.1 to the bi-framework (G_1, A, A_0, B, B_0) with F_1, ε_0 playing the roles of F, ε in order to obtain partitions $A_1, \dots, A_K$ and $B_1, \dots, B_K$ of A and B into sets of size m such that together with A_0 and B_0 the sets A_i and B_i form a $(K, m, \varepsilon_0, \varepsilon_1, \varepsilon_2)$-partition $\mathcal{P}$ for G_1.

Note that by Lemma 4.3.1(ii) and since F satisfies (WF5), for all $x \in A$ and $1 \leq j \leq K$, we have

$$\begin{aligned} d_{F_1}(x, B_j) &\geq \frac{d_{F_1}(x, B) - \varepsilon_1 n}{K} \overset{\text{(WF5)}}{\geq} \frac{d_{F_1}(x) - \varepsilon' n - |B_0| - \varepsilon_1 n}{K} \\ &\overset{(4.6.2)}{\geq} \frac{(1/2 - 3\varepsilon_{\rm ex}^{1/3})n - 2\varepsilon_1 n}{K} \geq (1 - 5\varepsilon_1)m. \end{aligned} \tag{4.6.3}$$

Similarly, $d_{F_1}(y, A_i) \geq (1 - 5\varepsilon_1)m$ for all $y \in B$ and $1 \leq i \leq K$.

Step 3: Choosing balanced exceptional systems for the almost decomposition. Apply Corollary 4.3.10 to the $(\varepsilon_0, \varepsilon', K, D_1)$-bi-framework (G_1,A,A_0,B,B_0) with F_1, G_1, ε_0, ε_0', D_1 playing the roles of F, G, ε, ε_0, D. Let $\mathcal{J}'$ be the union of the sets $\mathcal{J}_{i_1 i_2 i_3 i_4}$ guaranteed by Corollary 4.3.10. So $\mathcal{J}'$ consists of $K^4 t_K$ edge-disjoint balanced exceptional systems with parameter ε_0' in G_1 (with respect to $\mathcal{P}$). Let $\mathcal{C}_2$ denote the set of $10\varepsilon_4 D_1$ Hamilton cycles guaranteed by Corollary 4.3.10. Let F_2 be the subgraph obtained from F_1 by deleting all the Hamilton cycles in $\mathcal{C}_2$. Note that

$$D_2 := D_1 - 2|\mathcal{C}_2| = (1 - 20\varepsilon_4)D_1 = 2K^4 t_K = 2|\mathcal{J}'|. \tag{4.6.4}$$

Step 4: Finding the remaining Hamilton cycles. Our next aim is to apply Lemma 4.6.1 with F_2, $\mathcal{J}'$, ε' playing the roles of G, $\mathcal{J}$, ε_0.

Clearly, condition (c) of Lemma 4.6.1 is satisfied. In order to see that condition (a) is satisfied, let $\mu := 1/K$ and note that for all $w \in A$ we have

$$d_{F_2}(w, B_i) \geq d_{F_1}(w, B_i) - 2|\mathcal{C}_2| \overset{(4.6.3)}{\geq} (1 - 5\varepsilon_1)m - 20\varepsilon_4 D_1 \geq (1 - 1/K)m.$$

Similarly $d_{F_2}(v, A_i) \geq (1 - 1/K)m$ for all $v \in B$.

To check condition (b), note that

$$|\mathcal{J}'| \overset{(4.6.4)}{=} \frac{D_2}{2} \leq \frac{D}{2} \leq (1/2 - \alpha)\frac{n}{2} \leq (1/4 - \mu - \alpha/3)n.$$

Thus condition (b) of Lemma 4.6.1 holds with $\alpha/3$ playing the role of ρ. Since the edges in $\mathcal{J}'$ lie in G_1 and (G_1, A, A_0, B, B_0) is an $(\varepsilon_0, \varepsilon', K, D_1)$-bi-framework, (BFR5) implies that each $v \in A \cup B$ is incident with an edge in J for at most $\varepsilon' n + |V_0| \leq 2\varepsilon' n$ $J \in \mathcal{J}'$. (Recall that in a balanced exceptional system there are no edges between A and B.) So condition (d) of Lemma 4.6.1 holds with ε' playing the role of ε_0.

So we can indeed apply Lemma 4.6.1 to obtain a collection $\mathcal{C}_3$ of $|\mathcal{J}'|$ edge-disjoint Hamilton cycles in F_2 which cover all edges of $\bigcup \mathcal{J}'$. Then $\mathcal{C}_1 \cup \mathcal{C}_2 \cup \mathcal{C}_3$ is a set of edge-disjoint Hamilton cycles in F of size

$$|\mathcal{C}_1| + |\mathcal{C}_2| + |\mathcal{C}_3| \overset{(4.6.1),(4.6.4)}{=} \frac{D - D_1}{2} + \frac{D_1 - D_2}{2} + \frac{D_2}{2} = \frac{D}{2},$$

as required. □

4.7. Proof of Theorem 1.3.5

As mentioned earlier, the proof of Theorem 1.3.5 is similar to that of Theorem 1.3.8 except that we will also need to apply the robust decomposition lemma (Corollary 4.5.4). This means Steps 2–4 and Step 8 in the proof of Theorem 1.3.5 did not appear in the proof of Theorem 1.3.8. Steps 2–4 prepare the ground for the application of the robust decomposition lemma and in Step 8 we apply it to cover the leftover from the approximate decomposition step with Hamilton cycles. Steps 5–7 contain the approximate decomposition step, using Lemma 4.6.1.

In our proof of Theorem 1.3.5 it will be convenient to work with an undirected version of the bi-schemes introduced in Section 4.4.4. Given a graph G and partitions $\mathcal{P}$ and $\mathcal{P}'$ of a vertex set V, we call $(G, \mathcal{P}, \mathcal{P}')$ a $(K, L, m, \varepsilon_0, \varepsilon)$-*bi-scheme* if the following properties hold:

(BSch1) $(\mathcal{P}, \mathcal{P}')$ is a (K, L, m, ε_0)-partition of V. Moreover, $V(G) = A \cup B$.
(BSch2) Every edge of G joins some vertex in A to some vertex in B.
(BSch3) $d_G(v, A_{i,j}) \geq (1 - \varepsilon)m/L$ and $d_G(w, B_{i,j}) \geq (1 - \varepsilon)m/L$ for all $v \in B$, $w \in A$, $i \leq K$ and $j \leq L$.

We will also use the following proposition.

Proposition 4.7.1. *Suppose that $K, L, n, m/L \in \mathbb{N}$ and $0 < 1/n \ll \varepsilon, \varepsilon_0 \ll 1$. Let $(G, \mathcal{P}, \mathcal{P}')$ be a $(K, L, m, \varepsilon_0, \varepsilon)$-bi-scheme with $|G| = n$. Then there exists an orientation $G_{\rm dir}$ of G such that $(G_{\rm dir}, \mathcal{P}, \mathcal{P}')$ is a $[K, L, m, \varepsilon_0, 2\sqrt{\varepsilon}]$-bi-scheme.*

Proof. Randomly orient every edge in G to obtain an oriented graph $G_{\rm dir}$. (So given any edge xy in G with probability $1/2$, $xy \in E(G_{\rm dir})$ and with probability $1/2$, $yx \in E(G_{\rm dir})$.) (BSch1$'$) and (BSch2$'$) follow immediately from (BSch1) and (BSch2).

Note that Fact 1.4.3 and (BSch3) imply that $G[A_{i,j}, B_{i',j'}]$ is $[1, \sqrt{\varepsilon}]$-superregular with density at least $1 - \varepsilon$, for all $i, i' \leq K$ and $j, j' \leq L$. Using this, (BSch3$'$) follows easily from the large deviation bound in Proposition 1.4.4. (BSch4$'$) follows from Proposition 1.4.4 in a similar way. □

Proof of Theorem 1.3.5.
Step 1: Choosing the constants and a bi-framework. Define new constants such that

$$\begin{aligned}(4.7.1)\qquad 0 < 1/n_0 &\ll \varepsilon_{\rm ex} \ll \varepsilon_* \ll \varepsilon_0 \ll \varepsilon_0' \ll \varepsilon' \ll \varepsilon_1 \ll \varepsilon_2 \ll \varepsilon_3 \ll \varepsilon_4 \ll 1/K_2 \\ &\ll \gamma \ll 1/K_1 \ll \varepsilon'' \ll 1/L \ll 1/f \ll \gamma_1 \ll 1/g \ll \varepsilon \ll 1,\end{aligned}$$

where $K_1, K_2, L, f, g \in \mathbb{N}$ and both K_2, g are even. Note that we can choose the constants such that

$$\frac{K_1}{28fgL}, \frac{K_2}{4gLK_1}, \frac{4fK_1}{3g(g-1)} \in \mathbb{N}.$$

Let G and D be as in Theorem 1.3.5. By applying Dirac's theorem to remove a suitable number of edge-disjoint Hamilton cycles if necessary, we may assume that $D \leq n/2$. Apply Corollary 4.2.12 with G, $\varepsilon_{\rm ex}$, ε_*, ε_0, K_2 playing the roles of F, ε, ε^*, ε', K to find a set $\mathcal{C}_1$ of at most $\varepsilon_{\rm ex}^{1/3}n$ edge-disjoint Hamilton cycles in G so that the graph G_1 obtained from G by deleting all the edges in these Hamilton cycles forms part of an $(\varepsilon_*, \varepsilon_0, K_2, D_1)$-bi-framework (G_1, A, A_0, B, B_0), where

(4.7.2)
$$|A|+\varepsilon_0 n \geq n/2 \geq D_1 = D-2|\mathcal{C}_1| \geq D-2\varepsilon_{\rm ex}^{1/3}n \geq D-\varepsilon_0 n \geq n/2-2\varepsilon_0 n \geq |A|-2\varepsilon_0 n.$$

Note that G_1 is D_1-regular and that D_1 is even since D was even. Moreover, since $K_2/LK_1 \in \mathbb{N}$, (G_1, A, A_0, B, B_0) is also an $(\varepsilon_*, \varepsilon_0, K_1L, D_1)$-bi-framework and thus an $(\varepsilon_*, \varepsilon', K_1L, D_1)$-bi-framework.

Let

$$m_1 := \frac{|A|}{K_1} = \frac{|B|}{K_1}, \qquad r := \gamma m_1, \qquad r_1 := \gamma_1 m_1, \qquad r_2 := 192g^3K_1r,$$
$$r_3 := \frac{2rK_1}{L}, \qquad r^\diamond := r_1 + r_2 + r - (Lf-1)r_3,$$
$$D_4 := D_1 - 2(Lfr_3 + 7r^\diamond), \qquad t_{K_1L} := \frac{(1-20\varepsilon_4)D_1}{2(K_1L)^4}.$$

Note that (BFR4) implies $m_1/L \in \mathbb{N}$. Moreover,

$$(4.7.3) \qquad r_2, r_3 \leq \gamma^{1/2}m_1 \leq \gamma^{1/3}r_1, \qquad r_1/2 \leq r^\diamond \leq 2r_1.$$

Further, by changing $\gamma, \gamma_1, \varepsilon_4$ slightly, we may assume that $r/K_2^2, r_1, t_{K_1L} \in \mathbb{N}$. Since $K_1/L \in \mathbb{N}$ this implies that $r_3 \in \mathbb{N}$. Finally, note that

$$(4.7.4) \quad (1+3\varepsilon_*)|A| \geq D \geq D_4 \overset{(4.7.3)}{\geq} D_1 - \gamma_1 n \overset{(4.7.2)}{\geq} |A| - 2\gamma_1 n \geq (1-5\gamma_1)|A|.$$

Step 2: Choosing a $(K_1, L, m_1, \varepsilon_0)$-partition $(\mathcal{P}_1, \mathcal{P}_1')$. We now prepare the ground for the construction of the robustly decomposable graph $G^{\rm rob}$, which we will obtain via the robust decomposition lemma (Corollary 4.5.4) in Step 4.

Recall that (G_1, A, A_0, B, B_0) is an $(\varepsilon_*, \varepsilon', K_1L, D_1)$-bi-framework. Apply Lemma 4.3.1 with G_1, D_1, K_1L, ε_* playing the roles of G, D, K, ε to obtain partitions $A_1', \dots, A_{K_1L}'$ of A and $B_1', \dots, B_{K_1L}'$ of B into sets of size m_1/L such that together with A_0 and B_0 all these sets A_i' and B_i' form a $(K_1L, m_1/L, \varepsilon_*, \varepsilon_1, \varepsilon_2)$-partition $\mathcal{P}_1'$ for G_1. Note that $(1-\varepsilon_0)n \leq n - |A_0 \cup B_0| = 2K_1m_1 \leq n$ by (BFR4). For all $i \leq K_1$ and all $h \leq L$, let $A_{i,h} := A'_{(i-1)L+h}$. (So this is just a relabeling of the sets A_i'.) Define $B_{i,h}$ similarly and let $A_i := \bigcup_{h \leq L} A_{i,h}$ and $B_i := \bigcup_{h \leq L} B_{i,h}$. Let $\mathcal{P}_1 := \{A_0, B_0, A_1, \dots, A_{K_1}, B_1, \dots, B_{K_1}\}$ denote the corresponding $(K_1, m_1, \varepsilon_0)$-partition of $V(G)$. Thus $(\mathcal{P}_1, \mathcal{P}_1')$ is a $(K, L, m_1, \varepsilon_0)$-partition of $V(G)$, as defined in Section 4.4.2.

Let $G_2 := G_1[A, B]$. We claim that $(G_2, \mathcal{P}_1, \mathcal{P}'_1)$ is a $(K_1, L, m_1, \varepsilon_0, \varepsilon')$-bi-scheme. Indeed, clearly (BSch1) and (BSch2) hold. To verify (BSch3), recall that that (G_1, A, A_0, B, B_0) is an $(\varepsilon_*, \varepsilon_0, K_1L, D_1)$-bi-framework and so by (BFR5) for all $x \in B$ we have

$$d_{G_2}(x, A) \geq d_{G_1}(x) - d_{G_1}(x, B') - |A_0| \geq D_1 - \varepsilon_0 n - |A_0| \overset{(4.7.2)}{\geq} |A| - 4\varepsilon_0 n$$

and similarly $d_{G_2}(y, B) \geq |B| - 4\varepsilon_0 n$ for all $y \in A$. Since $\varepsilon_0 \ll \varepsilon'/K_1L$, this implies (BSch3).

Step 3: Balanced exceptional systems for the robustly decomposable graph. In order to apply Corollary 4.5.4, we first need to construct suitable balanced exceptional systems. Apply Corollary 4.3.10 to the $(\varepsilon_*, \varepsilon', K_1L, D_1)$-bi-framework (G_1, A, A_0, B, B_0) with G_1, K_1L, $\mathcal{P}'_1$, ε_* playing the roles of F, K, $\mathcal{P}$, ε in order to obtain a set $\mathcal{J}$ of $(K_1L)^4 t_{K_1L}$ edge-disjoint balanced exceptional systems in G_1 with parameter ε_0 such that for all $1 \le i'_1, i'_2, i'_3, i'_4 \le K_1L$ the set $\mathcal{J}$ contains precisely t_{K_1L} (i'_1, i'_2, i'_3, i'_4)-BES with respect to the partition $\mathcal{P}'_1$. (Note that F in Corollary 4.3.10 satisfies (WF5) since G_1 satisfies (BFR5).) So $\mathcal{J}$ is the union of all the sets $\mathcal{J}_{i'_1 i'_2 i'_3 i'_4}$ returned by Corollary 4.3.10. (Note that we will not use all the balanced exceptional systems in $\mathcal{J}$ and we do not need to consider the Hamilton cycles guaranteed by this result. So we do not need the full strength of Corollary 4.3.10 at this point.)

Our next aim is to choose two disjoint subsets $\mathcal{J}_{\rm CA}$ and $\mathcal{J}_{\rm PCA}$ of $\mathcal{J}$ with the following properties:

(a) In total $\mathcal{J}_{\rm CA}$ contains Lfr_3 balanced exceptional systems. For each $i \le f$ and each $h \le L$, $\mathcal{J}_{\rm CA}$ contains precisely r_3 (i_1, i_2, i_3, i_4)-BES of style h (with respect to the $(K, L, m_1, \varepsilon_0)$-partition $(\mathcal{P}_1, \mathcal{P}'_1)$) such that $i_1, i_2, i_3, i_4 \in \{(i-1)K_1/f + 2, \dots, iK_1/f\}$.

(b) In total $\mathcal{J}_{\rm PCA}$ contains $7r^\diamond$ balanced exceptional systems. For each $i \le 7$, $\mathcal{J}_{\rm PCA}$ contains precisely $r^\diamond$ (i_1, i_2, i_3, i_4)-BES (with respect to the partition $\mathcal{P}_1$) with $i_1, i_2, i_3, i_4 \in \{(i-1)K_1/7 + 2, \dots, iK_1/7\}$.

(Recall that we defined in Section 4.4.4 when an (i_1, i_2, i_3, i_4)-BES has style h with respect to a $(K, L, m_1, \varepsilon_0)$-partition $(\mathcal{P}_1, \mathcal{P}'_1)$.) To see that it is possible to choose $\mathcal{J}_{\rm CA}$ and $\mathcal{J}_{\rm PCA}$, split $\mathcal{J}$ into two sets $\mathcal{J}_1$ and $\mathcal{J}_2$ such that both $\mathcal{J}_1$ and $\mathcal{J}_2$ contain at least $t_{K_1L}/3$ (i'_1, i'_2, i'_3, i'_4)-BES with respect to $\mathcal{P}'_1$, for all $1 \le i'_1, i'_2, i'_3, i'_4 \le K_1L$. Note that there are $(K_1/f - 1)^4$ choices of 4-tuples (i_1, i_2, i_3, i_4) with $i_1, i_2, i_3, i_4 \in \{(i-1)K_1/f + 2, \dots, iK_1/f\}$. Moreover, for each such 4-tuple (i_1, i_2, i_3, i_4) and each $h \le L$ there is one 4-tuple (i'_1, i'_2, i'_3, i'_4) with $1 \le i'_1, i'_2, i'_3, i'_4 \le K_1L$ and such that any (i'_1, i'_2, i'_3, i'_4)-BES with respect to $\mathcal{P}'_1$ is an (i_1, i_2, i_3, i_4)-BES of style h with respect to $(\mathcal{P}_1, \mathcal{P}'_1)$. Together with the fact that

$$\frac{(K_1/f - 1)^4 t_{K_1L}}{3} \ge \frac{D_1}{7(Lf)^4} \ge \gamma^{1/2} n \overset{(4.7.3)}{\ge} r_3,$$

this implies that we can choose a set $\mathcal{J}_{\rm CA} \subseteq \mathcal{J}_1$ satisfying (a).

Similarly, there are $(K_1/7-1)^4$ choices of 4-tuples (i_1, i_2, i_3, i_4) with $i_1, i_2, i_3, i_4 \in \{(i-1)K_1/7 + 2, \dots, iK_1/7\}$. Moreover, for each such 4-tuple (i_1, i_2, i_3, i_4) there are L^4 distinct 4-tuples (i'_1, i'_2, i'_3, i'_4) with $1 \le i'_1, i'_2, i'_3, i'_4 \le K_1L$ and such that any (i'_1, i'_2, i'_3, i'_4)-BES with respect to $\mathcal{P}'_1$ is an (i_1, i_2, i_3, i_4)-BES with respect to $\mathcal{P}_1$. Together with the fact that

$$\frac{(K_1/7 - 1)^4 L^4 t_{K_1L}}{3} \ge \frac{D_1}{7^5} \ge \frac{n}{3 \cdot 7^5} \overset{(4.7.3)}{\ge} r^\diamond,$$

this implies that we can choose a set $\mathcal{J}_{\rm PCA} \subseteq \mathcal{J}_2$ satisfying (b).

Step 4: Finding the robustly decomposable graph. Recall that $(G_2, \mathcal{P}_1, \mathcal{P}'_1)$ is a $(K_1, L, m_1, \varepsilon_0, \varepsilon')$-bi-scheme. Apply Proposition 4.7.1 with G_2, $\mathcal{P}_1$, $\mathcal{P}'_1$, K_1, m_1, ε' playing the roles of G, $\mathcal{P}$, $\mathcal{P}'$, K, m, ε to obtain an orientation $G_{2,\rm dir}$ of G_2 such that $(G_{2,\rm dir}, \mathcal{P}_1, \mathcal{P}'_1)$ is a $[K_1, L, m_1, \varepsilon_0, 2\sqrt{\varepsilon'}]$-bi-scheme. Let $C = A_1B_1A_2 \dots A_{K_1}B_{K_1}$ be a spanning cycle on the clusters in $\mathcal{P}_1$.

Our next aim is to use Lemma 4.4.4 in order to extend the balanced exceptional systems in $\mathcal{J}_{\rm CA}$ into r_3 edge-disjoint balanced exceptional factors with parameters (L, f) for $G_{2,\rm dir}$ (with respect to C, $\mathcal{P}_1'$). For this, note that the condition on $\mathcal{J}_{CA}$ in Lemma 4.4.4 with r_3 playing the role of q is satisfied by (a). Moreover, $Lr_3/m_1 = 2rK_1/m_1 = 2\gamma K_1 \ll 1$. Thus we can indeed apply Lemma 4.4.4 to $(G_{2,\rm dir}, \mathcal{P}_1, \mathcal{P}_1')$ with $\mathcal{J}_{CA}$, $2\sqrt{\varepsilon'}$, K_1, r_3 playing the roles of $\mathcal{J}$, ε, K, q in order to obtain r_3 edge-disjoint balanced exceptional factors $BF_1, \dots, BF_{r_3}$ with parameters (L, f) for $G_{2,\rm dir}$ (with respect to C, $\mathcal{P}_1'$) such that together these balanced exceptional factors cover all edges in $\bigcup \mathcal{J}_{\rm CA}$. Let $\mathcal{BF}_{\rm CA} := BF_1 + \dots + BF_{r_3}$.

Note that $m_1/4g, m_1/L \in \mathbb{N}$ since $m_1 = |A|/K_1$ and $|A|$ is divisible by K_2 and thus m_1 is divisible by $4gL$ (since $K_2/4gLK_1 \in \mathbb{N}$ by our assumption). Furthermore, $4rK_1^2 = 4\gamma m_1 K_1^2 \leq \gamma^{1/2} m_1 \leq m_1$. Thus we can apply Corollary 4.5.4 to the $[K_1, L, m_1, \varepsilon_0, \varepsilon'']$-bi-scheme $(G_{2,\rm dir}, \mathcal{P}_1, \mathcal{P}_1')$ with K_1, ε'', g playing the roles of K, ε, ℓ' to obtain a spanning subgraph $CA(r)$ of G_2 as described there. (Note that G_2 equals the graph G' defined in Corollary 4.5.4.) In particular, $CA(r)$ is $2(r_1 + r_2)$-regular and edge-disjoint from $\mathcal{BF}_{\rm CA}$.

Let G_3 be the graph obtained from G_2 by deleting all the edges of $CA(r) + \mathcal{BF}_{\rm CA}$. Thus G_3 is obtained from G_2 by deleting at most $2(r_1 + r_2 + r_3) \leq 6r_1 = 6\gamma_1 m_1$ edges at every vertex in $A \cup B = V(G_3)$. Let $G_{3,\rm dir}$ be the orientation of G_3 in which every edge is oriented in the same way as in $G_{2,\rm dir}$. Then Proposition 1.4.1 implies that $(G_{3,\rm dir}, \mathcal{P}_1, \mathcal{P}_1)$ is still a $[K_1, 1, m_1, \varepsilon_0, \varepsilon]$-bi-scheme. Moreover,

$$\frac{r^\diamond}{m_1} \stackrel{(4.7.3)}{\leq} \frac{2r_1}{m_1} = 2\gamma_1 \ll 1.$$

Together with (b) this ensures that we can apply Lemma 4.4.4 to $(G_{3,\rm dir}, \mathcal{P}_1)$ with $\mathcal{P}_1$, $\mathcal{J}_{PCA}$, K_1, 1, 7, $r^\diamond$ playing the roles of $\mathcal{P}$, $\mathcal{J}$, K, L, f, q in order to obtain $r^\diamond$ edge-disjoint balanced exceptional factors $BF_1', \dots, BF_{r^\diamond}'$ with parameters $(1, 7)$ for $G_{3,\rm dir}$ (with respect to C, $\mathcal{P}_1$) such that together these balanced exceptional factors cover all edges in $\bigcup \mathcal{J}_{\rm PCA}$. Let $\mathcal{BF}_{\rm PCA} := BF_1' + \dots + BF_{r^\diamond}'$.

Apply Corollary 4.5.4 to obtain a spanning subgraph $PCA(r)$ of G_2 as described there. In particular, $PCA(r)$ is $10r^\diamond$-regular and edge-disjoint from $CA(r) + \mathcal{BF}_{\rm CA} + \mathcal{BF}_{\rm PCA}$.

Let $G^{\rm rob} := CA(r) + PCA(r) + \mathcal{BF}_{\rm CA} + \mathcal{BF}_{\rm PCA}$. Note that by (4.4.1) all the vertices in $V_0 := A_0 \cup B_0$ have the same degree $r_0^{\rm rob} := 2(Lfr_3 + 7r^\diamond)$ in $G^{\rm rob}$. So

$$7r_1 \stackrel{(4.7.3)}{\leq} r_0^{\rm rob} \stackrel{(4.7.3)}{\leq} 30r_1. \tag{4.7.5}$$

Moreover, (4.4.1) also implies that all the vertices in $A \cup B$ have the same degree $r^{\rm rob}$ in $G^{\rm rob}$, where $r^{\rm rob} = 2(r_1 + r_2 + r_3 + 6r^\diamond)$. So

$$r_0^{\rm rob} - r^{\rm rob} = 2\left(Lfr_3 + r^\diamond - (r_1 + r_2 + r_3)\right) = 2(Lfr_3 + r - (Lf - 1)r_3 - r_3) = 2r.$$

Step 5: Choosing a $(K_2, m_2, \varepsilon_0)$-partition $\mathcal{P}_2$. We now prepare the ground for the approximate decomposition step (i.e. to apply Lemma 4.6.1). For this, we need to work with a finer partition of $A \cup B$ than the previous one (this will ensure that the leftover from the approximate decomposition step is sufficiently sparse compared to $G^{\rm rob}$).

Let $G_4 := G_1 - G^{\rm rob}$ (where G_1 was defined in Step 1) and note that

$$D_4 = D_1 - r_0^{\rm rob} = D_1 - r^{\rm rob} - 2r. \tag{4.7.6}$$

So

(4.7.7) $d_{G_4}(x) = D_4 + 2r$ for all $x \in A \cup B$ and $d_{G_4}(x) = D_4$ for all $x \in V_0$.

(Note that D_4 is even since D_1 and $r_0^{\rm rob}$ are even.) So G_4 is D_4-balanced with respect to (A, A_0, B, B_0) by Proposition 4.2.1. Together with the fact that (G_1,A,A_0, B,B_0) is an $(\varepsilon_*, \varepsilon_0, K_2, D_1)$-bi-framework, this implies that (G_4,G_4,A,A_0,B,B_0) satisfies conditions (WF1)–(WF5) in the definition of an $(\varepsilon_*, \varepsilon_0, K_2, D_4)$-weak framework. However, some vertices in $A_0 \cup B_0$ might violate condition (WF6). (But every vertex in $A \cup B$ will still satisfy (WF6) with room to spare.) So we need to modify the partition of $V_0 = A_0 \cup B_0$ to obtain a new weak framework.

Consider a partition A_0^*, B_0^* of $A_0 \cup B_0$ which maximizes the number of edges in G_4 between $A_0^* \cup A$ and $B_0^* \cup B$. Then $d_{G_4}(v, A_0^* \cup A) \leq d_{G_4}(v)/2$ for all $v \in A_0^*$ since otherwise $A_0^* \setminus \{v\}, B_0^* \cup \{v\}$ would be a better partition of $A_0 \cup B_0$. Similarly $d_{G_4}(v, B_0^* \cup B) \leq d_{G_4}(v)/2$ for all $v \in B_0^*$. Thus (WF6) holds in G_4 (with respect to the partition $A \cup A_0^*$ and $B \cup B_0^*$). Moreover, Proposition 4.2.2 implies that G_4 is still D_4-balanced with respect to (A, A_0^*, B, B_0^*). Furthermore, with (BFR3) and (BFR4) applied to G_1, we obtain $e_{G_4}(A \cup A_0^*) \leq e_{G_1}(A \cup A_0) + |A_0^*||A \cup A_0^*| \leq \varepsilon_0 n^2$ and similarly $e_{G_4}(B \cup B_0^*) \leq \varepsilon_0 n^2$. Finally, every vertex in $A \cup B$ has internal degree at most $\varepsilon_0 n + |A_0 \cup B_0| \leq 2\varepsilon_0 n$ in G_4 (with respect to the partition $A \cup A_0^*$ and $B \cup B_0^*$). Altogether this implies that $(G_4, G_4, A, A_0^*, B, B_0^*)$ is an $(\varepsilon_0, 2\varepsilon_0, K_2, D_4)$-weak framework and thus also an $(\varepsilon_0, \varepsilon', K_2, D_4)$-weak framework.

Without loss of generality we may assume that $|A_0^*| \geq |B_0^*|$. Apply Lemma 4.2.11 to the $(\varepsilon_0, \varepsilon', K_2, D_4)$-weak framework $(G_4, G_4, A, A_0^*, B, B_0^*)$ to find a set $\mathcal{C}_2$ of $|\mathcal{C}_2| \leq \varepsilon_0 n$ edge-disjoint Hamilton cycles in G_4 so that the graph G_5 obtained from G_4 by deleting all the edges of these Hamilton cycles forms part of an $(\varepsilon_0, \varepsilon', K_2, D_5)$-bi-framework $(G_5, A, A_0^*, B, B_0^*)$, where

(4.7.8) $$D_5 = D_4 - 2|\mathcal{C}_2| \geq D_4 - 2\varepsilon_0 n.$$

Since D_4 is even, D_5 is even. Further,

(4.7.9)
$$d_{G_5}(x) \overset{(4.7.7)}{=} D_5 + 2r \text{ for all } x \in A \cup B \text{ and } d_{G_5}(x) \overset{(4.7.7)}{=} D_5 \text{ for all } x \in A_0^* \cup B_0^*.$$

Choose an additional constant ε_4' such that $\varepsilon_3 \ll \varepsilon_4' \ll 1/K_2$ and so that

$$t_{K_2} := \frac{(1 - 20\varepsilon_4')D_5}{2K_2^4} \in \mathbb{N}.$$

Now apply Lemma 4.3.1 to $(G_5, A, A_0^*, B, B_0^*)$ with D_5, K_2, ε_0 playing the roles of D, K, ε in order to obtain partitions $A_1, \dots, A_{K_2}$ and $B_1, \dots, B_{K_2}$ of A and B into sets of size

(4.7.10) $$m_2 := |A|/K_2$$

such that together with A_0^* and B_0^* the sets A_i and B_i form a $(K_2, m_2, \varepsilon_0, \varepsilon_1, \varepsilon_2)$-partition $\mathcal{P}_2$ for G_5. (Note that the previous partition of A and B plays no role in the subsequent argument, so denoting the clusters in $\mathcal{P}_2$ by A_i and B_i again will cause no notational conflicts.)

Step 6: Balanced exceptional systems for the approximate decomposition. In order to apply Lemma 4.6.1, we first need to construct suitable balanced exceptional systems. Apply Corollary 4.3.10 to the $(\varepsilon_0, \varepsilon', K_2, D_5)$-bi-framework $(G_5, A, A_0^*, B, B_0^*)$ with G_5, ε_0, ε_0', ε_4', K_2, D_5, $\mathcal{P}_2$ playing the roles of F, ε, ε_0, ε_4, K, D, $\mathcal{P}$. (Note that since we are letting G_5 play the role of F, condition (WF5) in the corollary immediately follows from (BFR5).) Let $\mathcal{J}'$ be the union of the sets $\mathcal{J}_{i_1 i_2 i_3 i_4}$ guaranteed by Corollary 4.3.10. So $\mathcal{J}'$ consists of $K_2^4 t_{K_2}$ edge-disjoint balanced exceptional systems with parameter ε_0' in G_5 (with respect to $\mathcal{P}_2$). Let $\mathcal{C}_3$ denote the set of Hamilton cycles guaranteed by Corollary 4.3.10. So $|\mathcal{C}_3| = 10\varepsilon_4' D_5$.

Let G_6 be the subgraph obtained from G_5 by deleting all those edges lying in the Hamilton cycles from $\mathcal{C}_3$. Set $D_6 := D_5 - 2|\mathcal{C}_3|$. So

(4.7.11)

$$d_{G_6}(x) \overset{(4.7.9)}{=} D_6 + 2r \text{ for all } x \in A \cup B \qquad \text{and} \qquad d_{G_6}(x) \overset{(4.7.9)}{=} D_6 \text{ for all } x \in V_0.$$

(Note that $V_0 = A_0 \cup B_0 = A_0^* \cup B_0^*$.) Let G_6' denote the subgraph of G_6 obtained by deleting all those edges lying in the balanced exceptional systems from $\mathcal{J}'$. Thus $G_6' = G^\diamond$, where $G^\diamond$ is as defined in Corollary 4.3.10(iv). In particular, V_0 is an isolated set in G_6' and G_6' is bipartite with vertex classes $A \cup A_0^*$ and $B \cup B_0^*$ (and thus also bipartite with vertex classes $A' = A \cup A_0$ and $B' = B \cup B_0$).

Consider any vertex $v \in V_0$. Then v has degree D_5 in G_5, degree two in each Hamilton cycle from $\mathcal{C}_3$, degree two in each balanced exceptional system from $\mathcal{J}'$ and degree zero in G_6'. Thus

$$D_6 + 2|\mathcal{C}_3| = D_5 \overset{(4.7.9)}{=} d_{G_5}(v) = 2|\mathcal{C}_3| + 2|\mathcal{J}'| + d_{G_6'}(v) = 2|\mathcal{C}_3| + 2|\mathcal{J}'|$$

and so

$$D_6 = 2|\mathcal{J}'|. \tag{4.7.12}$$

Step 7: Approximate Hamilton cycle decomposition. Our next aim is to apply Lemma 4.6.1 with G_6, $\mathcal{P}_2$, K_2, m_2, $\mathcal{J}'$, ε' playing the roles of G, $\mathcal{P}$, K, m, $\mathcal{J}$, ε_0. Clearly, condition (c) of Lemma 4.6.1 is satisfied. In order to see that condition (a) is satisfied, let $\mu := (r_0^{\rm rob} - 2r)/4K_2 m_2$ and note that

$$0 \le \frac{\gamma_1 m_1}{4K_2 m_2} \le \frac{7r_1 - 2r}{4K_2 m_2} \overset{(4.7.5)}{\le} \mu \overset{(4.7.5)}{\le} \frac{30 r_1}{4K_2 m_2} \le \frac{30\gamma_1}{K_1} \ll 1.$$

Recall that every vertex $v \in B$ satisfies

$$d_{G_5}(v) \overset{(4.7.9)}{=} D_5 + 2r \overset{(4.7.6),(4.7.8)}{=} D_1 - r_0^{\rm rob} + 2r \pm 2\varepsilon_0 n \overset{(4.7.2)}{=} |A| - r_0^{\rm rob} + 2r \pm 4\varepsilon_0 n.$$

Moreover,

$$d_{G_5}(v, A) = d_{G_5}(v) - d_{G_5}(v, B \cup B_0^*) - |A_0^*| \ge d_{G_5}(v) - 2\varepsilon' n,$$

where the last inequality holds since $(G_5, A, A_0^*, B, B_0^*)$ is an $(\varepsilon_0, \varepsilon', K_2, D_5)$-bi-framework (c.f. conditions (BFR4) and (BFR5)). Together with the fact that $\mathcal{P}_2$ is a $(K_2, m_2, \varepsilon_0, \varepsilon_1, \varepsilon_2)$-partition for G_5 (c.f. condition (P2)), this implies that

$$\begin{aligned} d_{G_5}(v, A_i) &= \frac{d_{G_5}(v, A) \pm \varepsilon_1 n}{K_2} = \frac{|A| - r_0^{\rm rob} + 2r \pm 2\varepsilon_1 n}{K_2} \\ &= \left(1 - \frac{r_0^{\rm rob} - 2r}{K_2 m_2} \pm 5\varepsilon_1\right) m_2 = (1 - 4\mu \pm 5\varepsilon_1) m_2 \\ &= (1 - 4\mu \pm 1/K_2) m_2. \end{aligned}$$

Recall that G_6 is obtained from G_5 by deleting all those edges lying in the Hamilton cycles in $\mathcal{C}_3$ and that

$$|\mathcal{C}_3| = 10\varepsilon_4' D_5 \le 10\varepsilon_4' D_4 \overset{(4.7.4)}{\le} 11\varepsilon_4' |A| \overset{(4.7.10)}{\le} m_2/K_2.$$

Altogether this implies that $d_{G_6}(v, A_i) = (1 - 4\mu \pm 4/K_2)m_2$. Similarly one can show that $d_{G_6}(w, B_j) = (1 - 4\mu \pm 4/K_2)m_2$ for all $w \in A$. So condition (a) of Lemma 4.6.1 holds.

To check condition (b), note that

$$|\mathcal{J}'| \stackrel{(4.7.12)}{=} \frac{D_6}{2} \leq \frac{D_4}{2} \stackrel{(4.7.6)}{\leq} \frac{D_1 - r_0^{\rm rob}}{2} \leq \frac{n}{4} - \mu \cdot 2K_2 m_2 - r \leq \left(\frac{1}{4} - \mu - \frac{\gamma}{3K_1}\right) n.$$

Thus condition (b) of Lemma 4.6.1 holds with $\gamma/3K_1$ playing the role of ρ.

Since the edges in $\mathcal{J}'$ lie in G_5 and $(G_5, A, A_0^*, B, B_0^*)$ is an $(\varepsilon_0, \varepsilon', K_2, D_5)$-bi-framework, (BFR5) implies that each $v \in A \cup B$ is incident with an edge in J for at most $\varepsilon' n + |V_0| \leq 2\varepsilon' n$ of the $J \in \mathcal{J}'$. (Recall that in a balanced exceptional system there are no edges between A and B.) So condition (d) of Lemma 4.6.1 holds with ε' playing the role of ε_0.

So we can indeed apply Lemma 4.6.1 to obtain a collection $\mathcal{C}_4$ of $|\mathcal{J}'|$ edge-disjoint Hamilton cycles in G_6 which cover all edges of $\bigcup \mathcal{J}'$.

Step 8: Decomposing the leftover and the robustly decomposable graph. Finally, we can apply the 'robust decomposition property' of $G^{\rm rob}$ guaranteed by Corollary 4.5.4 to obtain a Hamilton decomposition of the leftover from the previous step together with $G^{\rm rob}$.

To achieve this, let H' denote the subgraph of G_6 obtained by deleting all those edges lying in the Hamilton cycles from $\mathcal{C}_4$. Thus (4.7.11) and (4.7.12) imply that every vertex in V_0 is isolated in H' while every vertex $v \in A \cup B$ has degree $d_{G_6}(v) - 2|\mathcal{J}'| = D_6 + 2r - 2|\mathcal{J}'| = 2r$ in H' (the last equality follows from (4.7.12)). Moreover, $H'[A]$ and $H'[B]$ contain no edges. (This holds since H' is a spanning subgraph of $G_6 - \bigcup \mathcal{J}' = G_6'$ and since we have already seen that G_6' is bipartite with vertex classes A' and B'.) Now let $H := H'[A, B]$. Then Corollary 4.5.4(ii)(b) implies that $H + G^{\rm rob}$ has a Hamilton decomposition. Let $\mathcal{C}_5$ denote the set of Hamilton cycles thus obtained. Note that $H + G^{\rm rob}$ is a spanning subgraph of G which contains all edges of G which were not covered by $\mathcal{C}_1 \cup \mathcal{C}_2 \cup \mathcal{C}_3 \cup \mathcal{C}_4$. So $\mathcal{C}_1 \cup \mathcal{C}_2 \cup \mathcal{C}_3 \cup \mathcal{C}_4 \cup \mathcal{C}_5$ is a Hamilton decomposition of G. □

CHAPTER 5

Approximate decompositions

In this chapter we prove the approximate decomposition results, Lemmas 2.5.4 and 4.6.1. Recall that Lemma 2.5.4 gives an approximate Hamilton decomposition of our graph (with some additional properties) in the two cliques case whilst Lemma 4.6.1 gives an approximate Hamilton decomposition of our graph (with some additional properties) in the bipartite case. After introducing some tools in Section 5.1, we prove Lemma 2.5.4 in Sections 5.2–5.4. We then prove Lemma 4.6.1 using a similar approach in the final section. We remind the reader that many of the relevant definitions for Lemmas 2.5.4 and 4.6.1 are stated in Sections 2.3 and 4.3 respectively.

In this chapter it is convenient to view matchings as graphs (in which every vertex has degree precisely one).

5.1. Useful Results

5.1.1. Regular Spanning Subgraphs. The following lemma implies that any almost complete balanced bipartite graph has an approximate decomposition into perfect matchings. The proof is a straightforward application of the MaxFlowMinCut theorem.

LEMMA 5.1.1. *Suppose that $0 < 1/m \ll \varepsilon \ll \rho \ll 1$, that $0 \leq \mu \leq 1/4$ and that $m, \mu m, \rho m \in \mathbb{N}$. Suppose that Γ is a bipartite graph with vertex classes U and V of size m and with $(1-\mu-\varepsilon)m \leq \delta(\Gamma) \leq \Delta(\Gamma) \leq (1-\mu+\varepsilon)m$. Then Γ contains a spanning $(1-\mu-\rho)m$-regular subgraph Γ'. In particular, Γ contains at least $(1-\mu-\rho)m$ edge-disjoint perfect matchings.*

Proof. We first obtain a directed network N from Γ by adding a source s and a sink t. We add a directed edge su of capacity $(1-\mu-\rho)m$ for each $u \in U$ and a directed edge vt of capacity $(1-\mu-\rho)m$ for each $v \in V$. We give all the edges in Γ capacity 1 and direct them from U to V.

Our aim is to show that the capacity of any (s,t)-cut is at least $(1-\mu-\rho)m^2$. By the MaxFlowMinCut theorem this would imply that N admits an integer-valued flow of value $(1-\mu-\rho)m^2$ which by construction of N implies the existence of our desired subgraph Γ'.

Consider any (s,t)-cut $(S,\overline{S})$ where $S = \{s\} \cup S_1 \cup S_2$ with $S_1 \subseteq U$ and $S_2 \subseteq V$. Let $\overline{S}_1 := U \backslash S_1$ and $\overline{S}_2 := V \backslash S_2$. The capacity of this cut is

$$(1-\mu-\rho)m(m-|S_1|) + e(S_1,\overline{S}_2) + (1-\mu-\rho)m|S_2|$$

and therefore our aim is to show that

$$e(S_1,\overline{S}_2) \geq (1-\mu-\rho)m(|S_1|-|S_2|). \tag{5.1.1}$$

If $|S_1| \leq (1-\mu-\rho)m$, then

$$\begin{aligned} e(S_1, \overline{S}_2) &\geq ((1-\mu-\varepsilon)m - |S_2|)\,|S_1| \\ &= (1-\mu-\rho)m(|S_1|-|S_2|) + (\rho-\varepsilon)m|S_1| + |S_2|\,((1-\mu-\rho)m - |S_1|) \\ &\geq (1-\mu-\rho)m(|S_1|-|S_2|). \end{aligned}$$

Thus, we may assume that $|S_1| > (1-\mu-\rho)m$. Note that $|S_1|-|S_2| = |\overline{S}_2| - |\overline{S}_1|$. Therefore, by a similar argument, we may also assume that $|\overline{S}_2| > (1-\mu-\rho)m$ and so $|S_2| \leq (\mu+\rho)m$. This implies that

$$\begin{aligned} e(S_1, \overline{S}_2) &\geq \sum_{x\in S_1} d_\Gamma(x) - \sum_{y\in S_2} d_\Gamma(y) \geq (1-\mu-\varepsilon)m|S_1| - (1-\mu+\varepsilon)m|S_2| \\ &= (1-\mu-\rho)m(|S_1|-|S_2|) + \rho m(|S_1|-|S_2|) - \varepsilon m(|S_1|+|S_2|) \\ &> (1-\mu-\rho)m(|S_1|-|S_2|) + (1-2\mu-2\rho)\rho m^2 - (1+\mu+\rho)\varepsilon m^2 \\ &\geq (1-\mu-\rho)m(|S_1|-|S_2|). \end{aligned}$$

(Note that the last inequality follows as $\varepsilon \ll \rho \ll 1$ and $\mu \leq 1/4$.) So indeed (5.1.1) is satisfied, as desired. □

5.1.2. Hamilton Cyles in Robust Outexpanders. Recall that, given $0 < \nu \leq \tau < 1$, we say that a digraph G on n vertices is a *robust* (ν, τ)*-outexpander*, if for all $S \subseteq V(G)$ with $\tau n \leq |S| \leq (1-\tau)n$ the number of vertices that have at least νn inneighbours in S is at least $|S| + \nu n$. The following result was derived in [**20**] as a straightforward consequence of the result from [**25**] that every robust outexpander of linear minimum degree has a Hamilton cycle.

THEOREM 5.1.2. *Suppose that $0 < 1/n \ll \gamma \ll \nu \ll \tau \ll \eta < 1$. Let G be a digraph on n vertices with $\delta^+(G), \delta^-(G) \geq \eta n$ which is a robust (ν, τ)-outexpander. Let $y_1, \dots, y_p$ be distinct vertices in $V(G)$ with $p \leq \gamma n$. Then G contains a directed Hamilton cycle visiting $y_1, \dots, y_p$ in this order.*

5.1.3. A Regularity Concept for Sparse Graphs. We now formulate a concept of ε-superregularity which is suitable for 'sparse' graphs. Let G be a bipartite graph with vertex classes U and V, both of size m. Given $A \subseteq U$ and $B \subseteq V$, we write $d(A,B) := e(A,B)/|A||B|$ for the density of G between A and B. Given $0 < \varepsilon, d, d^*, c < 1$, we say that G is (ε, d, d^*, c)*-superregular* if the following conditions are satisfied:

(Reg1) Whenever $A \subseteq U$ and $B \subseteq V$ are sets of size at least εm, then $d(A,B) = (1 \pm \varepsilon)d$.
(Reg2) For all $u, u' \in V(G)$ we have $|N(u) \cap N(u')| \leq c^2 m$.
(Reg3) $\Delta(G) \leq cm$.
(Reg4) $\delta(G) \geq d^* m$.

Note that the above definitions also make sense if G is 'sparse' in the sense that $d < \varepsilon$ (which will be the case in our proofs). A bipartite digraph $G = G[U,V]$ is (ε, d, d^*, c)*-superregular* if this holds for the underlying undirected graph of G.

The following observation follows immediately from the definition.

PROPOSITION 5.1.3. *Suppose that $0 < 1/m \ll d^*, d, \varepsilon, \varepsilon', c \ll 1$ and $2\varepsilon' \leq d^*$. Let G be an (ε, d, d^*, c)-superregular bipartite graph with vertex classes U and V of size m. Let $U' \subseteq U$ and $V' \subseteq V$ with $|U'| = |V'| \geq (1-\varepsilon')m$. Then $G[U', V']$ is $(2\varepsilon, d, d^*/2, 2c)$-superregular.*

The following two simple observations were made in [**21**].

PROPOSITION 5.1.4. *Suppose that* $0 < 1/m \ll d^*, d, \varepsilon, c \ll 1$. *Let* G *be an* (ε, d, d^*, c)*-superregular bipartite graph with vertex classes* U *and* V *of size* m. *Suppose that* G' *is obtained from* G *by removing at most* $\varepsilon^2 dm$ *edges incident to each vertex from* G. *Then* G' *is* $(2\varepsilon, d, d^* - \varepsilon^2 d, c)$*-superregular.*

LEMMA 5.1.5. *Let* $0 < 1/m \ll \nu \ll \tau \ll d \leq \varepsilon \ll \mu, \zeta \leq 1/2$ *and let* G *be an* $(\varepsilon, d, \zeta d, d/\mu)$*-superregular bipartite graph with vertex classes* U *and* V *of size* m. *Let* $A \subseteq U$ *be such that* $\tau m \leq |A| \leq (1 - \tau)m$. *Let* $B \subseteq V$ *be the set of all those vertices in* V *which have at least* νm *neighbours in* A. *Then* $|B| \geq |A| + \nu m$.

5.2. Systems and Balanced Extensions

5.2.1. Sketch Proof of Lemma 2.5.4. Roughly speaking, the Hamilton cycles we find will have the following structure: let $A_1, \dots, A_K \subseteq A$ and $B_1, \dots, B_K \subseteq B$ be the clusters of the (K, m, ε_0)-partition $\mathcal{P}$ of $V(G)$ given in Lemma 2.5.4. So K is odd. Let $\mathcal{R}_A$ be the complete graph on $A_1, \dots, A_K$ and $\mathcal{R}_B$ be the complete graph on $B_1, \dots, B_K$. Since K is odd, Walecki's theorem [**26**] implies that $\mathcal{R}_A$ has a Hamilton decomposition $C_{A,1}, \dots, C_{A,(K-1)/2}$, and similarly $\mathcal{R}_B$ has a Hamilton decomposition $C_{B,1}, \dots, C_{B,(K-1)/2}$. Every Hamilton cycle C we construct in G will have the property that there is a j so that almost all edges of $C[A]$ wind around $C_{A,j}$ and almost all edges of $C[B]$ wind around $C_{B,j}$. Below, we describe the main ideas involved in the construction of the Hamilton cycles in more detail.

As indicated above, the first idea is that we can reduce the problem of finding the required edge-disjoint Hamilton cycles (and possibly perfect matchings) in G to that of finding appropriate Hamilton cycles on each of A and B separately.

More precisely, let $\mathcal{J}$ be a set of edge-disjoint exceptional systems as given in Lemma 2.5.4. By deleting some edges if necessary, we may further assume that $\mathcal{J}$ is an edge-decomposition of $G - G[A] - G[B]$. Thus, in order to prove Lemma 2.5.4, we have to find $|\mathcal{J}|$ suitable edge-disjoint subgraphs $H_{A,1}, \dots, H_{A,|\mathcal{J}|}$ of $G[A]$ and $|\mathcal{J}|$ suitable edge-disjoint subgraphs $H_{B,1}, \dots, H_{B,|\mathcal{J}|}$ of $G[B]$ such that $H_s := H_{A,s} + H_{B,s} + J_s$ are the desired spanning subgraphs of G. To prove this, for each $J \in \mathcal{J}$, we consider the two corresponding auxiliary subgraphs J^*_A and J^*_B defined at the beginning of Section 2.3. Thus J^*_A and J^*_B have the following crucial properties:

(α_1) J^*_A and J^*_B are matchings whose vertices are contained in A and B, respectively;

(α_2) the union of any Hamilton cycle C^*_A in $G[A] + J^*_A$ containing J^*_A (in some suitable order) and any Hamilton cycle C^*_B in $G[B] + J^*_B$ containing J^*_B (in some suitable order) corresponds to either a Hamilton cycle of G containing J or to the union of two edge-disjoint perfect matchings of G containing J.

Furthermore, J determines which of the cases in (α_2) holds: If J is a Hamilton exceptional system, then (α_2) will give a Hamilton cycle of G, while in the case when J is a matching exceptional system, (α_2) will give the union of two edge-disjoint perfect matchings of G. So roughly speaking, this allows us to work with multigraphs $G^*_A := G[A] + \sum_{J \in \mathcal{J}} J^*_A$ and $G^*_B := G[B] + \sum_{J \in \mathcal{J}} J^*_B$ rather than G in the two steps. Furthermore, the processes of finding Hamilton cycles in G^*_A and in G^*_B are independent (see Section 5.3.1 for more details).

By symmetry, it suffices to consider G^*_A in what follows. The second idea of the proof is that as an intermediate step, we decompose G^*_A into blown-up Hamilton cycles $G^*_{A,j}$. Roughly speaking, we will then find an approximate Hamilton decomposition of each $G^*_{A,j}$ separately.

More precisely, recall that $\mathcal{R}_A$ denotes the complete graph whose vertex set is $\{A_1, \dots, A_K\}$. As mentioned above, $\mathcal{R}_A$ has a Hamilton decomposition $C_{A,1}, \dots, C_{A,(K-1)/2}$. We decompose $G[A]$ into edge-disjoint subgraphs $G_{A,1}, \dots, G_{A,(K-1)/2}$ such that each $G_{A,j}$ corresponds to the 'blow-up' of $C_{A,j}$, i.e. $G_{A,j}[U, W] = G[U, W]$ for every edge $UW \in E(C_{A,j})$. (The edges of G lying inside one of the clusters $A_1, \dots, A_K$ are deleted.) We also partition the set $\{J^*_A : J \in \mathcal{J}\}$ into $(K-1)/2$ sets $\mathcal{J}^*_{A,1}, \dots, \mathcal{J}^*_{A,(K-1)/2}$ of roughly equal size. Set $G^*_{A,j} := G_{A,j} + \mathcal{J}^*_{A,j}$. Thus in order to prove Lemma 2.5.4, we need to find $|\mathcal{J}^*_{A,j}|$ edge-disjoint Hamilton cycles in $G^*_{A,j}$ (for each $j \le (K-1)/2$). Since $G^*_{A,j}$ is still close to being a blow-up of the cycle $C_{A,j}$, finding such Hamilton cycles seems feasible.

One complication is that in order to satisfy (α_2), we need to ensure that each Hamilton cycle in $G^*_{A,j}$ contains some $J^*_A \in \mathcal{J}^*_{A,j}$ (and it must traverse the edges of J^*_A in some given order). To achieve this, we will both orient and order the edges of J^*_A. So we will actually consider an ordered directed matching $J^*_{A,\mathrm{dir}}$ instead of J^*_A. (J^*_A itself will still be undirected and unordered). We orient the edges of $G_{A,j}$ such that the resulting oriented graph $G_{A,j,\mathrm{dir}}$ is a blow-up of the directed cycle $C_{A,j}$.

However, $J^*_{A,\mathrm{dir}}$ may not be 'locally balanced with respect to $C_{A,j}$'. This means that it is impossible to extend $J^*_{A,\mathrm{dir}}$ into a directed Hamilton cycle using only edges of $G_{A,j,\mathrm{dir}}$. For example, suppose that $G_{A,j,\mathrm{dir}}$ is a blow-up of the directed cycle $A_1A_2 \dots A_K$, i.e. each edge of $G_{A,j,\mathrm{dir}}$ joins A_i to A_{i+1} for some $1 \le i \le K$. If $J^*_{A,\mathrm{dir}}$ is non-empty and $V(J^*_{A,\mathrm{dir}}) \subseteq A_1$, then $J^*_{A,\mathrm{dir}}$ cannot be extended into a directed Hamilton cycle using edges of $G_{A,j,\mathrm{dir}}$ only. Therefore, each $J^*_{A,\mathrm{dir}}$ will first be extended into a 'locally balanced path sequence' PS. PS will have the property that it can be extended to a Hamilton cycle using only edges of $G_{A,j,\mathrm{dir}}$. We will call the set $\mathcal{BE}_j$ consisting of all such PS for all $J^*_A \in \mathcal{J}^*_{A,j}$ a balanced extension of $\mathcal{J}^*_{A,j}$. $\mathcal{BE}_j$ will be constructed in Section 5.3.3, using edges from a sparse graph H' on A (which is actually removed from $G[A]$ before defining $G_{A,1}, \dots, G_{A,(K-1)/2}$).

Finally, we find the required directed Hamilton cycles in $G_{A,j,\mathrm{dir}} + \mathcal{BE}_j$ in Section 5.4. We construct these by first extending the path sequences in $\mathcal{BE}_j$ into (directed) 1-factors, using edges which wind around the blow-up of $C_{A,j}$. These are then transformed into Hamilton cycles using a small set of edges set aside earlier (again the set of these edges winds around the blow-up of $C_{A,j}$).

5.2.2. Systems and Balanced Extensions. As mentioned above, the proof of Lemma 2.5.4 requires an edge-decomposition and orientation of $G[A]$ and $G[B]$ into blow-ups of directed cycles as well as 'balanced extensions'. These are defined in the current subsection.

Let $k, m \in \mathbb{N}$. Recall that a (k,m)-*equipartition* $\mathcal{Q}$ of a set V of vertices is a partition of V into sets $V_1, \dots, V_k$ such that $|V_i| = m$ for all $i \le k$. The V_i are called *clusters* of $\mathcal{Q}$. $(G, \mathcal{Q}, C)$ is a (k, m, μ, ε)-*cyclic system* if the following properties hold:

(Sys1) G is a digraph and $\mathcal{Q}$ is a (k,m)-equipartition of $V(G)$.

(Sys2) C is a directed Hamilton cycle on $\mathcal{Q}$ and G winds around C. Moreover, for every edge UW of C, we have $d^+_G(u,W) = (1-\mu\pm\varepsilon)m$ for every $u \in U$ and $d^-_G(w,U) = (1-\mu\pm\varepsilon)m$ for every $w \in W$.

So roughly speaking, such a cyclic system is a blown-up Hamilton cycle.

Let $\mathcal{Q}$ be a (k,m)-*equipartition* of V and let C be a directed Hamilton cycle on $\mathcal{Q}$. We say that a digraph H with $V(H) \subseteq V$ is *locally balanced with respect to* C if for every edge UW of C, the number of edges of H with initial vertex in U equals the number of edges of H with final vertex in W.

Recall that a path sequence is a digraph which is the union of vertex-disjoint directed paths (some of them might be trivial). Let M be a directed matching. We say a path sequence PS is a V_i-*extension of* M *with respect to* $\mathcal{Q}$ if each edge of M is contained in a distinct directed path in PS having its final vertex in V_i. Let $\mathcal{M} := \{M_1, \dots, M_q\}$ be a set of directed matchings. A set $\mathcal{BE}$ of path sequences is a *balanced extension of* $\mathcal{M}$ *with respect to* $(\mathcal{Q}, C)$ *and parameters* (ε, ℓ) if $\mathcal{BE}$ satisfies the following properties:

(BE1) $\mathcal{BE}$ consists of q path sequences $PS_1, \dots, PS_q$ such that $V(PS_i) \subseteq V$ for each $i \leq q$, each PS_i is locally balanced with respect to C and $PS_1 - M_1, \dots, PS_q - M_q$ are edge-disjoint from each other.

(BE2) Each PS_s is a V_{i_s}-extension of M_s with respect to $\mathcal{Q}$ for some $i_s \leq k$. Moreover, for each $i \leq k$ there are at most $\ell m/k$ indices $s \leq q$ such that $i_s = i$.

(BE3) $|V(PS_s) \cap V_i| \leq \varepsilon m$ for all $i \leq k$ and $s \leq q$. Moreover, for each $i \leq k$, there are at most $\ell m/k$ path sequences $PS_s \in \mathcal{BE}$ such that $V(PS_s) \cap V_i \neq \emptyset$.

Note that the 'moreover part' of (BE3) implies the 'moreover part' of (BE2).

Given an ordered directed matching $M = \{f_1, \dots, f_\ell\}$, we say that a directed cycle C' is *consistent with* M if C' contains M and visits the edges $f_1, \dots, f_\ell$ in this order. The following observation will be useful: suppose that PS is a V_i-extension of M and let x_j be the final vertex of the path in PS containing f_j. (So $x_1, \dots, x_\ell$ are distinct vertices of V_i.) Suppose also that C' is a directed cycle which contains PS and visits $x_1, \dots, x_\ell$ in this order. Then C' is consistent with M.

5.3. Finding Systems and Balanced Extensions for the Two Cliques Case

Let G be a graph, let $\mathcal{P}$ be a (K, m, ε_0)-partition of $V(G)$ and let $\mathcal{J}$ be a set of exceptional systems as given by Lemma 2.5.4. The aim of this section is to decompose $G[A] + G[B]$ into (k, m, μ, ε)-cyclic systems and to construct balanced extensions as described in Section 5.2.1. First we need to define $J^*_{A,\mathrm{dir}}$ and $J^*_{B,\mathrm{dir}}$ for each exceptional system $J \in \mathcal{J}$. Recall from Section 5.2.1 that these are introduced in order to be able to consider $G[A]$ and $G[B]$ separately (and thus to be able to ignore the exceptional vertices in $V_0 = A_0 \cup B_0$).

5.3.1. Defining the Graphs $J^*_{A,\mathrm{dir}}$ and $J^*_{B,\mathrm{dir}}$.

We recall the following definition of J^* from Section 2.3. Let A, A_0, B, B_0 be a partition of a vertex set V on n vertices and let J be an exceptional system with parameter ε_0. Since each maximal path in J has endpoints in $A \cup B$ and internal vertices in V_0, an exceptional system J naturally induces a matching J^*_{AB} on $A \cup B$. More precisely, if $P_1, \dots, P_{\ell'}$ are the non-trivial paths in J and x_i, y_i are the endpoints of P_i, then we define $J^*_{AB} := \{x_i y_i : i \leq \ell'\}$. Thus $e_{J^*_{AB}}(A, B)$ is equal to the number of AB-paths in J. In particular, $e_{J^*_{AB}}(A,B)$ is positive and even if J is a Hamilton exceptional system, while $e_{J^*_{AB}}(A,B)=0$ if J is a matching exceptional system. Without loss of generality we may assume that $x_1y_1,\dots,x_{2\ell}y_{2\ell}$ is an enumeration of the edges of $J^*_{AB}[A,B]$, where $x_i\in A$ and $y_i\in B$. Define

$$J^*_A := J^*_{AB}[A]\cup\{x_{2i-1}x_{2i}: 1\le i\le \ell\} \quad\text{and}\quad J^*_B := J^*_{AB}[B]\cup\{y_{2i}y_{2i+1}: 1\le i\le \ell\}$$

(with indices considered modulo 2ℓ). Let $J^*:=J^*_A+J^*_B$. So J^* is a matching and $e(J^*)=e(J^*_{AB})$. Moreover, by (EC2), (EC3) and (ES3) we have

$$e(J^*)=e(J^*_{AB})\le |V_0|+\sqrt{\varepsilon_0}n. \tag{5.3.1}$$

Recall that the edges in J^* are called *fictive* edges, and that if J_1 and J_2 are two edge-disjoint exceptional systems, then J^*_1 and J^*_2 may not be edge-disjoint.

Recall that we say that an (undirected) cycle C is *consistent with* J^*_A if C contains J^*_A and (there is an orientation of C which) visits the vertices $x_1,\dots,x_{2\ell}$ in this order. In a similar way we define when a cycle is consistent with J^*_B.

As mentioned in Section 5.2.1, we will orient and order the edges of J^*_A and J^*_B in a suitable way to obtain $J^*_{A,\mathrm{dir}}$ and $J^*_{B,\mathrm{dir}}$. Accordingly, we will need an oriented version of Proposition 2.3.1. For this, we first orient the edges of J^*_A by orienting the edge $x_{2i-1}x_{2i}$ from x_{2i-1} to x_{2i} for all $i\le\ell$ and the edges of $J^*_{AB}[A]$ arbitrarily. Next we order these directed edges as $f_1,\dots,f_{\ell_A}$ such that $f_i=x_{2i-1}x_{2i}$ for all $i\le\ell$, where $\ell_A:=e(J^*_A)$. Define $J^*_{A,\mathrm{dir}}$ to be the ordered directed matching $\{f_1,\dots,f_{\ell_A}\}$. Similarly, to define $J^*_{B,\mathrm{dir}}$, we first orient the edges of J^*_B by orienting the edge $y_{2i}y_{2i+1}$ from y_{2i} to y_{2i+1} for all $i\le\ell$ and the edges of $J^*_{AB}[B]$ arbitrarily. Next we order these directed edges as $f'_1,\dots,f'_{\ell_B}$ such that $f'_i=y_{2i}y_{2i+1}$ for all $i\le\ell$, where $\ell_B:=e(J^*_B)$. Define $J^*_{B,\mathrm{dir}}$ to be the ordered directed matching $\{f'_1,\dots,f'_{\ell_B}\}$. Note that if J is an (i,i')-ES, then $V(J^*_{A,\mathrm{dir}})\subseteq A_i$ and $V(J^*_{B,\mathrm{dir}})\subseteq B_{i'}$. Recall from Section 5.2.2 that a directed cycle $C_{A,\mathrm{dir}}$ is *consistent with* $J^*_{A,\mathrm{dir}}$ if $C_{A,\mathrm{dir}}$ contains $J^*_{A,\mathrm{dir}}$ and visits the edges $f_1,\dots,f_{\ell_A}$ in this order. The following proposition, which is similar to Proposition 2.8.1 follows easily from Proposition 2.3.1.

PROPOSITION 5.3.1. *Suppose that A,A_0,B,B_0 forms a partition of a vertex set V. Let J be an exceptional system. Let $C_{A,\mathrm{dir}}$ and $C_{B,\mathrm{dir}}$ be two directed cycles such that*

- *$C_{A,\mathrm{dir}}$ is a directed Hamilton cycle on A that is consistent with $J^*_{A,\mathrm{dir}}$;*
- *$C_{B,\mathrm{dir}}$ is a directed Hamilton cycle on B that is consistent with $J^*_{B,\mathrm{dir}}$.*

Then the following assertions hold, where C_A and C_B are the undirected cycles obtained from $C_{A,\mathrm{dir}}$ and $C_{B,\mathrm{dir}}$ by ignoring the directions of all the edges.

(i) *If J is a Hamilton exceptional system, then $C_A+C_B-J^*+J$ is a Hamilton cycle on V.*

(ii) *If J is a matching exceptional system, then $C_A+C_B-J^*+J$ is the union of a Hamilton cycle on A' and a Hamilton cycle on B'. In particular, if both $|A'|$ and $|B'|$ are even, then $C_A+C_B-J^*+J$ is the union of two edge-disjoint perfect matchings on V.*

5.3.2. Finding Systems. In this subsection, we will decompose (and orient) $G[A]$ into cyclic systems $(G_{A,j,\mathrm{dir}},\mathcal{Q}_A,C_{A,j})$, one for each $j\le(K-1)/2$. Roughly speaking, this corresponds to a decomposition into (oriented) blown-up Hamilton cycles. We will achieve this by considering a Hamilton decomposition of $\mathcal{R}_A$, where

$\mathcal{R}_A$ is the complete graph on $\{A_1, \ldots, A_K\}$. So each $C_{A,j}$ corresponds to one of the Hamilton cycles in this Hamilton decomposition. We will split the set $\{J^*_{A,\mathrm{dir}} : J \in \mathcal{J}\}$ into subsets $\mathcal{J}^*_{A,j}$ and assign $\mathcal{J}^*_{A,j}$ to the jth cyclic system. Moreover, for each $j \leq (K-1)/2$, we will also set aside a sparse spanning subgraph $H_{A,j}$ of $G[A]$, which is removed from $G[A]$ before the decomposition into cyclic systems. $H_{A,j}$ will be used later on in order to find a balanced extension of $\mathcal{J}^*_{A,j}$. We proceed similarly for $G[B]$.

LEMMA 5.3.2. *Suppose that $0 < 1/n \ll \varepsilon_0 \ll 1/K \ll \rho \ll 1$ and $0 \leq \mu \ll 1$, where $n, K \in \mathbb{N}$ and K is odd. Suppose that G is a graph on n vertices and $\mathcal{P}$ is a (K, m, ε_0)-partition of $V(G)$. Furthermore, suppose that the following conditions hold:*

- (a) *$d(v, A_i) = (1-4\mu \pm 4/K)m$ and $d(w, B_i) = (1-4\mu \pm 4/K)m$ for all $v \in A$, $w \in B$ and $1 \leq i \leq K$.*
- (b) *There is a set $\mathcal{J}$ which consists of at most $(1/4 - \mu - \rho)n$ edge-disjoint exceptional systems with parameter ε_0 in G.*
- (c) *$\mathcal{J}$ has a partition into K^2 sets $\mathcal{J}_{i,i'}$ (one for all $1 \leq i, i' \leq K$) such that each $\mathcal{J}_{i,i'}$ consists of precisely $|\mathcal{J}|/K^2$ (i, i')-ES with respect to $\mathcal{P}$.*
- (d) *If $\mathcal{J}$ contains matching exceptional systems then $|A'| = |B'|$ is even.*

Then for each $1 \leq j \leq (K-1)/2$, there is a pair of tuples $(G_{A,j}, \mathcal{Q}_A, C_{A,j}, H_{A,j}, \mathcal{J}^*_{A,j})$ and $(G_{B,j}, \mathcal{Q}_B, C_{B,j}, H_{B,j}, \mathcal{J}^*_{B,j})$ such that the following assertions hold:

- (a_1) *Each of $C_{A,1}, \ldots, C_{A,(K-1)/2}$ is a directed Hamilton cycle on $\mathcal{Q}_A := \{A_1, \ldots, A_K\}$ such that the undirected versions of these Hamilton cycles form a Hamilton decomposition of the complete graph on $\mathcal{Q}_A$.*
- (a_2) *$\mathcal{J}^*_{A,1}, \ldots, \mathcal{J}^*_{A,(K-1)/2}$ is a partition of $\{J^*_{A,\mathrm{dir}} : J \in \mathcal{J}\}$.*
- (a_3) *Each $\mathcal{J}^*_{A,j}$ has a partition into K sets $\mathcal{J}^*_{A,j,i}$ (one for each $1 \leq i \leq K$) such that $|\mathcal{J}^*_{A,j,i}| \leq (1 - 4\mu - 3\rho)m/K$ and each $J^*_{A,\mathrm{dir}} \in \mathcal{J}^*_{A,j,i}$ is an ordered directed matching with $e(J^*_{A,\mathrm{dir}}) \leq 5K\sqrt{\varepsilon_0}m$ and $V(J^*_{A,\mathrm{dir}}) \subseteq A_i$.*
- (a_4) *$G_{A,1}, \ldots, G_{A,(K-1)/2}, H_{A,1}, \ldots, H_{A,(K-1)/2}$ are edge-disjoint subgraphs of $G[A]$.*
- (a_5) *$H_{A,j}[A_i, A_{i'}]$ is a $10K\sqrt{\varepsilon_0}m$-regular graph for all $j \leq (K-1)/2$ and all $i, i' \leq K$ with $i \neq i'$.*
- (a_6) *For each $j \leq (K-1)/2$, there exists an orientation $G_{A,j,\mathrm{dir}}$ of $G_{A,j}$ such that $(G_{A,j,\mathrm{dir}}, \mathcal{Q}_A, C_{A,j})$ is a $(K, m, 4\mu, 5/K)$-cyclic system.*
- (a_7) *Analogous statements to (a_1)–(a_6) hold for $C_{B,j}, \mathcal{J}^*_{B,j}, G_{B,j}, H_{B,j}$ for all $j \leq (K-1)/2$, with $\mathcal{Q}_B := \{B_1, \ldots, B_K\}$.*

Proof. Since K is odd, by Walecki's theorem the complete graph on $\{A_1, \ldots, A_K\}$ has a Hamilton decomposition. (a_1) follows by orienting the edges of each of these Hamilton cycles to obtain directed Hamilton cycles $C_{A,1}, \ldots, C_{A,(K-1)/2}$.

For each $i, i' \leq K$, we partition $\mathcal{J}_{i,i'}$ into $(K-1)/2$ sets $\mathcal{J}_{i,i',j}$ (one for each $j \leq (K-1)/2$) whose sizes are as equal as possible. Note that if $J \in \mathcal{J}_{i,i',j}$, then J is an (i, i')-ES and so $V(J^*_{A,\mathrm{dir}}) \subseteq A_i$. Since $\mathcal{P}$ is a (K, m, ε_0)-partition of $V(G)$, $|V_0| \leq \varepsilon_0 n$ and $(1 - \varepsilon_0)n \leq 2Km$. Hence,

$$e(J^*_{A,\mathrm{dir}}) \leq e(J^*) \overset{(5.3.1)}{\leq} |V_0| + \sqrt{\varepsilon_0}n \leq 5\sqrt{\varepsilon_0}Km.$$

(a$_2$) is satisfied by setting $\mathcal{J}^*_{A,j,i} := \bigcup_{i' \leq K}\{J^*_{A,\mathrm{dir}} : J \in \mathcal{J}_{i,i',j}\}$ and $\mathcal{J}^*_{A,j} := \bigcup_{i \leq K} \mathcal{J}^*_{A,j,i}$. Note that

$$|\mathcal{J}^*_{A,j,i}| \leq \sum_{i' \leq K} \left(\frac{2|\mathcal{J}_{i,i'}|}{K-1} + 1 \right) \overset{\text{(c)}}{=} \frac{2|\mathcal{J}|}{K(K-1)} + K \overset{\text{(b)}}{\leq} \frac{(1/2 - 2\mu - 2\rho)n}{K(K-1)} + K$$
$$\leq (1 - 4\mu - 3\rho)m/K$$

as $2Km \geq (1-\varepsilon_0)n$ and $1/n \ll \varepsilon_0 \ll 1/K \ll \rho$. Hence (a$_3$) holds.

For $i, i' \leq K$ with $i \neq i'$, apply Lemma 5.1.1 with $G[A_i, A_{i'}], 4/K, \rho, 4\mu$ playing the roles of $\Gamma, \varepsilon, \rho, \mu$ to obtain a spanning $(1 - 4\mu - \rho)m$-regular subgraph $H_{i,i'}$ of $G[A_i, A_{i'}]$. Since $H_{i,i'}$ is a regular bipartite graph and $\varepsilon_0 \ll 1/K, \rho \ll 1$ and $0 \leq \mu \ll 1$, there exist $(K-1)/2$ edge-disjoint $10K\sqrt{\varepsilon_0}m$-regular spanning subgraphs $H_{i,i',1}, \dots, H_{i,i',(K-1)/2}$ of $H_{i,i'}$. Set $H_{A,j} := \sum_{1 \leq i,i' \leq K} H_{i,i',j}$ for each $j \leq (K-1)/2$. So (a$_5$) holds.

Define $G_A := G[A] - (H_{A,1} + \cdots + H_{A,(K-1)/2})$. Note that, as $\varepsilon_0 \ll 1/K$, (a) implies that $d_{G_A}(v, A_i) = (1 - 4\mu \pm 5/K)m$ for all $v \in A$ and all $i \leq K$. For each $j \leq (K-1)/2$, let $G_{A,j}$ be the graph on A whose edge set is the union of $G_A[A_i, A_{i'}]$ for each edge $A_iA_{i'} \in E(C_{A,j})$. Define $G_{A,j,\mathrm{dir}}$ to be the oriented graph obtained from $G_{A,j}$ by orienting every edge in $G_A[A_i, A_{i'}]$ from A_i to $A_{i'}$ (for each edge $A_iA_{i'} \in E(C_{A,j})$). Note that $(G_{A,j,\mathrm{dir}}, \mathcal{Q}_A, C_{A,j})$ is a $(K, m, 4\mu, 5/K)$-cyclic system for each $j \leq (K-1)/2$. Therefore, (a$_4$) and (a$_6$) hold. (a$_7$) can be proved by a similar argument. □

5.3.3. Constructing Balanced Extensions. Let $(G_{A,j}, \mathcal{Q}_A, C_{A,j}, H_{A,j}, \mathcal{J}^*_{A,j})$ be one of the 5-tuples obtained by Lemma 5.3.2. The next lemma will be applied to find a balanced extension of $\mathcal{J}^*_{A,j}$ with respect to $(\mathcal{Q}_A, C_{A,j})$, using edges of $H_{A,j}$ (after a suitable orientation of these edges). Consider any $J^*_{A,\mathrm{dir}} \in \mathcal{J}^*_{A,j,i}$. Lemma 5.3.2(a$_3$) guarantees that $V(J^*_{A,\mathrm{dir}}) \subseteq A_i$, and so $J^*_{A,\mathrm{dir}}$ is an A_i-extension of itself. Therefore, in order to find a balanced extension of $\mathcal{J}^*_{A,j}$, it is enough to extend each $J^*_{A,\mathrm{dir}} \in \mathcal{J}^*_{A,j}$ into a locally balanced path sequence by adding a directed matching which is vertex-disjoint from $J^*_{A,\mathrm{dir}}$ in such a way that (BE3) is satisfied as well.

LEMMA 5.3.3. *Suppose that $0 < 1/m \ll \varepsilon \ll 1$ and that $m, k \in \mathbb{N}$ with $k \geq 3$. Let $\mathcal{Q} = \{V_1, \dots, V_k\}$ be a (k, m)-equipartition of a set V of vertices and let $C = V_1 \dots V_k$ be a directed cycle. Suppose that there exist a set $\mathcal{M}$ and a graph H on V such that the following conditions hold:*

(i) *$\mathcal{M}$ can be partitioned into k sets $\mathcal{M}_1, \dots, \mathcal{M}_k$ such that $|\mathcal{M}_i| \leq m/k$ and each $M \in \mathcal{M}_i$ is an ordered directed matching with $e(M) \leq \varepsilon m$ and $V(M) \subseteq V_i$ (for all $i \leq k$).*

(ii) *$H[V_{i-1}, V_{i+1}]$ is a $2\varepsilon m$-regular graph for all $i \leq k$.*

Then there exist an orientation H_{dir} of H and a balanced extension $\mathcal{BE}$ of $\mathcal{M}$ with respect to $(\mathcal{Q}, C)$ and parameters $(2\varepsilon, 3)$ such that each path sequence in $\mathcal{BE}$ is obtained from some $M \in \mathcal{M}$ by adding edges of H_{dir}.

Proof. Fix $i \leq k$ and write $\mathcal{M}_i := \{M_1, \dots, M_{|\mathcal{M}_i|}\}$. We orient each edge in $H[V_{i-1}, V_{i+1}]$ from V_{i-1} to V_{i+1}. By (ii), $H[V_{i-1}, V_{i+1}]$ can be decomposed into $2\varepsilon m$ perfect matchings. Each perfect matching can be split into $1/2\varepsilon$ matchings, each containing at least εm edges. Recall from (i) that $|\mathcal{M}_i| \leq m/k$ and $e(M_j) \leq \varepsilon m$

for all $M_j \in \mathcal{M}_i$. Hence, $H[V_{i-1}, V_{i+1}]$ contains $|\mathcal{M}_i|$ edge-disjoint matchings $M'_1, \dots, M'_{|\mathcal{M}_i|}$ with $e(M'_j) = e(M_j)$ for all $j \leq |\mathcal{M}_i|$. Define $PS_j := M_j \cup M'_j$. Note that PS_j is locally balanced with respect to C. Also, PS_j is a V_i-extension of M_j (as $M_j \in \mathcal{M}_i$ and so $V(M_j) \subseteq V_i$ by (i)). Moreover,

$$(5.3.2)\quad |V(PS_j) \cap V_{i'}| = \begin{cases} |V(M_j)| = 2e(M_j) \leq 2\varepsilon m & \text{if } i' = i, \\ e(M_j) \leq \varepsilon m & \text{if } i' = i+1 \text{ or } i' = i-1, \\ 0 & \text{otherwise.} \end{cases}$$

For each $i \leq k$, set $\mathcal{PS}_i := \{PS_1, \dots, PS_{|\mathcal{M}_i|}\}$. Therefore, $\mathcal{BE} := \bigcup_{i \leq k} \mathcal{PS}_i$ is a balanced extension of $\mathcal{M}$ with respect to $(\mathcal{Q}, C)$ and parameters $(2\varepsilon, 3)$. Indeed, (BE3) follows from (5.3.2). As remarked after the definition of a balanced extension, this also implies the 'moreover part' of (BE2). Hence the lemma follows (by orienting the remaining edges of H arbitrarily). □

5.4. Constructing Hamilton Cycles via Balanced Extensions

Recall that a cyclic system can be viewed as a blow-up of a Hamilton cycle. Given a cyclic system $(G, \mathcal{Q}, C)$ and a balanced extension $\mathcal{BE}$ of a set $\mathcal{M}$ of ordered directed matchings, our aim is to extend each path sequence in $\mathcal{BE}$ into a Hamilton cycle using edges of G. Moreover, each Hamilton cycle has to be consistent with a distinct $M \in \mathcal{M}$. This is achieved by the following lemma, which is the key step in proving Lemma 2.5.4.

Lemma 5.4.1. *Suppose that $0 < 1/m \ll \varepsilon_0, \varepsilon', 1/k \ll 1/\ell, \rho \leq 1$, that $0 \leq \mu, \rho \ll 1$ and that $m, k, \ell, q \in \mathbb{N}$ with $q \leq (1 - \mu - \rho)m$. Let $(G, \mathcal{Q}, C)$ be a $(k, m, \mu, \varepsilon')$-cyclic system and let $\mathcal{M} = \{M_1, \dots, M_q\}$ be a set of q ordered directed matchings. Suppose that $\mathcal{BE} = \{PS_1, \dots, PS_q\}$ is a balanced extension of $\mathcal{M}$ with respect to $(\mathcal{Q}, C)$ and parameters (ε_0, ℓ) such that for each $s \leq q$, $M_s \subseteq PS_s$. Then there exist q Hamilton cycles $C_1, \dots, C_q$ in $G + \mathcal{BE}$ such that for all $s \leq q$, C_s contains PS_s and is consistent with M_s, and such that $C_1 - PS_1, \dots, C_q - PS_q$ are edge-disjoint subgraphs of G.*

Lemma 5.4.1 will be used both in the two cliques case (i.e. to prove Lemma 2.5.4) and in the bipartite case (i.e. to prove Lemma 4.6.1).

We now give an outline of the proof of Lemma 5.4.1, where for simplicity we assume that the path sequences in the balanced extension $\mathcal{BE}$ are edge-disjoint from each other. Our first step is to remove a sparse subdigraph H from G (see Lemma 5.4.2), and set $G' := G - H$. Next we extend each path sequence in $\mathcal{BE}$ into a (directed) 1-factor using edges of G' such that all these 1-factors are edge-disjoint from each other (see Lemma 5.4.3). Finally, we use edges of H to transform the 1-factors into Hamilton cycles (see Lemma 5.4.6).

The following lemma enables us to find a suitable sparse subdigraph H of G. Recall that (ε, d, d^*, c)-superregularity was defined in Section 5.1.3.

Lemma 5.4.2. *Suppose that $0 < 1/m \ll \varepsilon' \ll \gamma \ll \varepsilon \ll 1$ and $0 \leq \mu \ll \varepsilon$. Let G be a bipartite graph with vertex classes U and W of size m such that $d(v) = (1 - \mu \pm \varepsilon')m$ for all $v \in V(G)$. Then there is a spanning subgraph H of G which satisfies the following properties:*

(i) *H is $(\varepsilon, 2\gamma, \gamma, 3\gamma)$-superregular.*

(ii) *Let* $G' := G - H$. *Then* $d_{G'}(v) = (1 - \mu \pm 4\gamma)m$ *for all* $v \in V(G)$.

Proof. Note that $\delta(G) \geq (1 - \mu - \varepsilon')m \geq (1 - \varepsilon^3)m$ as $\varepsilon', \mu \ll \varepsilon \ll 1$. Thus, whenever $A \subseteq U$ and $B \subseteq W$ are sets of size at least εm, then

$$e_G(A, B) \geq (|B| - \varepsilon^3 m)|A| \geq (1 - \varepsilon^2)|A||B|. \tag{5.4.1}$$

Let H be a random subgraph of G which is obtained by including each edge of G with probability 2γ. (5.4.1) implies that whenever $A \subseteq U$ and $B \subseteq W$ are sets of size at least εm then

$$2\gamma(1 - \varepsilon^2)|A||B| \leq \mathbb{E}(e_H(A, B)) \leq 2\gamma|A||B|. \tag{5.4.2}$$

Further, for all $u, u' \in V(H)$,

$$\mathbb{E}(|N_H(u) \cap N_H(u')|) \leq 4\gamma^2 m \tag{5.4.3}$$

and

$$3\gamma m/2 \leq \mathbb{E}(\delta(H)), \mathbb{E}(\Delta(H)) \leq 2\gamma m. \tag{5.4.4}$$

Thus, (5.4.2)–(5.4.4) together with Proposition 1.4.4 imply that, with high probability, H is an $(\varepsilon, 2\gamma, \gamma, 3\gamma)$-superregular pair. Since $\Delta(H) \leq 3\gamma m$ by (Reg3) and $\varepsilon' \ll \gamma$, G' satisfies (ii). □

5.4.1. Transforming a Balanced Extension into 1-factors.

The next lemma will be used to extend each locally balanced path sequence PS belonging to a balanced extension $\mathcal{BE}$ into a (directed) 1-factor using edges from G'. We will select the edges from G' in such a way that (apart from the path sequences) the 1-factors obtained are edge-disjoint.

LEMMA 5.4.3. *Suppose that* $0 < 1/m \ll 1/k \leq \varepsilon \ll \rho, 1/\ell \leq 1$, *that* $\rho \ll 1$, *that* $0 \leq \mu \leq 1/4$ *and that* $q, m, k, \ell \in \mathbb{N}$ *with* $q \leq (1 - \mu - \rho)m$. *Let* $(G, \mathcal{Q}, C)$ *be a* (k, m, μ, ε)*-cyclic system, where* $C = V_1 \dots V_k$. *Suppose that there exists a set* $\mathcal{PS}$ *of* q *path sequences* $PS_1, \dots, PS_q$ *satisfying the following conditions:*

(i) *Each* $PS_s \in \mathcal{PS}$ *is locally balanced with respect to* C.
(ii) $|V(PS_s) \cap V_i| \leq \varepsilon m$ *for all* $i \leq k$ *and* $s \leq q$. *Moreover, for each* $i \leq k$, *there are at most* $\ell m/k$ PS_s *such that* $V(PS_s) \cap V_i \neq \emptyset$.

Then there exist q *directed* 1*-factors* $F_1, \dots, F_q$ *in* $G + \mathcal{PS}$ *such that for all* $s \leq q$ $PS_s \subseteq F_s$ *and* $F_1 - PS_1, \dots, F_q - PS_q$ *are edge-disjoint subgraphs of* G.

Proof. By changing the values of ρ, μ and ε slightly, we may assume that $\rho m, \mu m \in \mathbb{N}$. For each $s \leq q$ and each $i \leq k$, let $V_i^{s,-}$ (or $V_i^{s,+}$) be the set of vertices in V_i with indegree (or outdegree) one in PS_s. Since each PS_s is locally balanced with respect to C, $|V_i^{s,+}| = |V_{i+1}^{s,-}| \leq \varepsilon m$ for all $s \leq q$ and all $i \leq k$ (where the inequality follows from (ii)). To prove the lemma, it suffices to show that for each $i \leq k$, there exist edge-disjoint directed matchings $M_i^1, \dots, M_i^q$, so that each M_i^s is a perfect matching in $G[V_i \setminus V_i^{s,+}, V_{i+1} \setminus V_{i+1}^{s,-}]$. The lemma then follows by setting $F_s := PS_s + \sum_{i \leq k} M_i^s$ for each $s \leq q$.

Fix any $i \leq k$. Without loss of generality (by relabelling the PS_s if necessary) we may assume that there exists an integer s_0 such that $V_i^{s,+} \neq \emptyset$ for all $s \leq s_0$ and $V_i^{s,+} = \emptyset$ for all $s_0 < s \leq q$. By (ii), $s_0 \leq \ell m/k$. Suppose that for some s

with $1 \leq s \leq s_0$ we have already found our desired matchings $M_i^1, \dots, M_i^{s-1}$ in $G[V_i, V_{i+1}]$. Let

$$V_i' := V_i \setminus V_i^{s,+}, \quad V_{i+1}' := V_{i+1} \setminus V_{i+1}^{s,-} \quad \text{and} \quad G_s := G[V_i', V_{i+1}'] - \sum_{s'<s} M_i^{s'}.$$

Note that each $v \in V_i'$ satisfies

$$d_{G_s}^+(v) \geq d_G^+(v, V_{i+1}) - (|V_{i+1}^{s,-}| + s_0) \geq (1 - \mu - (2\varepsilon + \ell/k))m \geq (1 - \mu - \sqrt{\varepsilon})m$$

by (Sys2) and the fact that $1/k \leq \varepsilon \ll 1/\ell$. Similarly, each $v \in V_{i+1}'$ satisfies $d_{G_s}^-(v) \geq (1-\mu-\sqrt{\varepsilon})m$. Thus G_s contains a perfect matching M_i^s (this follows, for example, from Hall's theorem). So we can find edge-disjoint matchings $M_i^1, \dots, M_i^{s_0}$ in $G[V_i, V_{i+1}]$.

Let G' be the subdigraph of $G[V_i, V_{i+1}]$ obtained by removing all the edges in $M_i^1, \dots, M_i^{s_0}$. Since $V_i^{s,+} = \emptyset$ for all $s_0 < s \leq q$ (and thus also $V_{i+1}^{s,-} = \emptyset$ for all such s), in order to prove the lemma it suffices to find $q - s_0$ edge-disjoint perfect matchings in G'. Each $v \in V_i$ satisfies

$$d_{G'}^+(v) = d_G^+(v, V_{i+1}) \pm s_0 = d_G^+(v, V_{i+1}) \pm \ell m/k = (1 - \mu \pm \sqrt{\varepsilon})m$$

by (Sys2) and the fact that $1/k \leq \varepsilon \ll 1/\ell$. Similarly, each $v \in V_{i+1}$ satisfies $d_{G'}^-(v) = (1 - \mu \pm \sqrt{\varepsilon})m$. Set $\rho' := \rho + s_0/m$. Note that $\rho' m \in \mathbb{N}$ and $\rho \leq \rho' \leq \rho + \ell/k \leq 2\rho$ as $1/k \ll 1/\ell, \rho$. Hence, $\varepsilon \ll \rho' \ll 1$. Thus we can apply Lemma 5.1.1 with $G', \rho', \sqrt{\varepsilon}$ playing the roles of $\Gamma, \rho, \varepsilon$ to obtain $(1 - \mu - \rho')m$ edge-disjoint perfect matchings in G'. Since $(1 - \mu - \rho')m = (1 - \mu - \rho)m - s_0 \geq q - s_0$, there exists $q - s_0$ edge-disjoint perfect matchings $M_i^{s_0+1}, \dots, M_i^q$ in G'. This completes the proof of the lemma. □

5.4.2. Merging Cycles to Obtain Hamilton Cycles. Recall that we have removed a sparse subdigraph H from G and that $G' = G - H$. Our final step in the proof of Lemma 5.4.1 is to merge the cycles from each of the 1-factors F_s returned by Lemma 5.4.3 to obtain edge-disjoint (directed) Hamilton cycles. We will apply Lemma 5.4.4 to merge the cycles of each F_s, using the edges in H. However, the Hamilton cycles obtained in this way might not be consistent with the matching $M_s \in \mathcal{M}$ that lies in PS_s. Lemma 5.4.5 is designed to deal with this issue.

Lemma 5.4.4 was proved in [**21**] and was first used to construct approximate Hamilton decompositions in [**31**]. Roughly speaking, it asserts the following: suppose that we have a 1-factor F where most of the edges wind around a cycle $C = V_1 \dots V_k$. Suppose also that we have a digraph H which winds around C. (More precisely, H is the union of superregular pairs $H[V_i, V_{i+1}]$.) Then we can transform F into a Hamilton cycle C' by using a few edges of H. The crucial point is that when applying this lemma, the edges in $C' - F$ can be taken from a small number of the superregular pairs $H[V_i, V_{i+1}]$ (i.e. the set J in Lemma 5.4.4 will be very small compared to k). In this way, we can transform many 1-factors F into edge-disjoint Hamilton cycles without using any of the pairs $H[V_i, V_{i+1}]$ too often. This in turn means that we will be able to transform all of our 1-factors into edge-disjoint Hamilton cycles by using the edges of a single sparse graph H.

LEMMA 5.4.4. *Suppose that* $0 < 1/m \ll d' \ll \varepsilon \ll d \ll \zeta, 1/t \leq 1/2$. *Let* $V_1, \dots, V_k$ *be pairwise disjoint clusters, each of size* m, *and let* $C = V_1 \dots V_k$ *be a directed cycle on these clusters. Let* H *be a digraph on* $V_1 \cup \dots \cup V_k$ *and let*

$J \subseteq E(C)$. For each edge $V_iV_{i+1} \in J$, let $V_i^1 \subseteq V_i$ and $V_{i+1}^2 \subseteq V_{i+1}$ be such that $|V_i^1| = |V_{i+1}^2| \geq m/100$ and such that $H[V_i^1, V_{i+1}^2]$ is $(\varepsilon, d', \zeta d', td'/d)$-superregular. Suppose that F is a 1-regular digraph with $V_1 \cup \dots \cup V_k \subseteq V(F)$ such that the following properties hold:

(i) *For each edge $V_iV_{i+1} \in J$ the digraph $F[V_i^1, V_{i+1}^2]$ is a perfect matching.*
(ii) *For each cycle D in F there is some edge $V_iV_{i+1} \in J$ such that D contains a vertex in V_i^1.*
(iii) *Whenever $V_iV_{i+1}, V_jV_{j+1} \in J$ are such that J avoids all edges in the segment $V_{i+1}CV_j$ of C from V_{i+1} to V_j, then F contains a path P_{ij} joining some vertex $u_{i+1} \in V_{i+1}^2$ to some vertex $u_j' \in V_j^1$ such that P_{ij} winds around C.*

Then we can obtain a directed cycle on $V(F)$ from F by replacing $F[V_i^1, V_{i+1}^2]$ with a suitable perfect matching in $H[V_i^1, V_{i+1}^2]$ for each edge $V_iV_{i+1} \in J$.

LEMMA 5.4.5. *Suppose that $0 < 1/m \ll \gamma \ll d' \ll \varepsilon \ll d \ll \zeta, 1/t \leq 1/2$. Let $V_1, \dots, V_k$ be pairwise disjoint clusters, each of size m, and let $C = V_1 \dots V_k$ be a directed cycle on these clusters. Let $1 \leq i \leq k$ be fixed and let $V_i^1 \subseteq V_i$ and $V_{i+1}^2 \subseteq V_{i+1}$ be such that $|V_i^1| = |V_{i+1}^2| \geq m/100$. Suppose that $H = H[V_i^1, V_{i+1}^2]$ is an $(\varepsilon, d', \zeta d', td'/d)$-superregular bipartite digraph. Let $X = \{x_1, \dots, x_p\} \subseteq V_i^1$ with $|X| \leq \gamma m$. Suppose that C' is a directed cycle with $V_1 \cup \dots \cup V_k \subseteq V(C')$ such that $C'[V_i^1, V_{i+1}^2]$ is a perfect matching. Then we can obtain a directed cycle on $V(C')$ from C' that visits the vertices $x_1, \dots, x_p$ in order by replacing $C'[V_i^1, V_{i+1}^2]$ with a suitable perfect matching in $H[V_i^1, V_{i+1}^2]$.*

Proof. Pick ν and τ such that $\gamma \ll \nu \ll \tau \ll d'$. For every $u \in V_i^1$, starting at u we move along the cycle C' (but in the opposite direction to the orientation of the edges) and let $f(u)$ be the first vertex on C' in V_{i+1}^2. (Note that $f(u)$ exists since $C'[V_i^1, V_{i+1}^2]$ is a perfect matching. Moreover, $f(u) \neq f(v)$ if $u \neq v$.) Define an auxiliary digraph A on V_{i+1}^2 such that $N_A^+(f(u)) := N_H^+(u)$. So A is obtained by identifying each pair $(u, f(u))$ into one vertex with an edge from $(u, f(u))$ to $(v, f(v))$ if H has an edge from u to $f(v)$. So Lemma 5.1.5 applied with d', d/t playing the roles of d, μ implies that A is a robust (ν, τ)-outexpander. Moreover, $\delta^+(A), \delta^-(A) \geq \zeta d' |V_{i+1}^2| = \zeta d' |A|$ by (Reg4). Thus Theorem 5.1.2 implies that A has a Hamilton cycle visiting $f(x_1), \dots, f(x_p)$ in order, which clearly corresponds to a perfect matching M in H with the desired property. □

The above proof idea is actually quite similar to that for Lemma 5.4.4 itself. We now apply Lemmas 5.4.4 and 5.4.5 to each 1-factor F_s given by Lemma 5.4.3 and obtain edge-disjoint Hamilton cycles that are consistent with the M_s.

LEMMA 5.4.6. *Suppose that $0 < 1/m \ll \varepsilon_0, 1/k \ll \gamma \ll \varepsilon \ll 1$, that $\gamma \ll 1/\ell \leq 1$ and that $q, m, k, \ell \in \mathbb{N}$. Let $\mathcal{Q} = \{V_1, \dots, V_k\}$ be a (k, m)-equipartition of a vertex set V and let $C = V_1 \dots V_k$ be a directed cycle. Let $\mathcal{M} = \{M_1, \dots, M_q\}$ be a set of ordered directed matchings. Suppose that $\mathcal{BE} = \{PS_1, \dots, PS_q\}$ is a balanced extension of $\mathcal{M}$ with respect to $(\mathcal{Q}, C)$ and parameters (ε_0, ℓ). Furthermore, suppose that there exist 1-regular digraphs $F_1, \dots, F_q$ on V such that for each $s \leq q$, $PS_s \subseteq F_s$ and such that $F_s - PS_s$ winds around C. Let H be a digraph on V which is edge-disjoint from each of $F_1 - PS_1, \dots, F_q - PS_q$ and such that $H[V_i, V_{i+1}]$ is $(\varepsilon, 2\gamma, \gamma, 3\gamma)$-superregular for all $i \leq k$. Then there exist q Hamilton cycles $C_1, \dots, C_q$ in $F_1 + \dots + F_q + H$ such that C_s contains PS_s and is consistent* *with* M_s *for all* $s \leq q$ *and such that* $C_1 - F_1, \dots, C_q - F_q$ *are edge-disjoint subgraphs of* H.

Proof. Recall from (BE2) that for each $s \leq q$ there is some $i_s \leq k$ such that PS_s is a V_{i_s}-extension of M_s. In particular, $M_s \subseteq PS_s$. Let I_s be the set consisting of all $i \leq k$ such that $V_i \cap V(PS_s) \neq \emptyset$. Since $\mathcal{BE}$ is a balanced extension with parameters (ε_0, ℓ), (BE3) implies that for every $i \leq k$ we have

$$|\{s : i \in I_s\}| \leq \ell m/k. \tag{5.4.5}$$

For each $s \leq q$ in turn, we are going to show that there exist Hamilton cycles $C_1, \dots, C_s$ in $F_1 + \dots + F_s + H$ such that

(a_s) $PS_{s'} \subseteq C_{s'}$ and $C_{s'}$ is consistent with $M_{s'}$ for all $s' \leq s$,
(b_s) $E(C_{s'} - F_{s'}) \subseteq \bigcup_{i \in I_{s'}} E(H[V_i, V_{i+1}])$ for all $s' \leq s$,
(c_s) $C_1 - F_1, \dots, C_s - F_s$ are pairwise edge-disjoint.

So suppose that for some s with $1 \leq s \leq q$ we have already constructed $C_1, \dots, C_{s-1}$. We now construct C_s as follows. Let $H_s := H - \sum_{s'<s}(C_{s'} - F_{s'})$. Define a new constant d such that $\varepsilon \ll d \ll 1$.

Our first task is to apply Lemma 5.4.4 to F_s to merge all the cycles in F_s into a Hamilton cycle using only edges of H_s. For each $i \in I_s$, let V_i^- be the set of vertices in V_i with indegree one in PS_s and let V_i^+ be the set of vertices in V_i with outdegree one in PS_s. Set $V_i^1 := V_i \backslash V_i^+$ and $V_{i+1}^2 := V_{i+1} \backslash V_{i+1}^-$. Since PS_s is locally balanced, $|V_i^+| = |V_{i+1}^-| \leq \varepsilon_0 m$ for all $i \in I_s$ (where the inequality holds by (BE3)). By (b_{s-1}) and (5.4.5), $H_s[V_i, V_{i+1}]$ is obtained from $H[V_i, V_{i+1}]$ by removing at most $|\{s' < s : i \in I_{s'}\}| \leq \ell m/k \leq \varepsilon^2 \gamma m$ edges from each vertex (as $1/k \ll \varepsilon, \gamma, 1/\ell$). So by Proposition 5.1.4, $H_s[V_i, V_{i+1}]$ is still $(2\varepsilon, 2\gamma, \gamma/2, 3\gamma)$-superregular for each $i \in I_s$. Recall that $|V_i \setminus V_i^1| = |V_{i+1} \setminus V_{i+1}^2| \leq \varepsilon_0 m$. Hence $H_s[V_i^1, V_{i+1}^2]$ is $(4\varepsilon, 2\gamma, \gamma/4, 6\gamma)$-superregular by Proposition 5.1.3 and thus also $(4\varepsilon, 2\gamma, \gamma/4, 4\gamma/d)$-superregular.

Let $E_s := \{V_i V_{i+1} : i \in I_s\}$. Our aim is to apply Lemma 5.4.4 with F_s, E_s, H_s, 4ε, 2γ, 2, 1/8 playing the roles of F, J, H, ε, d', t, ζ. Our assumption that $F_s - PS_s$ winds around C implies that for each $i \in I_s$, $F_s[V_i^1, V_{i+1}^2]$ is a perfect matching. So Lemma 5.4.4(i) holds. Note that every final vertex of a nontrivial path in PS_s must lie in $\bigcup_{i \in I_s} V_i^1$, implying Lemma 5.4.4(ii). Finally, recall that $|V_i^1|, |V_{i+1}^2| \geq (1 - \varepsilon_0)m$ for all $i \in I_s$. Together with our assumption that $F_s - PS_s$ winds around C, this easily implies Lemma 5.4.4(iii). So we can apply Lemma 5.4.4 to obtain a Hamilton cycle C_s' which is constructed from F_s by replacing $F_s[V_i^1, V_{i+1}^2]$ with a suitable perfect matching in $H_s[V_i^1, V_{i+1}^2]$ for each $i \in I_s$. In particular, $PS_s \subseteq C_s'$.

Let $H_s' := H_s - (C_s' - F_s)$. Recall that M_s is an ordered directed matching, say $M_s = \{e_1, \dots, e_r\}$, and that PS_s is a V_{i_s}-extension of M_s. For each $j \leq r$, let P_j be the path in PS_s containing e_j and let x_j denote the final vertex of P_j. Hence $x_1, \dots, x_r$ are distinct and lie in $V_{i_s}^1$. Together with (BE3) this implies that $r \leq \varepsilon_0 m$. Note that $H_s'[V_{i_s}^1, V_{i_s+1}^2]$ is obtained from $H_s[V_{i_s}^1, V_{i_s+1}^2]$ by removing a perfect matching, namely $C_s'[V_{i_s}^1, V_{i_s+1}^2]$. So by Proposition 5.1.4, $H_s'[V_{i_s}^1, V_{i_s+1}^2]$ is still $(8\varepsilon, 2\gamma, \gamma/8, 4\gamma/d)$-superregular. Apply Lemma 5.4.5 with C_s', i_s, $H_s'[V_{i_s}^1, V_{i_s+1}^2]$, ε_0, 8ε, 2γ, 2, 1/16 playing the roles of C', i, H, γ, ε, d', t, ζ to obtain a Hamilton cycle C_s which visits $x_1, \dots, x_r$ in this order and is constructed from C_s' by replacing the perfect matching $C_s'[V_{i_s}^1, V_{i_s+1}^2]$ with a suitable perfect matching in $H_s'[V_{i_s}^1, V_{i_s+1}^2]$. In particular, $PS_s \subseteq C_s$.

Note that $E(C_s - F_s) \subseteq \bigcup_{i\in I_s} E(H_s[V_i, V_{i+1}])$, so ($b_s$) and ($c_s$) hold. Since $PS_s \subseteq C_s$ and x_j is the final vertex of P_j and since $e_j \in E(P_j)$, it follows that C_s visits the edges $e_1, \dots, e_r$ in order. So C_s is consistent with M_s, implying (a_s). □

Proof of Lemma 5.4.1. Let $\mathcal{Q} = \{V_1, \dots, V_k\}$. By relabeling the V_i if necessary, we may assume that $C = V_1 \dots V_k$. Define new constants γ and ε such that $\varepsilon_0, \varepsilon', 1/k \ll \gamma \ll \varepsilon, \rho, 1/\ell$ and $\mu \ll \varepsilon \ll 1$. For each $i \leq k$ we apply Lemma 5.4.2 to (the underlying undirected graph of) $G[V_i, V_{i+1}]$ in order to obtain a spanning subdigraph H of G which satisfies the following properties:

(i′) For each $i \leq k$, $H[V_i, V_{i+1}]$ is $(\varepsilon, 2\gamma, \gamma, 3\gamma)$-superregular.

(ii′) Let $G' := G - H$. Then $(G', \mathcal{Q}, C)$ is a $(k, m, \mu, 4\gamma)$-cyclic system.

Indeed, (ii′) follows easily from Lemma 5.4.2(ii) and the definition of a $(k, m, \mu, 4\gamma)$-cyclic system. Recall that $\mathcal{BE} = \{PS_1, \dots, PS_q\}$ with $M_s \subseteq PS_s$ for all $s \leq q$. Our next aim is to apply Lemma 5.4.3 with G', $\mathcal{BE}$, 4γ playing the roles of G, $\mathcal{PS}$, ε to obtain 1-factors F_s extending the PS_s. Note that (BE1) and (BE3) imply that conditions (i) and (ii) of Lemma 5.4.3 hold. So we can apply Lemma 5.4.3 to obtain q (directed) 1-factors $F_1, \dots, F_q$ in $G' + \mathcal{BE}$ such that $PS_s \subseteq F_s$ for all $s \leq q$ and $F_1 - PS_1, \dots, F_q - PS_q$ are edge-disjoint subgraphs of G'. Recall from (ii′) and (Sys2) that G' (and thus also $F_s - PS_s$) winds around C. So we can apply Lemma 5.4.6 to obtain q Hamilton cycles $C_1, \dots, C_q$ in $F_1 + \dots + F_q + H$ such that C_s contains PS_s and is consistent with M_s for all $s \leq q$, and such that $C_1 - F_1, \dots, C_q - F_q$ are edge-disjoint subgraphs of H. Since H and G' are edge-disjoint, $C_1 - PS_1, \dots, C_q - PS_q$ are edge-disjoint subgraphs of G. □

We can now put everything together to prove the approximate decomposition lemma in the two cliques case. First we apply Lemma 5.3.2 to obtain cyclic systems and sparse subgraphs $H_{A,j}$ and $H_{B,j}$. Then we apply Lemma 5.3.3 to balance out the exceptional systems into balanced extensions. Next, we apply Lemma 5.4.1 to A and B separately to extend the balanced extensions into Hamilton cycles.

Proof of Lemma 2.5.4. Apply Lemma 5.3.2 to $G, \mathcal{P}$ and $\mathcal{J}$ to obtain (for each $1 \leq j \leq (K-1)/2$) pairs of tuples $(G_{A,j}, \mathcal{Q}_A, C_{A,j}, H_{A,j}, \mathcal{J}^*_{A,j})$ and $(G_{B,j}, \mathcal{Q}_B, C_{B,j}, H_{B,j}, \mathcal{J}^*_{B,j})$ which satisfy (a_1)–(a_7). Fix $j \leq (K-1)/2$. Write $\mathcal{J}^*_{A,j} = \{J^*_{A,\mathrm{dir},1}, \dots, J^*_{A,\mathrm{dir},q}\}$, where

$$q := |\mathcal{J}^*_{A,j}| \leq (1 - 4\mu - 3\rho)m \tag{5.4.6}$$

by (a_3). We now apply Lemma 5.3.3 with $\mathcal{J}^*_{A,j}, \mathcal{Q}_A, C_{A,j}, H_{A,j}, K, 5K\sqrt{\varepsilon_0}$ playing the roles of $\mathcal{M}, \mathcal{Q}, C, H, k, \varepsilon$ to obtain an orientation $H_{A,j,\mathrm{dir}}$ of $H_{A,j}$ and a balanced extension $\mathcal{BE}_j$ of $\mathcal{J}^*_{A,j}$ with respect to $(\mathcal{Q}_A, C_{A,j})$ and parameters $(10K\sqrt{\varepsilon_0}, 3)$. (Note that ($a_3$) and ($a_5$) imply conditions (i) and (ii) of Lemma 5.3.3.) Write $\mathcal{BE}_j := \{PS_1, \dots, PS_q\}$ such that $J^*_{A,\mathrm{dir},s} \subseteq PS_s$ for all $s \leq q$. So (BE1) implies that $PS_1 - J^*_{A,\mathrm{dir},1}, \dots, PS_q - J^*_{A,\mathrm{dir},q}$ are edge-disjoint subgraphs of $H_{A,j,\mathrm{dir}}$. Since $(G_{A,j,\mathrm{dir}}, \mathcal{Q}_A, C_{A,j})$ is a $(K, m, 4\mu, 5/K)$-cyclic system by (a_6), (5.4.6) implies that we can apply Lemma 5.4.1 as follows:

	$G_{A,j,\mathrm{dir}}$	$\mathcal{Q}_A$	$C_{A,j}$	K	$\mathcal{J}^*_{A,j}$	q	4μ	3ρ	$10K\sqrt{\varepsilon_0}$	$5/K$	3
plays role of	G	$\mathcal{Q}$	C	k	$\mathcal{M}$	q	μ	ρ	ε_0	ε'	ℓ

In this way we obtain q directed Hamilton cycles $C'_{A,j,1},\dots,C'_{A,j,q}$ in $G_{A,j,\mathrm{dir}}+\mathcal{BE}_j$ such that $C'_{A,j,s}$ contains PS_s and is consistent with $J^*_{A,\mathrm{dir},s}$ for all $s\le q$. Moreover, $C'_1 - J^*_{A,\mathrm{dir},1},\dots,C'_q - J^*_{A,\mathrm{dir},q}$ are edge-disjoint subgraphs of $G_{A,j,\mathrm{dir}} + H_{A,j,\mathrm{dir}}$. Repeat this process for all $j\le (K-1)/2$.

Write $\mathcal{J}=\{J_1,\dots,J_{|\mathcal{J}|}\}$. Recall from (a$_2$) that the $\mathcal{J}^*_{A,1},\dots,\mathcal{J}^*_{A,(K-1)/2}$ partition $\{J^*_{A,\mathrm{dir}} : J\in\mathcal{J}\}$. Therefore, we have obtained $|\mathcal{J}|$ directed Hamilton cycles $C'_{A,1},\dots,C'_{A,|\mathcal{J}|}$ on vertex set A. Moreover, by relabelling the J_s if necessary, we may assume that $C'_{A,s}$ is consistent with $(J_s)^*_{A,\mathrm{dir}}$ for all $s\le|\mathcal{J}|$. Furthermore, (a$_4$) implies that the undirected versions of $C'_{A,1}-(J_1)^*_{A,\mathrm{dir}},\dots,C'_{A,|\mathcal{J}|}-(J_{|\mathcal{J}|})^*_{A,\mathrm{dir}}$ are edge-disjoint spanning subgraphs of $G[A]$.

Similarly we obtain directed Hamilton cycles $C'_{B,1},\dots,C'_{B,|\mathcal{J}|}$ on vertex set B so that $(J_s)^*_{B,\mathrm{dir}}\subseteq C'_{B,s}$ for all $s\le|\mathcal{J}|$. Let H_s be the undirected graph obtained from $C'_{A,s}+C'_{B,s}-J^*_s+J_s$ by ignoring all the orientations of the edges. Recall that $J_1,\dots,J_{|\mathcal{J}|}$ are edge-disjoint exceptional systems and that they are edge-disjoint from the $C'_{A,s}+C'_{B,s}-J^*_s$ by (EC3). So $H_1,\dots,H_{|\mathcal{J}|}$ are edge-disjoint spanning subgraphs of G. Finally, Proposition 5.3.1 implies that $H_1,\dots,H_{|\mathcal{J}|}$ are indeed as desired in Lemma 2.5.4. □

5.5. The Bipartite Case

Roughly speaking, the idea in this case is to reduce the problem of finding the desired edge-disjoint Hamilton cycles in G to that of finding suitable Hamilton cycles in an almost complete balanced bipartite graph. This is achieved by considering the graphs J^*_{dir}, whose definition we recall in the next subsection. The main steps are similar to those in the proof of Lemma 2.5.4 (in fact, we re-use several of the lemmas, in particular Lemma 5.4.1).

We will construct the graphs J^*_{dir}, which are based on balanced exceptional systems J, in Section 5.5.1. In Section 5.5.2 we describe a decomposition of G into blown-up Hamilton cycles. We will construct balanced extensions in Section 5.5.3 (this is more difficult than in the two cliques case). Finally, we obtain the desired Hamilton cycles using Lemma 5.4.1 (in the same way as in the two cliques case).

5.5.1. Defining the Graphs J^*_{dir} for the Bipartite Case. In this section we recall a number of definitions from Section 4.4.1. Let $\mathcal{P}$ be a (K,m,ε)-partition of a vertex set V and let J be a balanced exceptional system with respect to $\mathcal{P}$. Since each maximal path in J has endpoints in $A\cup B$ and internal vertices in V_0 by (BES1), a balanced exceptional system J naturally induces a matching J^*_{AB} on $A\cup B$. More precisely, if $P_1,\dots,P_{\ell'}$ are the non-trivial paths in J and x_i,y_i are the endpoints of P_i, then we define $J^*_{AB}:=\{x_iy_i : i\le\ell'\}$. Thus J^*_{AB} is a matching by (BES1) and $e(J^*_{AB})\le e(J)$. Moreover, J^*_{AB} and $E(J)$ cover exactly the same vertices in A. Similarly, they cover exactly the same vertices in B. So (BES3) implies that $e(J^*_{AB}[A]) = e(J^*_{AB}[B])$. We can write $E(J^*_{AB}[A]) = \{x_1x_2,\dots,x_{2s-1}x_{2s}\}$, $E(J^*_{AB}[B]) = \{y_1y_2,\dots,y_{2s-1}y_{2s}\}$ and $E(J^*_{AB}[A,B]) = \{x_{2s+1}y_{2s+1},\dots,x_{s'}y_{s'}\}$, where $x_i\in A$ and $y_i\in B$. Define $J^*:=\{x_iy_i : 1\le i\le s'\}$. Note that

$$e(J^*) = e(J^*_{AB}) \le e(J). \tag{5.5.1}$$

As before, all edges of J^* are called *fictive edges*. Recall that an (undirected) cycle D is *consistent with* J^* if D contains J^* and (there is an orientation of D which) visits the vertices $x_1, y_1, x_2, \dots, y_{s'-1}, x_{s'}, y_{s'}$ in this order.

We will need a directed version of Proposition 4.4.1(ii). This directed version immediately follows from Proposition 4.4.1(ii) and is similar to Proposition 4.4.2. For this, define $J^*_{\rm dir}$ to be the ordered directed matching $\{f_1, \dots, f_{s'}\}$ such that f_i is a directed edge from x_i to y_i for all $i \le s'$. So $J^*_{\rm dir}$ consists only of AB-edges. Similarly to the undirected case, we say that a directed cycle $D_{\rm dir}$ is *consistent with* $J^*_{\rm dir}$ if $D_{\rm dir}$ contains $J^*_{\rm dir}$ and visits the edges $f_1, \dots, f_{s'}$ in this order.

PROPOSITION 5.5.1. *Let $\mathcal{P}$ be a (K, m, ε)-partition of a vertex set V. Let G be a graph on V and let J be a balanced exceptional system with respect to $\mathcal{P}$ such that $J \subseteq G$. Suppose that $D_{\rm dir}$ is a directed Hamilton cycle on $A \cup B$ such that $D_{\rm dir}$ is consistent with $J^*_{\rm dir}$. Furthermore, suppose that $D - J^* \subseteq G$, where D is the cycle obtained from $D_{\rm dir}$ after ignoring the directions of all edges. Then $D - J^* + J$ is a Hamilton cycle of G.*

5.5.2. Finding Systems. The following lemma gives a decomposition of an almost complete bipartite graph G into blown-up Hamilton cycles (together with an associated decomposition of exceptional systems). Its proof is almost the same as that of Lemma 5.3.2, so we omit it here. The only difference is that instead of Walecki's theorem we use a result of Auerbach and Laskar [**1**] to decompose the complete bipartite graph $K_{K,K}$ into Hamilton cycles, where K is even.

LEMMA 5.5.2. *Suppose that $0 < 1/n \ll \varepsilon_0 \ll 1/K \ll \rho \ll 1$ and $0 \le \mu \ll 1$, where $n, K \in \mathbb{N}$ and K is even. Suppose that G is a graph on n vertices and $\mathcal{P} = \{A_0, A_1, \dots, A_K, B_0, B_1, \dots, B_K\}$ is a (K, m, ε_0)-partition of $V(G)$. Furthermore, suppose that the following conditions hold:*

- (a) *$d(v, B_i) = (1-4\mu \pm 4/K)m$ and $d(w, A_i) = (1-4\mu \pm 4/K)m$ for all $v \in A$, $w \in B$ and $1 \le i \le K$.*
- (b) *There is a set $\mathcal{J}$ which consists of at most $(1/4 - \mu - \rho)n$ edge-disjoint exceptional systems with parameter ε_0 in G.*
- (c) *$\mathcal{J}$ has a partition into K^4 sets $\mathcal{J}_{i_1,i_2,i_3,i_4}$ (one for all $1 \le i_1, i_2, i_3, i_4 \le K$) such that each $\mathcal{J}_{i_1,i_2,i_3,i_4}$ consists of precisely $|\mathcal{J}|/K^4$ (i_1, i_2, i_3, i_4)-BES with respect to $\mathcal{P}$.*

Then for each $1 \le j \le K/2$, there is a tuple $(G_j, \mathcal{Q}, C_j, H_j, \mathcal{J}_j)$ such that the following assertions hold, where $\mathcal{Q} := \{A_1, \dots, A_K, B_1, \dots, B_K\}$:

- (a_1) *Each of $C_1, \dots, C_{K/2}$ is a directed Hamilton cycle on $\mathcal{Q}$ such that the undirected versions of these cycles form a Hamilton decomposition of the complete bipartite graph whose vertex classes are $\{A_1, \dots, A_K\}$ and $\{B_1, \dots, B_K\}$.*
- (a_2) *$\mathcal{J}_1, \dots, \mathcal{J}_{K/2}$ is a partition of $\mathcal{J}$.*
- (a_3) *Each $\mathcal{J}_j$ has a partition into K^4 sets $\mathcal{J}_{j,i_1,i_2,i_3,i_4}$ (one for all $1 \le i_1, i_2, i_3, i_4 \le K$) such that $\mathcal{J}_{j,i_1,i_2,i_3,i_4}$ consists of (i_1, i_2, i_3, i_4)-BES with respect to $\mathcal{P}$ and $|\mathcal{J}_{j,i_1,i_2,i_3,i_4}| \le (1 - 4\mu - 3\rho)m/K^4$.*
- (a_4) *$G_1, \dots, G_{K/2}, H_1, \dots, H_{K/2}$ are edge-disjoint subgraphs of $G[A, B]$.*
- (a_5) *$H_j[A_i, B_{i'}]$ is a $(11K + 248/K)\varepsilon_0 m$-regular graph for all $j \le K/2$ and all $i, i' \le K$.*
- (a_6) *For each $j \le K/2$, there exists an orientation $G_{j,\rm dir}$ of G_j such that $(G_{j,\rm dir}, \mathcal{Q}, C_j)$ is a $(2K, m, 4\mu, 5/K)$-cyclic system.*

5.5.3. Constructing Balanced Extensions. Let $\mathcal{P} = \{A_0, A_1, \dots, A_K, B_0, B_1, \dots, B_K\}$ be a (K, m, ε)-partition of a vertex set V, let $\mathcal{Q} := \{A_1, \dots, A_K, B_1, \dots, B_K\}$ and let $C = A_1B_1A_2B_2 \dots A_KB_K$ be a directed cycle. Given a set $\mathcal{J}$ of balanced exceptional systems with respect to $\mathcal{P}$, we write $\mathcal{J}^*_{\rm dir} := \{J^*_{\rm dir} : J \in \mathcal{J}\}$. So $\mathcal{J}^*_{\rm dir}$ is a set of ordered directed matchings and thus it makes sense to construct a balanced extension of $\mathcal{J}^*_{\rm dir}$ with respect to $(\mathcal{Q}, C)$. (Recall that balanced extensions were defined in Section 5.2.2.)

Now consider any of the tuples $(G_j, \mathcal{Q}, C_j, H_j, \mathcal{J}_j)$ guaranteed by Lemma 5.5.2. We will apply the following lemma to find a balanced extension of $(\mathcal{J}_j)^*_{\rm dir}$ with respect to $(\mathcal{Q}, C_j)$, using edges of H_j (after a suitable orientation of these edges). So the lemma is a bipartite analogue of Lemma 5.3.3. However, the proof is more involved than in the two cliques case.

LEMMA 5.5.3. *Suppose that* $0 < 1/n \ll \varepsilon \ll 1/K \ll 1$, *where* $n, K \in \mathbb{N}$. *Let* $\mathcal{P} = \{A_0, A_1, \dots, A_K, B_0, B_1, \dots, B_K\}$ *be a* (K, m, ε)*-partition of a set* V *of* n *vertices. Let* $\mathcal{Q} := \{A_1, \dots, A_K, B_1, \dots, B_K\}$ *and let* $C := A_1B_1A_2B_2 \dots A_KB_K$ *be a directed cycle. Suppose that there exist a set* $\mathcal{J}$ *of edge-disjoint balanced exceptional systems with respect to* $\mathcal{P}$ *and parameter* ε *and a graph* H *such that the following conditions hold:*

(i) *$\mathcal{J}$ can be partitioned into K^4 sets $\mathcal{J}_{i_1,i_2,i_3,i_4}$ (one for all $1 \le i_1, i_2, i_3, i_4 \le K$) such that $\mathcal{J}_{i_1,i_2,i_3,i_4}$ consists of (i_1, i_2, i_3, i_4)-BES with respect to $\mathcal{P}$ and $|\mathcal{J}_{i_1,i_2,i_3,i_4}| \le m/K^4$.*
(ii) *For each $v \in A \cup B$ the number of all those $J \in \mathcal{J}$ for which v is incident to an edge in J is at most $2\varepsilon n$.*
(iii) *$H[A_i, B_{i'}]$ is a $(11K + 248/K)\varepsilon m$-regular graph for all $i, i' \le K$.*

Then there exist an orientation $H_{\rm dir}$ of H and a balanced extension $\mathcal{BE}$ of $\mathcal{J}^*_{\rm dir}$ with respect to $(\mathcal{Q}, C)$ and parameters $(12\varepsilon K, 12)$ such that each path sequence in $\mathcal{BE}$ is obtained from some $J^*_{\rm dir} \in \mathcal{J}^*_{\rm dir}$ by adding edges of $H_{\rm dir}$.

The proof proceeds roughly as follows. Consider any $J \in \mathcal{J}_{i_1,i_2,i_3,i_4}$. We extend $J^*_{\rm dir}$ into a locally balanced path sequence in two steps. For this, recall that $J^*_{\rm dir}$ consists only of edges from $A_{i_1} \cup A_{i_2}$ to $B_{i_3} \cup B_{i_4}$. In the first step, we construct a path sequence PS that is an A_{i_1}-extension of $J^*_{\rm dir}$ by adding suitable $B_{i_3}A_{i_1}$- and $B_{i_4}A_{i_1}$-edges from H to $J^*_{\rm dir}$. In the second step, we locally balance PS in such a way that (BE1)–(BE3) are satisfied.

Proof. First we decompose H into H' and H'' such that $H'[A_i, B_{i'}]$ is a $11\varepsilon Km$-regular graph for all $i, i' \le K$ and $H'' := H - H'$. Hence $H''[A_i, B_{i'}]$ is a $248\varepsilon m/K$-regular graph for all $i, i' \le K$.

Write $\mathcal{J} := \{J_1, \dots, J_{|\mathcal{J}|}\}$. For each $s \le |\mathcal{J}|$, we will extend $J^*_{s,{\rm dir}} := (J_s)^*_{\rm dir}$ into a path sequence PS_s satisfying the following conditions:

(α_s) Suppose that $J_s \in \mathcal{J}_{i_1,i_2,i_3,i_4}$. Then PS_s is an A_{i_1}-extension of $J^*_{s,{\rm dir}}$ consisting of precisely $e(J^*_s)$ vertex-disjoint directed paths of length two.
(β_s) $V(PS_s) = V(J^*_{s,{\rm dir}}) \cup A'_s$, where $A'_s \subseteq A_{i_1} \setminus V(J^*_{s,{\rm dir}})$ and $|A'_s| = e(J^*_s)$.
(γ_s) $PS_s - J^*_{s,{\rm dir}}$ is a matching of size $e(J^*_s)$ from B'_s to A'_s, where $B'_s := V(J^*_{s,{\rm dir}}) \cap (B_{i_3} \cup B_{i_4})$.
(δ_s) Let M_s be the set of undirected edges obtained from $PS_s - J^*_{s,{\rm dir}}$ after ignoring all the orientations. Then $M_1, \dots, M_s$ are edge-disjoint matchings in H'.

(ε_s) PS_s consists only of edges from $A_{i_1} \cup A_{i_2}$ to $B_{i_3} \cup B_{i_4}$, and from $B_{i_3} \cup B_{i_4}$ to A_{i_1}.

Note that (β_s) and (γ_s) together imply (ε_s). Suppose that for some s with $1 \leq s \leq |\mathcal{J}|$ we have already constructed $PS_1, \dots, PS_{s-1}$. We will now construct PS_s as follows. Let i_1, i_2, i_3, i_4 be such that $J_s \in \mathcal{J}_{i_1,i_2,i_3,i_4}$ and let $H'_s := H' - (M_1 + \dots + M_{s-1})$. (BES4) implies that

(5.5.2)
$$e(J^*_{s,\mathrm{dir}}) = e(J^*_s) \overset{(5.5.1)}{\leq} e(J_s) \leq \varepsilon n \leq 3\varepsilon K m \quad \text{and} \quad |V(J^*_{s,\mathrm{dir}}) \cap A_{i_1}| \leq 3\varepsilon K m.$$

Consider any $s' < s$. Recall from the definition of $J^*_{s',\mathrm{dir}}$ that $V(J^*_{s',\mathrm{dir}})$ is the set of all those vertices in $A \cup B$ which are covered by edges of $J_{s'}$. Together with ($\beta_{s'}$) and ($\gamma_{s'}$) this implies that a vertex $v \in B$ is covered by $M_{s'}$ if and only if v is incident to an edge of $J_{s'}$. Together with (ii) this in turn implies that for all $v \in B$ we have

$$\begin{aligned} d_{H'_s}(v, A_{i_1}) &\geq d_{H'}(v, A_{i_1}) - \sum_{s'<s} d_{M_{s'}}(v) \geq 11K\varepsilon m - 2\varepsilon n \\ &\geq 11K\varepsilon m - 5K\varepsilon m \overset{(5.5.2)}{\geq} |V(J^*_{s,\mathrm{dir}}) \cap A_{i_1}| + e(J^*_s). \end{aligned}$$

Note that $e(J^*_s) = |V(J^*_{s,\mathrm{dir}}) \cap (B_{i_3} \cup B_{i_4})| = |B'_s|$. So we can greedily find a matching M_s of size $e(J^*_s)$ in $H'_s[A_{i_1} \setminus V(J^*_{s,\mathrm{dir}}), B'_s]$ (which therefore covers all vertices in B'_s). Orient all edges of M_s from B'_s to A_{i_1} and call the resulting directed matching $M_{s,\mathrm{dir}}$. Set

$$PS_s := J^*_{s,\mathrm{dir}} + M_{s,\mathrm{dir}}.$$

Note that PS_s consists of precisely $e(J^*_s)$ directed paths of length two whose final vertices lie in A_{i_1}, so (α_s)–(ε_s) hold by our construction. This shows that we can obtain path sequences $PS_1, \dots, PS_{|\mathcal{J}|}$ satisfying (α_s)–(ε_s) for all $s \leq |\mathcal{J}|$.

The following claim provides us with a 'reservoir' of edges which we will use to balance out the edges of each PS_s and thus extend each PS_s into a path sequence PS'_s which is locally balanced with respect to C.

Claim. *H'' contains $|\mathcal{J}|$ subgraphs $H''_1, \dots, H''_{|\mathcal{J}|}$ satisfying the following properties for all $s \leq |\mathcal{J}|$ and all $i, i' \leq K$:*

(a_1) *If PS_s contains an $A_iB_{i'}$-edge, then H''_s contains a matching between $A_{i'}$ and B_i of size $30\varepsilon Km$.*

(a_2) *If PS_s contains a $B_iA_{i'}$-edge, then H''_s contains a matching between A_{i+1} and $B_{i'-1}$ of size $30\varepsilon Km$.*

(a_3) *$H''_1, \dots, H''_{|\mathcal{J}|}$ are edge-disjoint and for all $s \leq |\mathcal{J}|$ the matchings guaranteed by (a_1) and (a_2) are edge-disjoint.*

So if PS_s contains both an $A_iB_{i'}$-edge and a $B_{i'-1}A_{i+1}$-edge, then H''_s contains a matching between $A_{i'}$ and B_i of size $60\varepsilon Km$.

To prove the claim, first recall that $H''[A_i, B_{i'}]$ is a $248\varepsilon m/K$-regular graph for all $i, i' \leq K$. So $H''[A_i, B_{i'}]$ can be decomposed into $248\varepsilon m/K$ perfect matchings. Each perfect matching can be split into $1/(31\varepsilon K)$ matchings, each of size at least $30\varepsilon Km$. Therefore $H''[A_i, B_{i'}]$ contains $8m/K^2$ edge-disjoint matchings, each of size at least $30\varepsilon Km$. (i) and (ε_s) together imply that for any $i, i' \leq K$, the number of PS_s containing an $A_iB_{i'}$-edge is at most

$$\sum_{(i_1,i_2,i_3,i_4)\ :\ i\in\{i_1,i_2\},\ i'\in\{i_3,i_4\}} |\mathcal{J}_{i_1,i_2,i_3,i_4}| \leq 4m/K^2.$$

Recall that $H''[A_{i'}, B_i]$ contains $8m/K^2$ edge-disjoint matchings, each of size at least $30\varepsilon m$. Thus we can assign a distinct matching in $H''[A_{i'}, B_i]$ of size $30\varepsilon m$ to each PS_s that contains an $A_iB_{i'}$-edge. Additionally, we can also assign a distinct matching in $H''[A_{i+1}, B_{i'-1}]$ of size $30\varepsilon m$ to each PS_s that contains a $B_iA_{i'}$-edge. For all $s \leq |\mathcal{J}|$, let H''_s be the union of all those matchings assigned to PS_s. Then $H''_1, \dots, H''_{|\mathcal{J}|}$ are as desired in the claim.

For each $s \leq |\mathcal{J}|$, we will now add suitable edges from H''_s to PS_s in order to obtain a path sequence PS'_s which is locally balanced with respect to C. So fix $s \leq |\mathcal{J}|$ and let $e_1, \dots, e_\ell$ denote the edges of PS_s. Note that $\ell = 2e(J^*_s) \leq 6K\varepsilon m$ by (γ_s) and (5.5.2). For each $r \leq \ell$, we will find a directed edge f_r satisfying the following conditions:

(b_1) If e_r is an $A_iB_{i'}$-edge, then f_r is an $A_{i'}B_i$-edge.
(b_2) If e_r is a $B_iA_{i'}$-edge, then f_r is a $B_{i'-1}A_{i+1}$-edge.
(b_3) The undirected version of $\{f_1, \dots, f_\ell\}$ is a matching in H''_s and vertex-disjoint from $V(PS_s)$.

Suppose that for some $r \leq \ell$ we have already constructed $f_1, \dots, f_{r-1}$. Suppose that e_r is an $A_iB_{i'}$-edge. (The argument for the other case is similar.) By (a_1), $H''_s[A_{i'}, B_i]$ contains a matching of size $30K\varepsilon m$. Note by (α_s) and (b_3) that

$$|V(PS_s \cup \{f_1, \dots, f_{r-1}\})| \leq 3e(J^*_s) + 2(r-1) < 5\ell \leq 30K\varepsilon m.$$

Hence there exists an edge in $H''_s[A_{i'}, B_i]$ that is vertex-disjoint from $PS_s \cup \{f_1, \dots, f_{r-1}\}$. Orient one such edge from $A_{i'}$ to B_i and call it f_r. In this way, we can construct $f_1, \dots, f_\ell$ satisfying (b_1)–(b_3).

Let PS'_s be digraph obtained from PS_s by adding all the edges $f_1, \dots, f_\ell$. Note that PS'_s is a locally balanced path sequence with respect to C. (Indeed, PS'_s is locally balanced since $\{e_r, f_r\}$ is locally balanced for each $r \leq \ell$.) Let i_1, i_2, i_3, i_4 be such that $J \in \mathcal{J}_{i_1,i_2,i_3,i_4}$. Then the following properties hold:

(c_1) PS'_s is an A_{i_1}-extension of $J^*_{s,\mathrm{dir}}$.
(c_2) $|V(PS'_s) \cap A_i|, |V(PS'_s) \cap B_i| \leq 12\varepsilon Km$ for all $i \leq K$.
(c_3) If $V(PS'_s) \cap A_i \neq \emptyset$, then $i \in \{i_1, i_2, i_3, i_4, i_3+1, i_4+1\}$.
(c_4) If $V(PS'_s) \cap B_i \neq \emptyset$, then $i \in \{i_1-1, i_1, i_2, i_3, i_4\}$.

Indeed, (c_1) is implied by (α_s) and the definition of PS'_s. Since $e(PS'_s) = 2e(PS_s) = 4e(J^*_s)$, ($c_2$) holds by (5.5.2). Finally, (c_3) and (c_4) are implied by (ε_s), (b_1) and (b_2) as $J_s \in \mathcal{J}_{i_1,i_2,i_3,i_4}$.

Note that $PS'_1 - J^*_{1,\mathrm{dir}}, \dots, PS'_{|\mathcal{J}|} - J^*_{|\mathcal{J}|,\mathrm{dir}}$ are pairwise edge-disjoint and let $\mathcal{BE} := \{PS'_1, \dots, PS'_{|\mathcal{J}|}\}$. We claim that $\mathcal{BE}$ is a balanced extension of $\mathcal{J}^*_{\mathrm{dir}}$ with respect to $(\mathcal{Q}, C)$ and parameters $(12\varepsilon K, 12)$. To see this, recall that $\mathcal{Q} = \{A_1, \dots, A_K, B_1, \dots, B_K\}$ is a $(2K, m)$-equipartition of $V' := V \setminus (A_0 \cup B_0)$. Clearly, (BE1) holds with V' playing the role of V. (c_3) and (i) imply that for every $i \leq K$ there are at most $6m/K$ $PS'_s \in \mathcal{BE}$ such that $V(PS'_s) \cap A_i \neq \emptyset$. A similar statement also holds for each B_i. So together with (c_2), this implies (BE3), where $2K$ plays the role of k in (BE3). As remarked after the definition of a balanced extension, this implies the 'moreover part' of (BE2). So (BE2) holds too. Therefore $\mathcal{BE}$ is a balanced extension, so the lemma follows (by orienting the remaining edges of H arbitrarily). $\square$

Proof of Lemma 4.6.1. Apply Lemma 5.5.2 to obtain tuples $(G_j, \mathcal{Q}, C_j, H_j, \mathcal{J}_j)$ for all $j \le K/2$ satisfying (a$_1$)–(a$_6$). Fix $j \le K/2$ and write $\mathcal{J}_j := \{J_{j,1} \dots, J_{j,|\mathcal{J}_j|}\}$. Next, apply Lemma 5.5.3 with $\mathcal{J}_j, C_j, H_j, \varepsilon_0$ playing the roles of $\mathcal{J}, C, H, \varepsilon$ to obtain an orientation $H_{j,\mathrm{dir}}$ of H_j and a balanced extension $\mathcal{BE}_j$ of $\mathcal{J}_j$ with respect to $(\mathcal{Q}, C_j)$ and parameters $(12\varepsilon_0 K, 12)$. (Note that (a$_3$) and (a$_5$) imply conditions (i) and (iii) of Lemma 5.5.3. Condition (ii) follows from Lemma 4.6.1(d).) So we can write $\mathcal{BE}_j := \{PS_{j,1} \dots, PS_{j,|\mathcal{J}_j|}\}$ such that $(J_{j,s})^*_{\mathrm{dir}} \subseteq PS_{j,s}$ for all $s \le |\mathcal{J}_j|$. Each path sequence in $\mathcal{BE}_j$ is obtained from some $(J_{j,s})^*_{\mathrm{dir}}$ by adding edges of $H_{j,\mathrm{dir}}$. Since $(G_{j,\mathrm{dir}}, \mathcal{Q}, C_j)$ is a $(2K, m, 4\mu, 5/K)$-cyclic system by Lemma 5.5.2(a$_6$), we can apply Lemma 5.4.1 as follows:

	$G_{j,\mathrm{dir}}$	$\mathcal{Q}$	C_j	$2K$	$\mathcal{J}^*_{j,\mathrm{dir}}$	$\lvert\mathcal{J}_j\rvert$	4μ	3ρ	$12\varepsilon_0 K$	$5/K$	12
plays role of	G	$\mathcal{Q}$	C	k	$\mathcal{M}$	q	μ	ρ	ε_0	ε'	ℓ

This gives us $|\mathcal{J}_j|$ directed Hamilton cycles $C'_{j,1}, \dots, C'_{j,|\mathcal{J}_j|}$ in $G_{j,\mathrm{dir}} + \mathcal{BE}_j$ such that each $C'_{j,s}$ contains $PS_{j,s}$ and is consistent with $(J_{j,s})^*_{\mathrm{dir}}$. Moreover, (a$_4$) implies that $C'_{j,1} - (J_{j,1})^*_{\mathrm{dir}}, \dots, C'_{j,|\mathcal{J}_j|} - (J_{j,|\mathcal{J}_j|})^*_{\mathrm{dir}}$ are edge-disjoint subgraphs of $G_{j,\mathrm{dir}} + H_{j,\mathrm{dir}}$. Repeat this process for all $j \le K/2$.

Recall from Lemma 5.5.2(a$_2$) that $\mathcal{J}_1, \dots, \mathcal{J}_{K/2}$ is a partition of $\mathcal{J}$. Thus we have obtained $|\mathcal{J}|$ directed Hamilton cycles $C'_1, \dots, C'_{|\mathcal{J}|}$ on $A \cup B$ such that each C'_s is consistent with $(J_s)^*_{\mathrm{dir}}$ for some $J_s \in \mathcal{J}$ (and $J_s \neq J_{s'}$ whenever $s \neq s'$). Let H_s be the undirected graph obtained from $C'_s - J^*_s + J_s$ by ignoring all the orientations of the edges. Since $J_1, \dots, J_{|\mathcal{J}|}$ are edge-disjoint exceptional systems, $H_1, \dots, H_{|\mathcal{J}|}$ are edge-disjoint spanning subgraphs of G. Finally, Proposition 5.5.1 implies that $H_1, \dots, H_{|\mathcal{J}|}$ are indeed as desired in Lemma 4.6.1. $\square$

Acknowledgement

We are grateful to Katherine Staden for drawing Figure 2.8.1.

Bibliography

[1] Bruce Auerbach and Renu Laskar, *On decomposition of r-partite graphs into edge-disjoint Hamilton circuits*, Discrete Math. **14** (1976), no. 3, 265–268. MR0389662 (52 #10493)

[2] D. Cariolaro, The 1-factorization problem and some related conjectures, Ph. D. Thesis, University of Reading, 2004.

[3] A. G. Chetwynd and A. J. W. Hilton, *Regular graphs of high degree are* 1*-factorizable*, Proc. London Math. Soc. (3) **50** (1985), no. 2, 193–206, DOI 10.1112/plms/s3-50.2.193. MR772711 (86d:05093)

[4] A. G. Chetwynd and A. J. W. Hilton, *Star multigraphs with three vertices of maximum degree*, Math. Proc. Cambridge Philos. Soc. **100** (1986), no. 2, 303–317, DOI 10.1017/S030500410006610X. MR848854 (87i:05094)

[5] A. G. Chetwynd and A. J. W. Hilton, 1*-factorizing regular graphs of high degree—an improved bound*, Discrete Math. **75** (1989), no. 1-3, 103–112, DOI 10.1016/0012-365X(89)90082-4. Graph theory and combinatorics (Cambridge, 1988). MR1001390 (90g:05080)

[6] Demetres Christofides, Daniela Kühn, and Deryk Osthus, *Edge-disjoint Hamilton cycles in graphs*, J. Combin. Theory Ser. B **102** (2012), no. 5, 1035–1060, DOI 10.1016/j.jctb.2011.10.005. MR2959389

[7] A. Ferber, M. Krivelevich and B. Sudakov, Counting and packing Hamilton cycles in dense graphs and oriented graphs, *J. Combin. Theory B*, to appear.

[8] A. Ferber, M. Krivelevich and B. Sudakov, Counting and packing Hamilton ℓ-cycles in dense hypergraphs, *Journal of Combinatorics* **7** (2016), 135–157.

[9] Alan Frieze and Michael Krivelevich, *On packing Hamilton cycles in ϵ-regular graphs*, J. Combin. Theory Ser. B **94** (2005), no. 1, 159–172, DOI 10.1016/j.jctb.2004.12.003. MR2130284 (2005m:05171)

[10] S.G. Hartke, R. Martin and T. Seacrest, Relating minimum degree and the existence of a k-factor, research manuscript.

[11] Stephen G. Hartke and Tyler Seacrest, *Random partitions and edge-disjoint Hamiltonian cycles*, J. Combin. Theory Ser. B **103** (2013), no. 6, 742–766, DOI 10.1016/j.jctb.2013.09.004. MR3127592

[12] Ian Holyer, *The NP-completeness of edge-coloring*, SIAM J. Comput. **10** (1981), no. 4, 718–720, DOI 10.1137/0210055. MR635430 (83b:68050)

[13] Bill Jackson, *Edge-disjoint Hamilton cycles in regular graphs of large degree*, J. London Math. Soc. (2) **19** (1979), no. 1, 13–16, DOI 10.1112/jlms/s2-19.1.13. MR527728 (80k:05078)

[14] Svante Janson, Tomasz Łuczak, and Andrzej Rucinski, *Random graphs*, Wiley-Interscience Series in Discrete Mathematics and Optimization, Wiley-Interscience, New York, 2000. MR1782847 (2001k:05180)

[15] P. Katerinis, *Minimum degree of a graph and the existence of k-factors*, Proc. Indian Acad. Sci. Math. Sci. **94** (1985), no. 2-3, 123–127, DOI 10.1007/BF02880991. MR844252 (87f:05132)

[16] Jeong Han Kim and Nicholas C. Wormald, *Random matchings which induce Hamilton cycles and Hamiltonian decompositions of random regular graphs*, J. Combin. Theory Ser. B **81** (2001), no. 1, 20–44, DOI 10.1006/jctb.2000.1991. MR1809424 (2001m:05235)

[17] F. Knox, D. Kühn and D. Osthus, Edge-disjoint Hamilton cycles in random graphs, *Random Structures Algorithms* **46** (2015), 397–445.

[18] Michael Krivelevich and Wojciech Samotij, *Optimal packings of Hamilton cycles in sparse random graphs*, SIAM J. Discrete Math. **26** (2012), no. 3, 964–982, DOI 10.1137/110849171. MR3022117

[19] Daniela Kühn, John Lapinskas, and Deryk Osthus, *Optimal packings of Hamilton cycles in graphs of high minimum degree*, Combin. Probab. Comput. **22** (2013), no. 3, 394–416, DOI 10.1017/S0963548312000569. MR3053854

[20] D. Kühn, A. Lo, D. Osthus and K. Staden, *The robust component structure of dense regular graphs and applications*, Proc. London Math. Soc. **110** (2015), 19–56.

[21] Daniela Kühn and Deryk Osthus, *Hamilton decompositions of regular expanders: a proof of Kelly's conjecture for large tournaments*, Adv. Math. **237** (2013), 62–146, DOI 10.1016/j.aim.2013.01.005. MR3028574

[22] Daniela Kühn and Deryk Osthus, *Hamilton decompositions of regular expanders: applications*, J. Combin. Theory Ser. B **104** (2014), 1–27, DOI 10.1016/j.jctb.2013.10.006. MR3132742

[23] D. Kühn and D. Osthus, Hamilton cycles in graphs and hypergraphs: an extremal perspective, *Proceedings of the ICM 2014*, Seoul, Korea, Vol. 4, 381–406.

[24] Daniela Kühn, Deryk Osthus, and Andrew Treglown, *Hamilton decompositions of regular tournaments*, Proc. Lond. Math. Soc. (3) **101** (2010), no. 1, 303–335, DOI 10.1112/plms/pdp062. MR2661248 (2012e:05159)

[25] Daniela Kühn, Deryk Osthus, and Andrew Treglown, *Hamiltonian degree sequences in digraphs*, J. Combin. Theory Ser. B **100** (2010), no. 4, 367–380, DOI 10.1016/j.jctb.2009.11.004. MR2644240 (2011e:05143)

[26] E. Lucas, *Récréations Mathématiques*, Vol. 2, Gautheir-Villars, 1892.

[27] C.St.J.A. Nash-Williams, Valency sequences which force graphs to have Hamiltonian circuits, University of Waterloo Research Report, Waterloo, Ontario, 1969.

[28] C. St. J. A. Nash-Williams, *Hamiltonian lines in graphs whose vertices have sufficiently large valencies*, Combinatorial theory and its applications, III (Proc. Colloq., Balatonfüred, 1969), North-Holland, Amsterdam, 1970, pp. 813-819. MR0299517 (45 #8565)

[29] C. St. J. A. Nash-Williams, *Edge-disjoint Hamiltonian circuits in graphs with vertices of large valency*, Studies in Pure Mathematics (Presented to Richard Rado), Academic Press, London, 1971, pp. 157–183. MR0284366 (44 #1594)

[30] C. St. J. A. Nash-Williams, *Hamiltonian arcs and circuits*, Recent Trends in Graph Theory (Proc. Conf., New York, 1970), Lecture Notes in Mathematics, Vol. 186, Springer, Berlin, 1971, pp. 197–210. MR0277417 (43 #3150)

[31] Deryk Osthus and Katherine Staden, *Approximate Hamilton decompositions of robustly expanding regular digraphs*, SIAM J. Discrete Math. **27** (2013), no. 3, 1372–1409, DOI 10.1137/120880951. MR3090152

[32] L. Perkovic and B. Reed, *Edge coloring regular graphs of high degree*, Discrete Math. **165/166** (1997), 567–578, DOI 10.1016/S0012-365X(96)00202-6. Graphs and combinatorics (Marseille, 1995). MR1439301 (97k:05084)

[33] Michael J. Plantholt and Shailesh K. Tipnis, *All regular multigraphs of even order and high degree are 1-factorable*, Electron. J. Combin. **8** (2001), no. 1, Research Paper 41, 8 pp. (electronic). MR1877660 (2002k:05197)

[34] Michael Stiebitz, Diego Scheide, Bjarne Toft, and Lene M. Favrholdt, *Graph edge coloring*, Wiley Series in Discrete Mathematics and Optimization, John Wiley & Sons, Inc., Hoboken, NJ, 2012. Vizing's theorem and Goldberg's conjecture; With a preface by Stiebitz and Toft. MR2975974

[35] E. R. Vaughan, *An asymptotic version of the multigraph 1-factorization conjecture*, J. Graph Theory **72** (2013), no. 1, 19–29, DOI 10.1002/jgt.21629. MR2993074

[36] V. G. Vizing, *Critical graphs with given chromatic class* (Russian), Diskret. Analiz No. **5** (1965), 9–17. MR0200202 (34 #101)

[37] D.B. West, Introduction to Graph Theory (2nd Edition), Pearson 2000.

Editorial Information

To be published in the *Memoirs*, a paper must be correct, new, nontrivial, and significant. Further, it must be well written and of interest to a substantial number of mathematicians. Piecemeal results, such as an inconclusive step toward an unproved major theorem or a minor variation on a known result, are in general not acceptable for publication.

Papers appearing in *Memoirs* are generally at least 80 and not more than 200 published pages in length. Papers less than 80 or more than 200 published pages require the approval of the Managing Editor of the Transactions/Memoirs Editorial Board. Published pages are the same size as those generated in the style files provided for AMS-LaTeX or AMS-TeX.

Information on the backlog for this journal can be found on the AMS website starting from `http://www.ams.org/memo`.

A Consent to Publish is required before we can begin processing your paper. After a paper is accepted for publication, the Providence office will send a Consent to Publish and Copyright Agreement to all authors of the paper. By submitting a paper to the *Memoirs*, authors certify that the results have not been submitted to nor are they under consideration for publication by another journal, conference proceedings, or similar publication.

Information for Authors

Memoirs is an author-prepared publication. Once formatted for print and on-line publication, articles will be published as is with the addition of AMS-prepared frontmatter and backmatter. Articles are not copyedited; however, confirmation copy will be sent to the authors.

Initial submission. The AMS uses Centralized Manuscript Processing for initial submissions. Authors should submit a PDF file using the Initial Manuscript Submission form found at `www.ams.org/submission/memo`, or send one copy of the manuscript to the following address: Centralized Manuscript Processing, MEMOIRS OF THE AMS, 201 Charles Street, Providence, RI 02904-2294 USA. If a paper copy is being forwarded to the AMS, indicate that it is for *Memoirs* and include the name of the corresponding author, contact information such as email address or mailing address, and the name of an appropriate Editor to review the paper (see the list of Editors below).

The paper must contain a *descriptive title* and an *abstract* that summarizes the article in language suitable for workers in the general field (algebra, analysis, etc.). The *descriptive title* should be short, but informative; useless or vague phrases such as "some remarks about" or "concerning" should be avoided. The *abstract* should be at least one complete sentence, and at most 300 words. Included with the footnotes to the paper should be the 2010 *Mathematics Subject Classification* representing the primary and secondary subjects of the article. The classifications are accessible from `www.ams.org/msc/`. The Mathematics Subject Classification footnote may be followed by a list of *key words and phrases* describing the subject matter of the article and taken from it. Journal abbreviations used in bibliographies are listed in the latest *Mathematical Reviews* annual index. The series abbreviations are also accessible from `www.ams.org/msnhtml/serials.pdf`. To help in preparing and verifying references, the AMS offers MR Lookup, a Reference Tool for Linking, at `www.ams.org/mrlookup/`.

Electronically prepared manuscripts. The AMS encourages electronically prepared manuscripts, with a strong preference for AMS-LaTeX. To this end, the Society has prepared AMS-LaTeX author packages for each AMS publication. Author packages include instructions for preparing electronic manuscripts, samples, and a style file that generates the particular design specifications of that publication series. Though AMS-LaTeX is the highly preferred format of TeX, author packages are also available in AMS-TeX.

Authors may retrieve an author package for *Memoirs of the AMS* from `www.ams.org/journals/memo/memoauthorpac.html` or via FTP to `ftp.ams.org` (login as `anonymous`, enter your complete email address as password, and type `cd pub/author-info`). The

AMS Author Handbook and the *Instruction Manual* are available in PDF format from the author package link. The author package can also be obtained free of charge by sending email to tech-support@ams.org or from the Publication Division, American Mathematical Society, 201 Charles St., Providence, RI 02904-2294, USA. When requesting an author package, please specify $\mathcal{A}\mathcal{M}\mathcal{S}$-LaTeX or $\mathcal{A}\mathcal{M}\mathcal{S}$-TeX and the publication in which your paper will appear. Please be sure to include your complete mailing address.

After acceptance. The source files for the final version of the electronic manuscript should be sent to the Providence office immediately after the paper has been accepted for publication. The author should also submit a PDF of the final version of the paper to the editor, who will forward a copy to the Providence office.

Accepted electronically prepared files can be submitted via the web at www.ams.org/submit-book-journal/, sent via FTP, or sent on CD to the Electronic Prepress Department, American Mathematical Society, 201 Charles Street, Providence, RI 02904-2294 USA. TeX source files and graphic files can be transferred over the Internet by FTP to the Internet node ftp.ams.org (130.44.1.100). When sending a manuscript electronically via CD, please be sure to include a message indicating that the paper is for the *Memoirs*.

Electronic graphics. Comprehensive instructions on preparing graphics are available at www.ams.org/authors/journals.html. A few of the major requirements are given here.

Submit files for graphics as EPS (Encapsulated PostScript) files. This includes graphics originated via a graphics application as well as scanned photographs or other computer-generated images. If this is not possible, TIFF files are acceptable as long as they can be opened in Adobe Photoshop or Illustrator.

Authors using graphics packages for the creation of electronic art should also avoid the use of any lines thinner than 0.5 points in width. Many graphics packages allow the user to specify a "hairline" for a very thin line. Hairlines often look acceptable when proofed on a typical laser printer. However, when produced on a high-resolution laser imagesetter, hairlines become nearly invisible and will be lost entirely in the final printing process.

Screens should be set to values between 15% and 85%. Screens which fall outside of this range are too light or too dark to print correctly. Variations of screens within a graphic should be no less than 10%.

Any graphics created in color will be rendered in grayscale for the printed version unless color printing is authorized by the Managing Editor and the Publisher. In general, color graphics will appear in color in the online version.

Inquiries. Any inquiries concerning a paper that has been accepted for publication should be sent to memo-query@ams.org or directly to the Electronic Prepress Department, American Mathematical Society, 201 Charles St., Providence, RI 02904-2294 USA.

Editors

This journal is designed particularly for long research papers, normally at least 80 pages in length, and groups of cognate papers in pure and applied mathematics. Papers intended for publication in the *Memoirs* should be addressed to one of the following editors. The AMS uses Centralized Manuscript Processing for initial submissions to AMS journals. Authors should follow instructions listed on the Initial Submission page found at www.ams.org/memo/memosubmit.html.

Algebra, to MICHAEL LARSEN, Department of Mathematics, Rawles Hall, Indiana University, 831 E 3rd Street, Bloomington, IN 47405, USA; e-mail: mjlarsen@indiana.edu

Algebraic geometry, to LUCIA CAPORASO, Department of Mathematics and Physics, Roma Tre University, Largo San Leonardo Murialdo, I-00146 Roma, Italy; e-mail: LCedit@mat.uniroma3.it

Algebraic topology, to SOREN GALATIUS, Department of Mathematics, Stanford University, Stanford, CA 94305 USA; e-mail: transactions@lists.stanford.edu

Arithmetic geometry, to TED CHINBURG, Department of Mathematics, University of Pennsylvania, Philadelphia, PA 19104-6395; e-mail: math-tams@math.upenn.edu

Automorphic forms, representation theory and combinatorics, to DANIEL BUMP, Department of Mathematics, Stanford University, Building 380, Sloan Hall, Stanford, California 94305; e-mail: bump@math.stanford.edu

Combinatorics and discrete geometry, to IGOR PAK, Department of Mathematics, University of California, Los Angeles, California 90095; e-mail: pak@math.ucla.edu

Commutative and homological algebra, to LUCHEZAR L. AVRAMOV, Department of Mathematics, University of Nebraska, Lincoln, NE 68588-0130; e-mail: avramov@math.unl.edu

Differential geometry, to CHIU-CHU MELISSA LIU, Department of Mathematics, Columbia University, New York, NY 10027; e-mail: ccliu@math.columbia.edu

Dynamical systems and ergodic theory, to VIVIANE BALADI, Analyse Algébrique, Institut de Mathématiques de Jussieu-Paris Rive Gauche, U.P.M.C., B.C. 247, 4 Place Jussieu, F 75252 Paris cedex 05, France; e-mail: viviane.baladi@imj-prg.fr

Ergodic theory and combinatorics, to VITALY BERGELSON, Ohio State University, Department of Mathematics, 231 W. 18th Ave, Columbus, OH 43210; e-mail: vitaly@math.ohio-state.edu

Functional analysis and operator algebras, to STEFAAN VAES, KU Leuven, Department of Mathematics, Celestijnenlaan 200B, B-3001 Leuven, Belgium; e-mail: stefaan.vaes@wis.kuleuven.be

Geometric analysis, to TATIANA TORO, Department of Mathematics, University of Washington, Box 354350; e-mail: toro@uw.edu

Harmonic analysis, complex analysis, to MALABIKA PRAMANIK, Department of Mathematics, 1984 Mathematics Road, University of British Columbia, Vancouver, BC, Canada V6T 1Z2; e-mail: malabika@math.ubc.ca

Harmonic analysis, representation theory, and Lie theory, to E. P. VAN DEN BAN, Department of Mathematics, Utrecht University, P.O. Box 80 010, 3508 TA Utrecht, The Netherlands; e-mail: E.P.vandenBan@uu.nl

Logic, to NOAM GREENBERG, School of Mathematics and Statistics, Victoria University of Wellington, Wellington 6140, New Zealand; e-mail: greenberg@msor.vuw.ac.nz

Low-dimensional topology and geometric structures, to RICHARD CANARY, Department of Mathematics, University of Michigan, Ann Arbor, MI 48109-1043; e-mail: canary@umich.edu

Number theory, to HENRI DARMON, Department of Mathematics, McGILL University, Montreal, Quebec H3A 0G4, Canada; e-mail: darmon@math.mcgill.ca

Partial differential equations, to MARKUS KEEL, School of Mathematics, University of Minnesota, Minneapolis, MN 55455; e-mail: keel@math.umn.edu

Partial differential equations and functional analysis, to ALEXANDER KISELEV, Department of Mathematics, MS-136, Rice University, 6100 Main Street, Houston, TX 77005; e-mail: kisilev@rice.edu

Probability and statistics, to PATRICK FITZSIMMONS, Department of Mathematics, University of California, San Diego, 9500 Gilman Drive, La Jolla, CA 92093-0112; e-mail: pfitzsim@math.ucsd.edu

Real analysis and partial differential equations, to WILHELM SCHLAG, Department of Mathematics, The University of Chicago, 5734 South University Avenue, Chicago, IL 60615; e-mail: schlag@math.uchicago.edu

All other communications to the editors, should be addressed to the Managing Editor, ALEJANDRO ADEM, Department of Mathematics, The University of British Columbia, Room 121, 1984 Mathematics Road, Vancouver, B.C., Canada V6T 1Z2; e-mail: adem@math.ubc.ca

SELECTED PUBLISHED TITLES IN THIS SERIES

1151 **Reiner Hermann,** Monoidal Categories and the Gerstenhaber Bracket in Hochschild Cohomology, 2016

1150 **J. P. Pridham,** Real Non-Abelian Mixed Hodge Structures for Quasi-Projective Varieties: Formality and Splitting, 2016

1149 **Ariel Barton and Svitlana Mayboroda,** Layer Potentials and Boundary-Value Problems for Second Order Elliptic Operators with Data in Besov Spaces, 2016

1148 **Joseph Hundley and Eitan Sayag,** Descent Construction for GSpin Groups, 2016

1147 **U. Meierfrankenfeld, B. Stellmacher, and G. Stroth,** The Local Structure Theorem for Finite Groups With a Large p-Subgroup, 2016

1146 **Genni Fragnelli and Dimitri Mugnai,** Carleman Estimates, Observability Inequalities and Null Controllability for Interior Degenerate Nonsmooth Parabolic Equations, 2016

1145 **Bart Bories and Willem Veys,** Igusa's p-Adic Local Zeta Function and the Monodromy Conjecture for Non-Degenerate Surface Singularities, 2016

1144 **Atsushi Moriwaki,** Adelic Divisors on Arithmetic Varieties, 2016

1143 **Wen Huang, Song Shao, and Xiangdong Ye,** Nil Bohr-Sets and Almost Automorphy of Higher Order, 2015

1142 **Roland Donninger and Joachim Krieger,** A Vector Field Method on the Distorted Fourier Side and Decay for Wave Equations with Potentials, 2015

1141 **Su Gao, Steve Jackson, and Brandon Seward,** Group Colorings and Bernoulli Subflows, 2015

1140 **Michael Aschbacher,** Overgroups of Root Groups in Classical Groups, 2015

1139 **Mingmin Shen and Charles Vial,** The Fourier Transform for Certain HyperKähler Fourfolds, 2015

1138 **Volker Bach and Jean-Bernard Bru,** Diagonalizing Quadratic Bosonic Operators by Non-Autonomous Flow Equations, 2015

1137 **Michael Skeide,** Classification of E_0-Semigroups by Product Systems, 2015

1136 **Kieran G. O'Grady,** Moduli of Double EPW-Sextics, 2015

1135 **Vassilios Gregoriades,** Classes of Polish Spaces Under Effective Borel Isomorphism, 2015

1134 **Hongzi Cong, Jianjun Liu, and Xiaoping Yuan,** Stability of KAM Tori for Nonlinear Schrödinger Equation, 2015

1133 **P. Cannarsa, P. Martinez, and J. Vancostenoble,** Global Carleman Estimates for Degenerate Parabolic Operators with Applications, 2015

1132 **Weiwei Ao, Chang-Shou Lin, and Juncheng Wei,** On Non-Topological Solutions of the A_2 and B_2 Chern-Simons System, 2015

1131 **Bob Oliver,** Reduced Fusion Systems over 2-Groups of Sectional Rank at Most 4, 2015

1130 **Timothy C. Burness, Soumaïa Ghandour, and Donna M. Testerman,** Irreducible Geometric Subgroups of Classical Algebraic Groups, 2015

1129 **Georgios Daskalopoulos and Chikako Mese,** On the Singular Set of Harmonic Maps into DM-Complexes, 2015

1128 **M. Dickmann and F. Miraglia,** Faithfully Quadratic Rings, 2015

1127 **Svante Janson and Sten Kaijser,** Higher Moments of Banach Space Valued Random Variables, 2015

1126 **Toshiyuki Kobayashi and Birgit Speh,** Symmetry Breaking for Representations of Rank One Orthogonal Groups, 2015

1125 **Tetsu Mizumachi,** Stability of Line Solitons for the KP-II Equation in $\mathbb{R}^2$, 2015

For a complete list of titles in this series, visit the
AMS Bookstore at **www.ams.org/bookstore/memoseries/**.